IISE Annual Conference and Expo 2023 (Late Papers)

New Orleans, Louisiana, USA
21-23 May 2023

ISBN: 978-1-7138-9249-6

Printed from e-media with permission by:

Curran Associates, Inc.
57 Morehouse Lane
Red Hook, NY 12571

Some format issues inherent in the e-media version may also appear in this print version.

Printed with permission by Curran Associates, Inc. (2024)

For permission requests, please contact Institute of Industrial & Systems Engineers (IISE) at the address below.

Institute of Industrial & Systems Engineers (IISE)
3577 Parkway Lane, Suite 200
Norcross, GA 30092
USA

Phone: (770) 449-0460
Fax: (770) 441-3295

www.iise.org

Additional copies of this publication are available from:

Curran Associates, Inc.
57 Morehouse Lane
Red Hook, NY 12571 USA
Phone: 845-758-0400
Fax: 845-758-2633
Email: curran@proceedings.com
Web: www.proceedings.com

TABLE OF CONTENTS

https://doi.org/10.21872/2023IISE_1518

Proceedings of the IISE Annual Conference & Expo 2023
K. Babski-Reeves, B. Eksioglu, D. Hampton, eds.

Deep Neural Network and Mixed-Integer Programming Framework for Scheduling Textual Construction Workplace Audits

Farouq Halawa, Szu-Kai Hayden, Raashid Mohammed
Global Engineering Technical Services
Amazon.com

Abstract

Construction auditing is a vital step in designing new sites in fulfillment centers (FCs). The audits include inspection processes for newly launched buildings which ensure that the buildings meet workplace standards. In current practice the process of scheduling construction audits and prioritizing the checklists when launching new buildings is highly manual, and requires going over lengthy textual data to determine dates of implementation and sequence. The lack of visibility on actual equipment installation dates result in conducting more auditing visits than needed and significant traveling and labor costs. Natural Language Processing (NLP) has been recently proposed to automate the auditing processes in construction sites. In this research, a hybrid machine learning and mathematical optimization framework is proposed to schedule and sequence the auditing checks. The proposed model bundles the checklists and assigns visit months based on historical probability of occurrence using Mixed Integer Linear Programming (MILP). A deep neural network with fine-tuned BERT-TSDAE architecture is proposed to generate embeddings that capture semantic similarities between checklist and checklist types and predict the occurrence time of the checks. The proposed model was validated on a subset of 49 warehouses, contributing to a potential reduction of 51.3 % of visits per site due to optimal assignment of visits. The proposed Sentence BERT fined-tuned with TSDAE model over performed CNN-Glove architecture, proposed in the literature for construction auditing, and led to the highest accuracy and F1 score.

Keywords
Natural Language Processing, Deep Learning, Machine Learning, Data Science

1. Introduction

Construction inspection is a vital step to ensure that the new established buildings meet guidelines before the hand-over to operations. In this current practice, engineers and designers go over a large set of textual specifications in a manual fashion [1]. Lack of prioritization of checklists can create ambiguity and delays impacting commissioning timelines and targets. Natural language processing (NLP) has recently emerged as a promising tool to automate the construction inspection. NLP utilizes machine learning to extract semantic relationships and knowledge from textual datasets. The field of NLP in construction and design is still in its infancy due to limited algorithms and frameworks tested on such a dataset [2]. In the recent few years, several algorithms were proposed to automate the construction auditing [1–4, 9, 10], with multiple focuses including selection of relevant texts, clustering checklists, and duplicate detection. Nevertheless, the concept of checklist audit scheduling has not been tackled yet. Recognizing manually the relevant checklists in construction consumes at least one-third of an inspector time [2]. Significant time is also spent by firms to plan for the inspector visits, create schedules, and determine the optimal sequence of checks. Traditionally, the field of mathematical programming has been used to find optimal schedules of resources, staffing, or machinery [5, 7, 11]. In the context of NLP, sequencing textual data is a challenging task as the auditing checks are different depending on the type of site, type of asset, and the time-line of the construction plan. Often, the schedule is created based on the construction milestones. Thus, there are multiple factors that can influence the type of checks that are implemented. A fusion of machine learning and mathematical optimization is possible when having a large historical dataset of checklists at a warehousing network and a software to display the checks. With the advent of Industry 4.0, engineers can conduct those auditing checks using software applications. The integration of intelligent systems in those apps is significant as it enables the engineers to view the optimal lists of audits and select the ones that are relevant in an automated fashion. The proposed framework, can also be utilized in maintenance checks or any other application that is textual in nature. The contribution of this work is a novel deep learning & mixed-integer linear model to schedule checklist requirements and optimize the number of visits required by the designers and engineers,

https://doi.org/10.21872/2023IISE_1518

focusing on auditing requirements in the workplace. The remainder of this paper details the research methodology in Section 2, provides experimental results in Section 3, and indicates the research conclusion and future works in Section 4.

2. Methodology

The methodology of scheduling audits and prioritizing checklists is divided into four main phases, including data preparation, feature generation, checklist selection, and finally scheduling. Data preparation is performed to combine the checklist data for all sites from a software app environment, and transform the dataset into a format that also represents the checklist occurrence over time. Feature generation is performed through Kernel Density Estimation and word embeddings, followed by a deep neural network to predict checklist occurrence, and a mixed-integer linear programming model to assign checklists to visits. Figure 1 provides a high level illustration of the proposed framework. Detailed explanation of the methodology is shown in sections 2.1-2.4.

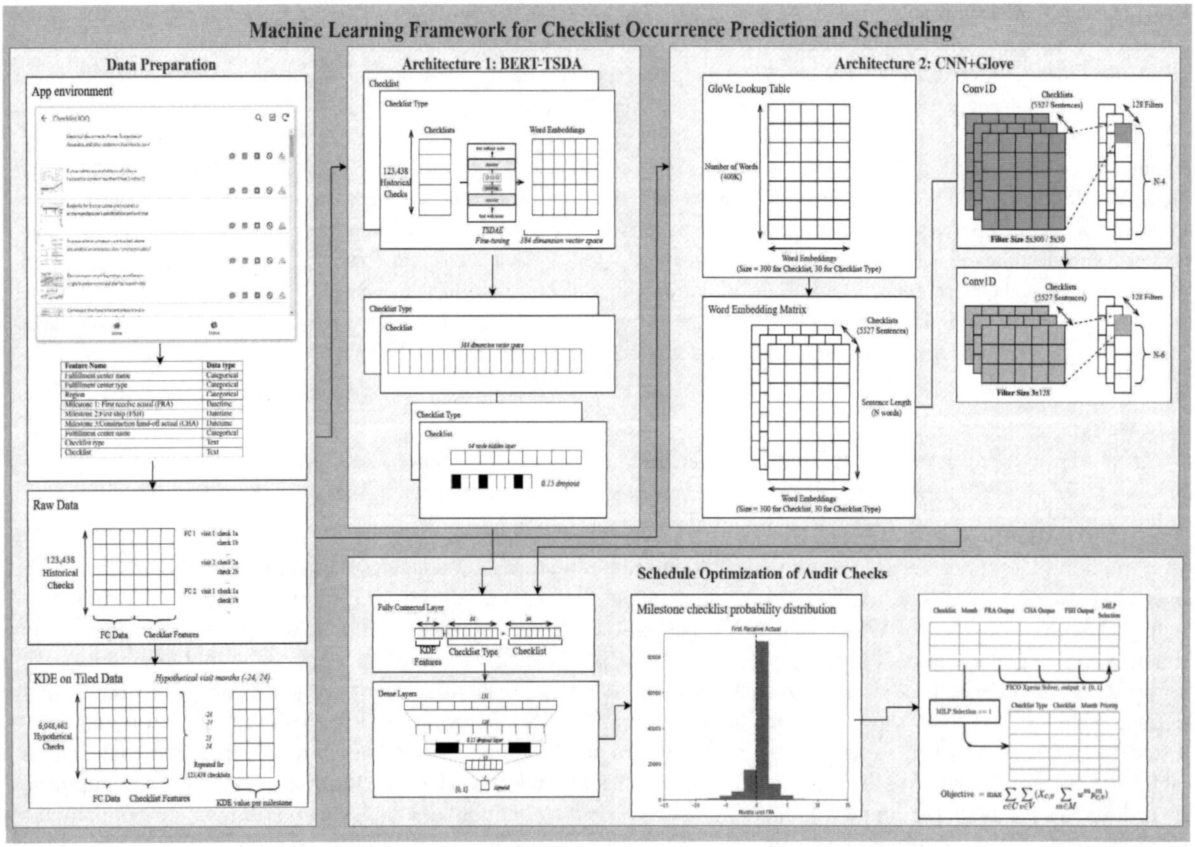

Figure 1: Proposed hybrid ML and MILP framework for scheduling checklist visits, showing the vectors representation and sequence of steps

2.1 Data Preparation

Two main types of data are used in the framework, fulfillment center data and checklist data. Fulfillment center data contains several construction milestones, assumed in this work as construction actual time (CHA) (i.e., time when the construction ends), facility first-received actual time (FRA) (i.e., time when the facility receives first items to store), facility first ship actual time (FSH) (i.e., time when the facility starts shipping items and packages). The number of milestones can vary depending on the site, however, those are the basic ones in warehouses. Checklist data includes a list of all textual auditing checks that were historically performed at a site or fullfilment center (FC). The model

https://doi.org/10.21872/2023IISE_1518

features are outlined in Figure 1, after being extracted from a live software application. The dataframes are merged on fulfillment center name to produce a dataset containing both fulfillment center and checklist level features. Each checklist has a specific datetime at which it was performed. The difference between each milestone date and the checklist historical date is then calculated and is expressed in number of months. To avoid over-fitting, the dataframe was tiled by duplicating the features for all available months in a time-line of 2 years. Figure 1 further visualizes the data structure discussed in this phase. The model labels are encoded as "1", for the month in which the check was performed historically, and "0" on all other months. As a result, the dataset to train the model will be highly imbalanced since the non-occurrences will exceed the occurrences. The model was trained on a dataset of 6,048,462 data points, which is highly imbalanced with 97.96% of rows representing checklist non-occurrences and only 2.04% of rows representing checklist occurrences. The number of FCs is randomly split into train, test, and validation sets of size 316, 25, and 24. The split resulted in datasets of length 5,153,085, length 463,932, and length 431,445, respectively.The existing dataframe includes a combination of features with different types: categorical, text, and time specific. The feature generation part aims to transform the text features into continuous values and to extract additional features that supports the prediction of checklist occurrence, as explained in the coming sections.

2.2 Feature Generation Using KDE

Checklist occurrence counts in relation to each milestone take on discrete values which may not reflect the true probability of occurrence. For example, if a checklist was historically only performed within one month of the FRA milestone, using discrete counts would imply that a checklist has a 0% probability of occurrence in any other month other than the FRA month. To tackle this, KDE (Kernel density estimation) [6] was implemented to use historical data to generate probability of checklist occurrence over an interval of two years before and after each milestone. Let n be the number of data points, σ is the standard deviation or bandwidth of the Gaussian distribution, x_i is the value of a particular data point i and $\hat{x}$ is the function input or the value that the kernel density is evaluated. Equation (1) represents the KDE function. KDE was performed 3 times for each of the milestones.

$$f(\hat{x}) = \frac{1}{n\sigma\sqrt{2\pi}} \sum_{i=0}^{n} e^{-\frac{1}{2}(\frac{x_i-\hat{x}}{\sigma})^2} \tag{1}$$

2.3 Checklist Occurrence Prediction

Two state-of-the-art architectures were tested, and the results produced by training each architecture on the 3 milestones were used to select the best model. The purpose of each neural network is to output the probability of the two classes [0,1], as a proxy for the probability of occurrence of a given checklist at a given number of months away from a milestone. For each model, each individual checklist will have 48 predictions indicating the probability of occurrence from 24 months before to 24 months after each milestone.

BERT & TSDAE: In this work, we implement the framework proposed by Wang et al. 2021 [8] for text embeddings. BERT allows for the production of fixed size vectors of sentence embeddings given variable length input sentences. A pre-trained all-MiniLM-L6-V2 BERT model was initialized, and fine-tuned over 18 epochs of training using TS-DAE (Transformers and Sequential Denoising Auto-Encoder) with a dataset consisting of 5,290 unique checklists and checklist types with a word length greater than 10. During the training process, the sentences are damaged through the removal of words and then encoded into fixed-size vectors, while the model attempts to reconstruct the vectors into the original, undamaged inputs. The embeddings are then passed into 3 neural networks used in this work, each consisting of an input layer receiving two sets of size 384 sentence embeddings, followed by two size 64 hidden layers. These hidden states are then concatenated with each other and the 3 milestone-specific KDE features, followed by 3 hidden layers of size 128, 128, 32 and output a single value between 0 and 1 using a sigmoid activation function, as they aim to predict the likelihood of checklist occurrence.Both models were trained using a binary cross-entropy loss function.
Glove & CNN: This architecture was proposed by Jeon et al. 2021 [2] for checklist auditing applications. In this work, we adapted the architecture as the following. The network consists of a size 300 embedding layer for a checklist and a size 30 embedding layer for checklist type, which each take a vectorized representation of words, followed by a convolutional layer that scans across the word embeddings to output a hidden layer of sentence embeddings. These hidden states are then concatenated with each other and the 3 milestone-specific KDE features, followed by 3 hidden layers of size 128, 128, 32. For the two tested architectures, the loss function is represented in equation 2 using Binary Cross-Entropy, where y_i is the target label of the data sample, $\hat{y}$ is the model output and n is the number of samples in

https://doi.org/10.21872/2023IISE_1518

the training dataset (equation 2). Due to the imbalance in the data caused by having more months where the check is not implemented (being labeled as 0), we are giving more importance to the minority class. Given a set of classes C, the observations of each class is x_i, and the weight of a class is shown in equation 3.

$$\text{Loss} = -\frac{1}{n}\sum_{i=1}^{n}[y_i \cdot \log(\hat{y}_i) + (1-y_i)\cdot \log(1-\hat{y}_i)] \tag{2}$$

$$W_i = \frac{1}{|x_i|} * \frac{\sum_{i\in C}|x_i|}{2} \tag{3}$$

2.4 Mixed Integer Linear Program (MILP) for Visits Optimization

Given the results of section 2.3, the MILP model bundles the checklists and schedules them into months. Let $i \in 1,2,..$ represents the number of checks, j $\in 1,2,..$ represents the number of visits, and $m \in 1,2,..$ represents the construction milestones. The objective function is to assign each checklist to a specific visit so that the total assignment probability coming from ML selection model is maximized for a construction milestone $p^m_{i,j}$. The variable $x_{i,j}$ is a binary variable that is 1 if the check is assigned to a visit j, and w^m is a weight given to each milestone, as shown in equation (4).

$$\text{Objective } = \max \sum_{i=1}\sum_{j=1} X_{i,j} \sum_{m=1} w^m p^m_{i,j} \tag{4}$$

The model is subject to multiple constraints. Equation 5, makes sure that each check will be assigned to a visit, and thus all checks will be assigned.

$$\sum_{j=1} X_{i,j} = 1, \forall i \in N \tag{5}$$

Let T_i be the time it takes for one check to be performed. Let J_j be the available hours per visit j. The following equation state that the total number of checks in a visit should not exceed the available time, as shown in equation 6.

$$\sum_{i=1} X_{i,j} \leq T_i J_j, \forall j \in N \tag{6}$$

To avoid the case that we have one visit with only few checks, let us define V_j as a semi-continuous variable, where l is the lower threshold, explained in equation 7.

$$V_j \geq l \text{ or } V_j = 0, \forall j \in N \tag{7}$$

We then set the sum of checks per visit equal to V_j to ensure that each visit is assigned to a pre-defined number, as explained in equation 8.

$$\sum_{i=1} X_{i,j} = V_j, \forall j \in N \tag{8}$$

3. Experimental Results

This section examines the results of implementing the proposed models on a subset of 25 fulfillment centers.

3.1 Checklist Occurrence Model Selection

Two model architectures were tested in Table 1 on three construction milestones.

Table 1: Selection of Machine learning architecture using different accuracy metrics

Milestone	GloVe with Convolutional Neural Network						Sentence BERT & TSDAE Deep Neural Network					
	Accuracy	Precision	Recall	F1	ROC AUC	Time/epoch	Accuracy	Precision	Recall	F1	ROC AUC	Time/epoch
FRA	0.96	0.97	0.96	0.97	0.95	674s	0.97	0.98	0.98	0.98	0.95	22s
CHA	0.96	0.96	0.95	0.96	0.95	680s	0.97	0.98	0.97	0.97	0.96	23s
FSH	0.97	0.98	0.95	0.97	0.94	675s	0.97	0.98	0.98	0.98	0.96	23s
Average	0.96	0.97	0.95	0.97	0.95	676s	0.97	0.98	0.98	0.98	0.96	23s

https://doi.org/10.21872/2023IISE_1518

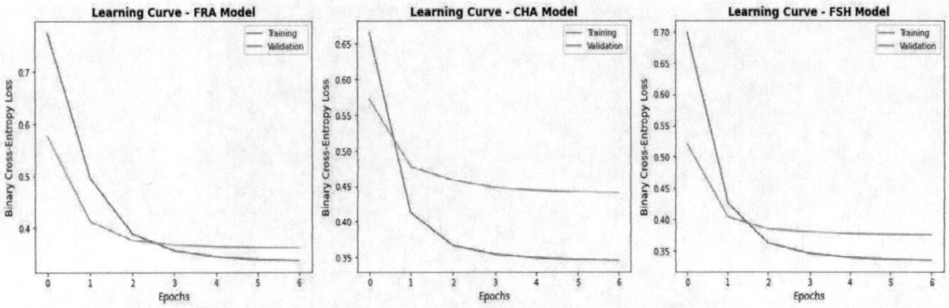

Figure 2: Model learning curves using 3 construction milestones (FRA, CHA, FSH)

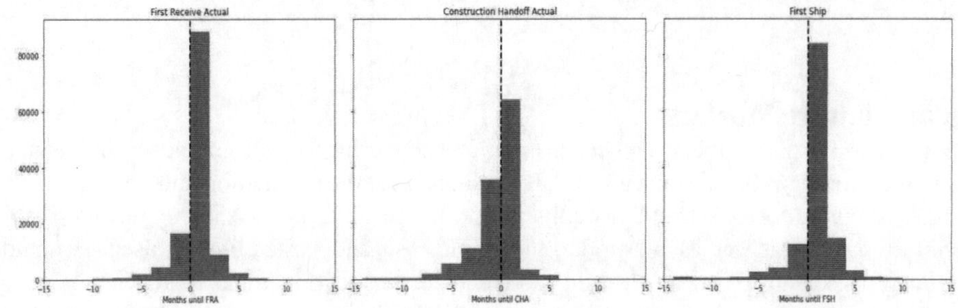

Figure 3: Distribution of checklists using 3 construction milestones (FRA, CHA, FSH)

As shown in the comparison tables, the BERT & TSDAE Deep Neural Network model was selected for use in the final implementation, based on a slightly improved performance and a significantly faster run time over the GloVe & CNN model. The loss curves for the BERT & TSDAE deep neural network are represented in Figure 2. Figure 3 visualizes the distribution of all checklists across a time-line of 2 years using one of the construction milestones (FRA) as obtained from the model. The classification probabilities resulted from the machine learning models were used to create the histogram shown in Figure 2. However, this is not enough as the current results would not provide an estimation on how many visits are required, which is the focus of the MILP phase.

3.2 Mixed-Integer Linear Programming (MILP) Model Results

MILP model was applied to 49 FC's. Number of visits ranged between 1 to 11 . Overall, we find that the average number of visits per fulfillment center was reduced by 51.3%, shown in Figure 4 for a sample of 10 sites. Figure 4 shows that the model is more beneficial for sites that included high number of visits (>5 visits).

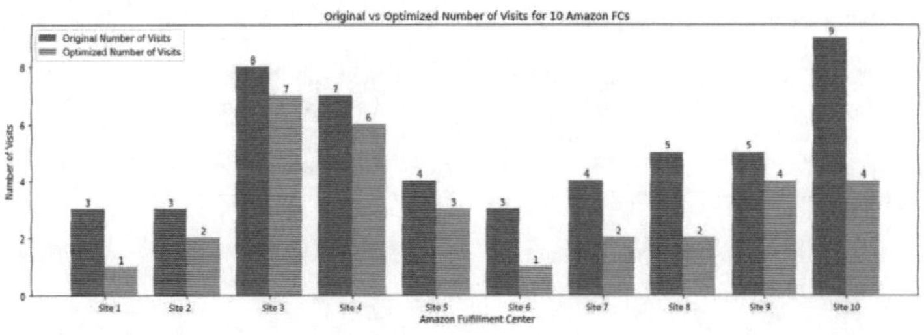

Figure 4: Comparison of number of auditing visits at 10 sites before and after implementing the proposed algorithm

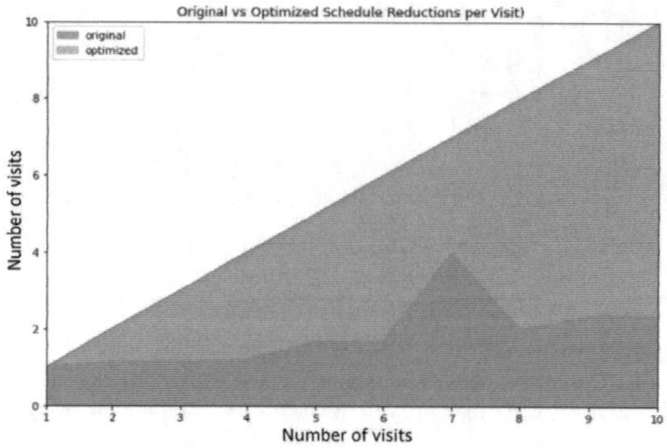

Figure 5: Reduction rate in number of visits: Simulation of 49 sites in 2022

4. Conclusion and Future Works

Optimal scheduling of checklists is a crucial step in construction and can increase the efficiency and reduce the number of auditing visits in the auditing process while ensure all facilities meet the installation and operational standards and specifications. The existing process of scheduling audits and prioritizing checklists is highly manual, leading to a high monetary cost from extra visits, as well as many other intangible costs and wasted time. This study demonstrated the value of applying the proposed novel machine learning and mathematical optimization approach to generate schedules of checklists for each fulfillment center visit, which can result in significant cost avoidance and improvement of workforce efficiency and work life harmony without adversely impacting existing operations. The proposed framework does not currently consider the time taken to complete checklists, and uses a naïve approach of setting a maximum number of checks per visit. Future work should consider the time aspect in the model. Future work should also consider other machine learning architectures to handle imbalanced data.

References

[1] Halawa, F., Abdul, M., Mohammed, R., 2022. "Applying Machine Learning for Duplicate Detection, Throttling and Prioritization of Equipment Commissioning Audits at Fulfillment Network". IISE EXPO Proceedings.

[2] Jeon, J., Xu, X., Zhang, Y., Yang, L., Cai, H., 2021. "Extraction of Construction Quality Requirements from Textual Specifications via Natural Language Processing". Transportation Research Record, 2675(9), 222-237.

[3] Yuan, C., Park, J., Xu, X., Cai, H., Abraham, D. M., Bowman, M. D., 2018. "Risk-based Prioritization of Construction Inspection. Transportation Research Record", 2672(26), 96-105.

[4] Goh, Y. M., Ubeynarayana, C. U., 2017, " Construction Accident Narrative Classification: An Evaluation of Text Mining Techniques. Accident Analysis Prevention, 108, 122-130.

[5] Hou, G. F. C., Thekumparampil, K., Oh, S., 2021. Multistage Stepsize Schedule in Federated Learning: Bridging Theory and Practice. In ICML Workshop (Vol. 12).

[6] Fan, J, James S., 1994. "Fast Implementations of Nonparametric Curve Estimators". Journal of Computational and Graphical Statistics 3(1)

[7] Rohaninejad, M., Janota, M., Hanzálek, Z. (2023). "Integrated Lot-sizing and Scheduling: Mitigation of uncertainty in Demand and Processing Time by Machine Learning". Engineering Applications of Artificial Intelligence, 118, 105676.

[8] Wang, K., Reimers, N., Gurevych, I., 2021. "Tsdae: Using Transformer-based Sequential Denoising Auto-encoder for Unsupervised Sentence Embedding Learning". arXiv preprint arXiv:2104.06979.

[9] Chokor, A., N, H., Chong, W. K., El Asmar, M., 2016. "Analyzing Arizona OSHA Injury Reports Using Unsupervised Machine Learning". Procedia engineering, 145, 1588-1593.

[10] Zhang, J., El-Gohary, N. M., 2016. "Semantic NLP-based Information Extraction From Construction Regulatory Documents for Automated Compliance Checking". Journal of Computing in Civil Engineering, 30(2), 04015014.

[11] Erekat, A., Sweidan, H., Mdathil, S. C., Khasawneh, M., 2019. "Staffing and Scheduling Framework of Merged IR and CathLab Departments". IISE EXPO Proceedings, 360-365.

https://doi.org/10.21872/2023IISE_1676

Turning Managers into Continuous Improvement Coaches

Abstract ID: 1676

Dr. Jill Speece, California Polytechnic University, San Luis Obispo, San Luis Obispo, CA

Abstract

Many organizations invest in training programs to help employees learn the basics of continuous improvement (CI) but miss the opportunity to develop those same employees' managers as CI coaches. This common problem was addressed by an outpatient imaging center through development of a specific "train the coaches" series of classes that addresses six key topics in developing confident CI coaches.

The purpose of the first class, "How to Explain Continuous Improvement," is to help managers identify what their staff needs to understand better and know how to effectively teach them that. The purpose of the second class, "Learn by Doing," is to help managers create opportunities for building CI muscle memory. The purpose of the third class, "Speaking with Data," is to help managers coach their staff to be proactive in collecting and communicating data to independently understand the problems facing the organization. The fourth class, "Making It Visual," focuses on learning how to get the right information to the right place at the right time. The fifth class, "The Benefits of Standardization," reviews the importance of providing a consistent customer experience and establishing a baseline for improvement using standardized processes. The sixth class, aptly named "Forever Aiming Higher," teaches the managers how to use an X-matrix to cultivate a culture of rising to the occasion. This course series fills the training gap for developing managers as CI coaches and the course content can be utilized successfully across multiple industries.

Keywords

Continuous Improvement, Manager Training, Coaching

1. Introduction

The ability to provide clear and timely coaching of continuous improvement (CI) tools and concepts was a knowledge gap realized by the management team at an outpatient imaging center on the central coast of California. This management team was fully trained in continuous improvement tools and concepts, and the majority earned a Lean Six Sigma Green or Black Belt. However, the traditional medical adage "see one, do one, teach one" can also be applied to developing a manager's continuous improvement skillset and, in this case, the knowledge gap identified was how to "teach one." In organizations dedicated to continuous improvement, middle managers "face a triangle of roles and challenges: leading change, implementing change, and behavioral change – together forming a so-called Bermuda triangle" [1]. Knowing this and knowing that the consistent presence of in-house continuous improvement champions is one of the factors that distinguishes successful continuous improvement systems from the unsuccessful ones [2], support for these middle managers became a strategic initiative for the organization in 2022. Middle managers are perfectly positioned to be the needed CI champions, but the organization must provide them with the necessary resources to navigate the aforementioned "Bermuda triangle." This paper provides an overview of the training program developed to address the "how to teach one" gap, a summary of the participants' feedback, and a recommended path forward for all management teams with the same gap.

2. Training Approach

A modular training course was developed with six focus areas. The learning objectives and the why for each training topic relevant to the middle manager position were established collaboratively with the whole management team. See Table 1 below. Each session was made up of six training modules scheduled for two modules a week in three-hour increments each. Two training sessions were conducted, and the managers were split evenly between sessions. Five managers attended each session for a total of ten managers that received the full training in the summer of 2022.

https://doi.org/10.21872/2023IISE_1676

Table 1: Training topics, learning objectives, and the why for each

Training Module	Learning Objective	Why
How to explain continuous improvement	Be able to quickly explain and provide context for the elements of continuous improvement	Identify what your staff needs to understand better and know how to effectively teach them that
Learn by doing	Learn how to create opportunities for building continuous improvement "muscle memory"	Help your staff learn, retain, and recall CI concepts and tools
Speaking with data	Master the use of check sheets to collect data for questions who answers are not immediately obvious and to communicate the data effectively	Help your staff be proactive in collecting and communicating data to independently understand the problems facing the organization
Making it visual	Learn how to get the right information to the right place at the right time	Create a self-ordering, self-explaining, self-regulating, and self-improving work environment
The benefits of standardization	Distinguish when it is necessary to standardize a process and when it is not	Provide a consistent customer experience and establish a baseline for improvement
Forever aiming higher	Learn how to use an x-matrix to cultivate a culture of forever aiming higher	Set a high bar that encourages people to rise to the occasion

3. Training Module Specifics

The goal of the first module "How to explain continuous improvement" is to equip managers with a method to explain more clearly difficult to understand CI concepts. As connecting leaders [3], middle managers in an organization dedicated to continuous improvement need to maintain a consistent attitude towards change and employee empowerment [4]. To support middle managers in this manner, the first training module focuses on the power of metaphors. Managers start the session by reflecting on which CI tools or concepts they are having a tough time explaining and/or their staff is having a tough time understanding. Metaphors are a well-known method for effectively communicating complicated information. During the training, managers work through an exercise to identify a list of abstract concepts and a list of concrete objects and are then randomly assigned one of those concepts with one of their concrete objects and are asked to create a meaningful metaphor what would help the listener better understand the abstract concept. They are then tasked with picking one of the "hard to explain or hard to understand" CI concepts and creating a metaphor that they can share with their team. An example of a metaphor created during the training is as follows:

> *Doing a Gemba walk is management gardening. You want to nurture and support your staff just like you would a garden. You need a balance of attentiveness and time away from them to develop and grow. Tending to them is being present in each department, asking about any gaps in their process, sharing any current information, asking them to address a concern or even acquiring information or concerns from them. This feedback you give and receive is the nutrient-rich soil needed for staff to flourish. A Gemba can have "just do it" moments where there is a quick fix to an identified problem, just like pulling a weed. Micromanaging or not being available for feedback could be detrimental to staff just like over or underwatering your garden.*

The metaphors created by the managers were shared with one another.

The goal of the second module "Learn by doing" was to help managers create opportunities for their staff to build CI muscle memory. The steps to improve CI muscle memory parallel the steps used to improve physical muscle memory. See Figure 1.

https://doi.org/10.21872/2023IISE_1676

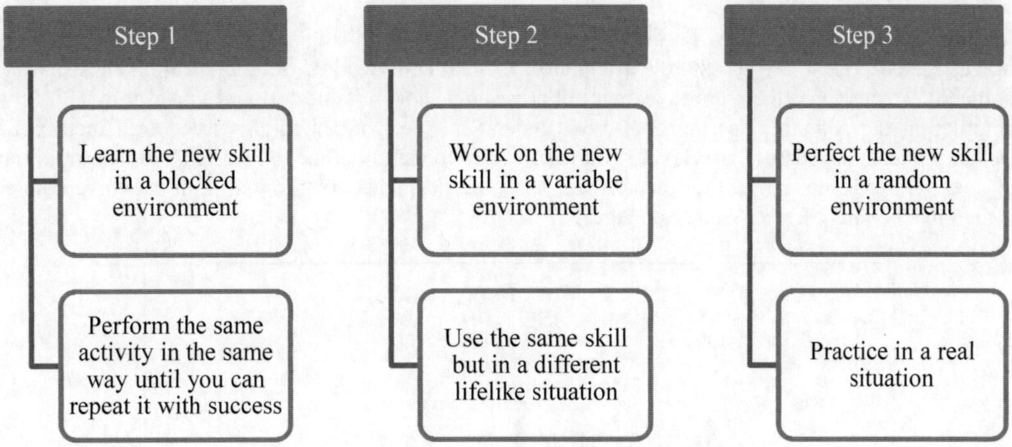

Figure *1*: Steps to Improving Muscle Memory *[5]*

During the training session, to demonstrate each step, participants worked through creating process flow maps. In the first step of a blocked environment, a prompt for making a process flow map was provided and all participants had to create an accurate and well-formatted process flow chart before they could move to the next step. The second step involved creating a process flow chart for planning a party. This gave them the opportunity to use the same skill but in a more realistic and variable situation. Finally, participants were asked to identify a process within their work environment that should be visually represented, create a process flow chart, and then review the chart with stakeholders in that environment to validate it. For homework, the managers had to identify a CI concept they needed to help their staff understand and produce a training plan that incorporates all three steps.

The goal of the third module "Speaking with data" was to review the importance of check sheets for helping staff be more proactive in collecting and communicating data to independently understand the problems facing the organization. The American Society for Quality (ASQ) outlined a check sheet procedure [6] that served as the basis for this training module. Participants were asked to follow this process for analyzing a hundred emails in their inbox. They created a check sheet, collected data, adjusted the check sheet, and then analyzed the results. This activity helped participants review and refine a structured method for collecting data. The second part of the training was to work on their ability to communicate the data. This was done through reviewing the meaningful order of elements for effective data storytelling [7] and having participants present a data story from their email inbox analysis check sheet.

The goal of the fourth module "Making it visual" was to remind managers of tools that help them get the right information to the right people at the right time. A visual workplace is a self-ordering, self-explaining, self-regulating, and self-improving work environment [8]. A simple tool for visual management is the checklist. Checklists help verify if a required set of steps has been performed or not and help identify defects. Performing a Gemba walk is an opportunity for managers to connect with their staff on a regular basis, but these walks can be prone to distractions so in this activity the managers were tasked with creating a checklist for their Gemba walk and then testing the goodness of the checklist by performing an actual Gemba walk. The process taught for creating a checklist was pulled from The Checklist Manifesto by Atul Gawande – keep the checklist short and include five to nine items only, keep the wording simple and precise, and make sure only the most important items are on the list [9].

The goal of the fifth module "The benefits of standardization" was twofold - to reiterate the importance of establishing standard processes to provide a consistent customer experience and to provide a structure for allowing customized processes. Standardization versus customization is a significant tension for healthcare organizations that are balancing the patient's need for a personalized healthcare experience and their need to operate to a certain standard [10]. Healthcare professionals know that what works best for patients is a hybrid approach [11]. Managers can best support their staff in executing this hybrid approach by clearly identifying the elements of the process that must be done the same way every time by everybody and why, in conjunction with identifying the elements of the process that can be

https://doi.org/10.21872/2023IISE_1676

customized by the staff and are up to their discretion. Participants were asked to do this with a recent standard work they created and challenged to use a metaphor to retrain their staff on that standard work.

The goal of the sixth and final module "Forever aiming higher" was to utilize a process that sets a high bar for staff and encourages them to rise to the occasion. Participants systematically worked through creating an X matrix to link corporate objectives with department improvement initiatives and short-term performance measures [12]. In working through completing the X matrix, participants must understand their organization's true north metrics (TNMs), identify big hairy audacious goals (BHAGs) [13] for their value stream (aka functional area), brainstorm initiatives to accomplish these BHAGs, determine their department-level metrics, and show how all four areas are connected by shading in the intersections between each. See Figure 2.

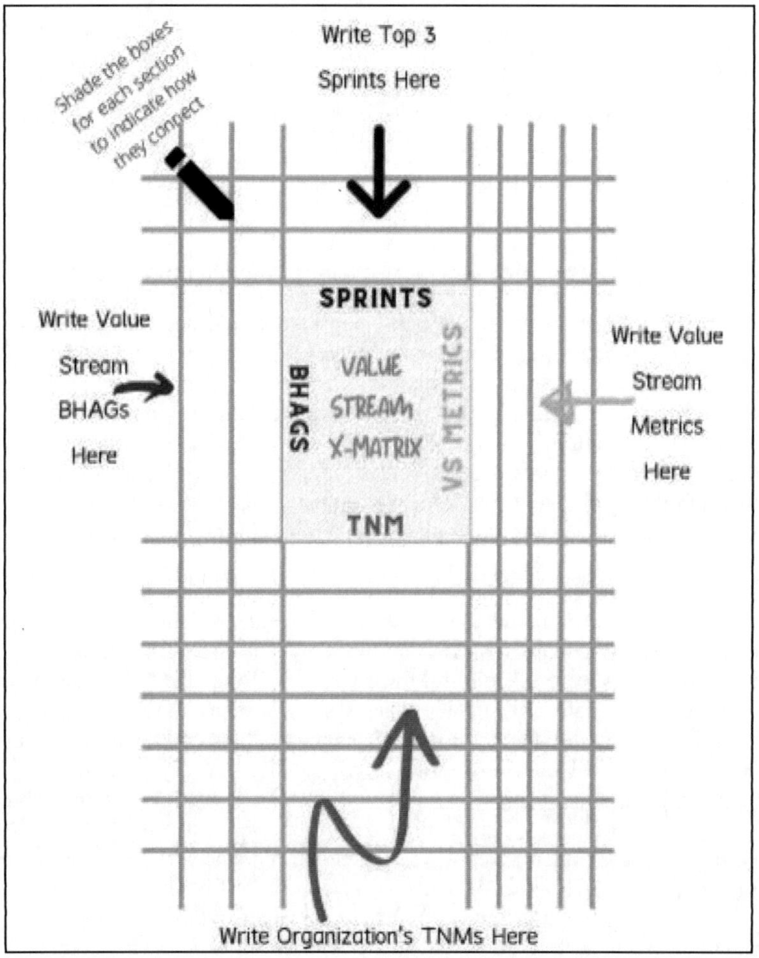

Figure 2: Strategic X Matrix Template

4. Conclusions and Final Thoughts

At the end of each training module, a lessons learned session was performed. Each participant had to identify at least one thing that went well during training and at least one thing that did not go well. This feedback helped shape the structure of subsequent modules and the second session. Not all feedback was relevant to the training course. For example, sometimes participants felt that what did not go well was getting pulled away from the training to address a patient issue. This is sometimes unavoidable in healthcare so for evaluating the training, only relevant feedback is included in Table 2.

Table *2*: Summarized Training Feedback

Training Module	What Went Well	What Did Not Go Well
How to explain continuous improvement	Enjoyed the metaphor teach one activity. Like how the class was interactive and got everyone involved. Metaphors are helpful. Challenging class.	Not enough time. Should have brushed up on Lean concepts prior to training. Brain block on metaphors. I have not used metaphors in ages.
Learn by doing	Good refresher of flowcharts and swim lane diagrams.	I'm stressed about time to complete activities.
Speaking with data	Email reality check. Storytelling exercise was effective. Feel empowered to train, teach, and mentor. Relevant to what I can use often in my position. Check sheets are a straightforward way to show data.	Overwhelmed with homework.
Making it visual	Learned a lot about visual workplaces. Checklists are my jam. Getting back into Gemba walks. It would be great training for all management. Created a checklist I will actually use.	Wanted more time to focus on overall visual management training.
The benefits of standardization	Like the standardization versus customization, helps explain the why so much more. Like how the activity built upon itself. Better understanding of the benefits of standardization.	Struggled with standard work comparison.
Forever aiming higher	Helps focus on how to get good work done with the staff. Loved this series of classes.	Needed more time.

Participants consolidated all their work into a slideshow and presented at a monthly operations meeting. The presentations included a summary of how the training was utilized by each manager. Some examples are as follows:

- Determine a strategic plan for the customer navigation department for the first time
- Train staff effectively on a process for routing reports back to a radiologist for an addendum
- Develop and execute a training plan for adding items correctly to a problem countermeasure board
- Teach staff the value of a Pareto chart and how to create one themselves
- Create an effective metaphor to explain the 8 wastes
- Establish a standard process for radiologist to technologist notes

This training was a first step towards filling the gap identified by the management team. The training course helped managers feel more empowered to teach, train, and mentor. Time constraints were an issue and experimentation with determining the optimal dedicated training time is recommended. The topics covered in this training course would benefit middle managers in any organization dedicated to continuous improvement.

Acknowledgements

The author thanks the management team at Radiology Associates for commissioning the training, participating actively in each module, and providing feedback that effectively shaped each training module.

References

[1] F. Hermkens and . R. A. Georges L., "The Role of Middle Management in Continuous Improvement: The Bermuda Triangle of Leadership, Implementation and Behavioral Change," *Journal of Management Policies and Practices,* vol. 8, no. 1, pp. 24-35, 2020.

[2] L. Perry, "Effective facilitators - a key element in successful continuous improvement processes," *Training for Quality,* vol. 03, no. 4, pp. 9-14, 1995.

[3] Z. Jaser, "The Real Value of Middle Managers," Harvard Business Review, 2021.

[4] "Continuous Improvement," Encyclopedia of Management.

[5] S. Milano, *How to Build Muscle Memory,* Livestrong.com.

[6] ASQ.org, "Check Sheet," [Online]. Available: https://asq.org/QUALITY-RESOURCES/CHECK-SHEET. [Accessed 5 February 2023].

[7] A. Ojo and B. Heravi, "Patterns in Award Winning Data Storytelling," *Digital Journalism,* vol. 6, no. 6, pp. 693-718, 2018.

[8] G. Galsworth, Visual Workplace Visual Thinking: Creating Enterprise Excellence Through the Technologies of the Visual Workplace, 2nd ed., Productivity Press, 2017.

[9] A. Gawande, The Checklist Manifesto: How to Get Things Right, Picador, 2011.

[10] T. J. Vogus, A. Gallan, C. Rathert, D. El-Manstrly and A. Strong, "Whose experience is it anyway? Toward a constructive engagement of tensions in patient-centered health care," *Journal of Service Management,* vol. 31, no. 5, pp. 979-1013, 2020.

[11] Wolters Kluwer, "How does standardizing care affect quality? Can standardization and personalized care co-exist?," 9 November 2017. [Online]. Available: https://www.wolterskluwer.com/EN/EXPERT-INSIGHTS/HOW-DOES-STANDARDIZING-CARE-AFFECT-QUALITY. [Accessed 8 February 2023].

[12] A. Chiarini, "Corporate social responsibility strategies using the TQM: Hoshin kanri as an alternative system to the balanced scorecard," *TQM Journal,* vol. 28, no. 3, pp. 360-376, 2016.

[13] J. Collins and J. I. Porras, Built to Last: Successful Habits of Visionary Companies, Harper Business, 1994.

https://doi.org/10.21872/2023IISE_1720

Proceedings of the IISE Annual Conference & Expo 2023
K. Babski-Reeves, B. Eksioglu, D. Hampton, eds.

Integrated Hurricane Relief Logistics and Evacuation Planning under Forecast Uncertainty: A Case Study for Hurricane Florence

Sudhan Bhattarai and Yongjia Song
Clemson University
Clemson, SC, USA

Abstract

In this paper, we study an integrated hurricane relief logistics and evacuation planning (HRLEP) problem. We propose stochastic optimization models and methods that integrate the hurricane relief item pre-positioning problem and the hurricane evacuation planning problem, which are often treated as separate problems in the literature, by incorporating the forecast information as well as the forecast errors (FE). Specifically, we fit historical FE data into an auto-regressive model of order one (AR-1), from which we generate FE realizations to create evacuation demand scenarios. We compare a static decision policy based on the proposed stochastic optimization model with a dynamic policy obtained by applying this model in a rolling-horizon (RH) procedure. We conduct a preliminary numerical experiment based on real-world data to validate the value of stochastic optimization and the value of the dynamic policy based on the RH procedure.

Keywords
Hurricane evacuation and relief logistics, stochastic programming, rolling horizon, auto-regressive model

1. Introduction

Hurricanes are among the deadliest disasters in the US. Hurricane Katrina in 2005 alone caused about 1800 deaths and an estimated 108 billion US dollars worth of infrastructure damage [1]. Population under the threat of a hurricane evacuates to safe shelters because of personal safety concerns, power outages, and flooding [2]. While most evacuees shelter in hotels or resort to families and friends, around 11% evacuees from the potential affected areas, which we refer to as demand points (DPs), are evacuated to the shelter points (SPs) [2]. In the meantime, relief items (such as food, water, and medical kits) are shipped from distribution centers (DCs) to the SPs. Clearly, the operations of evacuation and the relief item pre-positioning at the SPs must be coordinated to ensure the well-being of the evacuees at the shelters. In this paper, we study an integrated hurricane relief logistics and evacuation planning (HRLEP) problem.

Evacuation is a multi-period operation, which typically starts approximately two to three days before the hurricane's predicted landfall time, depending on the hurricane's severity [3]. A timely evacuation plan is crucial to ensure sufficient time for the evacuation operation. Every six hours after the hurricane's formation, the National Hurricane Center (NHC) issues a forecast of the hurricane attributes up to the next five days including the hurricane's projected track (longitude, latitude) and intensity (wind speed). The projected landfall location of an impending hurricane determines the risk zone. Inside the risk zone, the evacuation demand from the DPs, i.e., the number of evacuees, depends on the affected population's behaviors and the hurricane's attributes [4], which we assume to be characterized by the hurricane's intensity and location. In the case of a deterministic hurricane forecast, one may perform a deterministic demand estimation and create a deterministic optimization model for the HRLEP problem. However, the hurricane forecast is imperfect and subject to FEs, which imposes uncertainty on the hurricane's future attributes and hence the demand realization. A deterministic optimization model may be insufficient to address the challenges brought by the various sources of uncertainty in the hurricane event, and one typically resort to tools and models in optimization under uncertainty, such as stochastic optimization.

Given a stochastic process that characterizes the uncertain hurricane attributes over time with a known probability distribution, a fully adaptive multi-stage stochastic programming (MSSP) model can be used to create dynamic decision policies. However, MSSP models are complex and computationally expensive to solve [5]. Two-stage stochastic programming (2SSP) models are more desirable in this situation as they are computationally less demanding. Despite its efficiency, 2SSP models only produce static decisions that may suffer from lack of adaptability. This issue may

https://doi.org/10.21872/2023IISE_1720

be partially addressed by the so-called rolling horizon (RH) approach [6]. There is an extensive set of literature on hurricane evacuation and relief logistics operation, which are typically treated as two separate problems. For example, the pre-positioning of emergency relief items in a logistics network has been studied both in a single-period setting and in a dynamic multi-period setting with uncertainty in the supply and demand of relief items [7, 8]. For evacuation, a dynamic forecast-driven Markov decision process model has been used to determine the evacuation timing decisions [9]. The integrated problem of evacuation and relief logistics planning has not appeared in the literature until recently, where the problem is studied in a single-period setting [10]. There is a lack of research that incorporates evacuation into disaster relief logistics planning in an integrated network under a multi-stage setting.

In this paper, we aim to address this issue by studying both a static policy given by a 2SSP model and a dynamic policy by using the 2SSP model in the RH procedure for multi-stage integrated evacuation and disaster relief logistics planning. In this procedure, we sequentially solve a stochastic look-ahead model (SLAM) at every period, which corresponds to a 2SSP model defined for the remaining of the planning horizon, and only implements the decisions that apply to the current period. The remainder of the paper is structured as follows: Section 2 describes the mathematical formulation of HRLEP problem as a 2SSP model. Section 3 presents the experimental setup, a case study, and the corresponding results. Section 4 summarizes the paper with some concluding remarks.

2. Problem formulation

This section discusses the model formulation for the proposed 2SSP model for the HRLEP problem. We consider a planning horizon of T periods : $\mathcal{T} = \{1,...,T\}$. At any time period $t' \in \mathcal{T}$, we define a SLAM at t' for the remaining planning horizon of $\mathcal{T}' = \{t', t'+1, ..., T\}$. Let I, J, K represent the sets of DPs, SPs, and DCs, respectively, and let S represent the set of hurricane scenarios for 2SSP model. For simplicity, we make the following assumptions in our model: (i) The DCs and SPs can be opened at any time $t \in \mathcal{T}$, which is available for use immediately, and once opened, they will stay open until the end of the planning horizon; (ii) The evacuees evacuated to an SP cannot be moved to other SPs; (iii) We treat the relief items as a single-commodity package that contains all essential items needed by an evacuee. The problem parameters are defined as follows:

D_i^{POP}	The total population (maximum potential demand for evacuation) of DP $i \in I$
q_j/q_k	Capacity for the number of evacuees/relief items at SP $j \in J$ / DC $k \in K$
c_j^F/c_k^F	One-time fixed setup cost of SP $j \in J$ / DC $k \in K$ upon activation
c_k^P	Unit procurement cost of relief items at DC $k \in K$
c_{ij}^E	Unit evacuation cost from DP $i \in I$ to SP $j \in J$
$c_{kj}^R/c_{j'j}^R$	Unit transportation cost of relief items from DC $k \in K$ or SP $j' \in J$ to SP $j \in J$
c^{PE}	Unit penalty cost for not evacuating a person needing evacuation at a period
c_j^G/c_j^H	Unit cost for *emergency*/*unused* relief items shipped to/from SP $j \in J$
c_j^{InvR}/c_k^{InvR}	Unit inventory cost of relief items in SP $j \in J$ / DC $k \in K$ per period
c_j^{InvE}	Unit operating cost of SP $j \in J$ (per evacuee) per period
p_s	Probability of occurrence for scenario s $\in$ S

At each period t', the hurricane's attributes in the remaining planning horizon, $t'+1, ..., T$ are uncertain. To come up with a 2SSP model, we represent the uncertain demand at periods $t \in \{t'+1, ..., T\}$ in the form of scenarios. Each scenario $s \in S$ is a sample path that represents the evolution of a hurricane's attributes from $t'+1$ until T and corresponding demand realization vector $\{[D_{is}^{t'+1}]_{i \in I}, ..., [D_{is}^{T}]_{i \in I}\}$, which represents the fraction of the remaining population at $i \in I$ to evacuate each time t. For simplicity, the demand realization is computed via a deterministic mapping from the realization of the hurricane's attributes (see Section 3.1 for details). This captures the empirical behavior analysis results for hurricane evacuation [4].

The first-stage decisions are defined as follows:

z_j^t/z_k^t	Whether or not $j \in J/k \in K$ is open during time period $t \in \mathcal{T}'$
ℓ_j^t/ℓ_k^t	Inventory level of relief items at $j \in J/k \in K$ at the end of period $t \in \mathcal{T}'$
$y_{ij}^{t'}$	Number of people to evacuate from $i \in I$ to $j \in J$ at time t'
$u_i^{t'}$	Number of people whose evacuation need/demand is not met at $i \in I$ at time t'
$e_i^{t'}/e_j^{t'}$	Number of evacuees at $i \in I/j \in J$ at the end of period t'
$x_k^{t'}$	Amount of relief items procured at $k \in K$ in time period t'
$x_{kj}^{t'}/x_{j'j}^{t'}$	Amount of relief items shipped from $k \in K/j' \in J \setminus \{j\}$ to SP $j \in J$ at time t'

https://doi.org/10.21872/2023IISE_1720

$g_j^{t'}/h_j^{t'}$ Amount of *emergency/unused* relief items shipped to/from $j \in J$ in time t'

The first-stage decisions can be categorized by *local* variables and *state* variables. Specifically, z_j^t, z_k^t, ℓ_j^t, and ℓ_k^t are the state variables that characterize the state of the logistics system over time. All other first-stage variables are local variables for period t' only.

The second-stage decision variables are scenario-dependent, which are defined as follows:

y_{ijs}^t Number of people to evacuate from $i \in I$ to $j \in J$ at time $t \in \mathcal{T}' \setminus \{t'\}$ in $s \in S$

u_{is}^t Number of people whose evacuation need/demand is not met at $i \in I$ at time $t \in \mathcal{T}' \setminus \{t'\}$ in $s \in S$

e_{is}^t/e_{js}^t Number of evacuees at $i \in I / j \in J$ at the end of period $t \in \mathcal{T}'$ in $s \in S$

x_{ks}^t Amount of relief items procured at $k \in K$ in time period $t \in \mathcal{T}' \setminus \{t'\}$ in scenario $s \in S$

$x_{kjs}^t/x_{j'js}^t$ Number of relief items shipped from $k \in K / j' \in J \setminus \{j\}$ to $j \in J$ at time $t \in \mathcal{T}' \setminus \{t'\}$ in $s \in S$

g_{js}^t/h_{js}^t Amount of *emergency/unused* relief items shipped to/from $j \in J$ in time $t \in \mathcal{T}' \setminus \{t'\}$ in scenario $s \in S$

At time t', the values of the state variables from $t'-1$, i.e., $\hat{\ell}^{t'-1}$ and $\hat{z}^{t'-1}$, become input parameters to the SLAM. At $t'=1$, we assume the initial conditions are $\ell_j^0, z_j^0, e_j^0 = 0,\ \forall j \in J,\ \ell_k^0, z_k^0 = 0,\ \forall k \in K$, and $e_i^0 = D_i^{POP},\ \forall i \in I$. The SLAM at t' is a 2SSP model that combines the first-and second-stage models defined as follows:

The first-stage model

$$\min \sum_{j\in J} c_j^F \left(z_j^T - \hat{z}_j^{t'-1}\right) + \sum_{k\in K} c_k^F \left(z_k^T - \hat{z}_k^{t'-1}\right) + \sum_{k\in K} c_k^P x_k^{t'} + \sum_{j\in J}\sum_{j'\in J} c_{jj'}^R x_{jj'}^{t'} + \sum_{j\in J}\sum_{k\in K} c_{kj}^R x_{kj}^{t'} + \sum_{i\in I}\sum_{j\in J} c_{ij}^E y_{ij}^{t'} +$$
$$\sum_{j\in J} c_j^G g_j^{t'} + \sum_{j\in J} c_j^H h_j^{t'} + c_{t'}^{PE} \sum_{i\in I} u_i^{t'} + \sum_{j\in J} c_j^{invE} e_j^{t'} + \sum_{t\in\mathcal{T}'}\sum_{j\in J} c_j^{InvR} \ell_j^t + \sum_{t\in\mathcal{T}'}\sum_{k\in K} c_k^{InvR} \ell_k^t \tag{1a}$$

$$\text{s.t. } \hat{z}_k^{t'-1} \le z_k^{t'},\ \forall k \in K;\ z_k^{t-1} \le z_k^t,\ \forall k \in K, \forall t \in \mathcal{T}' \setminus \{t'\} \tag{1b}$$
$$\hat{z}_j^{t'-1} \le z_j^{t'},\ \forall j \in J;\ z_j^{t-1} \le z_j^t,\ \forall j \in J, \forall t \in \mathcal{T}' \setminus \{t'\} \tag{1c}$$
$$\ell_k^t \le q_k z_k^t,\ \forall k \in K, \forall t \in \mathcal{T}' \tag{1d}$$
$$\ell_j^t \le q_j z_j^t,\ \forall j \in J, \forall t \in \mathcal{T}' \tag{1e}$$
$$\sum_{j\in J} y_{ij}^{t'} + u_i^{t'} = D_i^{t'} \hat{e}_i^{t'-1},\ \forall i \in I \tag{1f}$$
$$e_i^{t'} = \hat{e}_i^{t'-1} - \sum_{j\in J} y_{ij}^{t'},\ \forall i \in I \tag{1g}$$
$$\sum_{i\in I} y_{ij}^{t'} + \hat{e}_j^{t'-1} = e_j^{t'},\ \forall j \in J \tag{1h}$$
$$e_j^{t'} \le q_j z_j^{t'},\ \forall j \in J \tag{1i}$$
$$\hat{\ell}_j^{t'-1} + \sum_{k\in K} x_{kj}^{t'} + \sum_{j'\in J, j\neq j'} x_{j'j}^{t'} - \sum_{j'\in J, j\neq j'} x_{jj'}^{t'} - e_j^{t'} + g_j^{t'} - h_j^{t'} = \ell_j^{t'},\ \forall j \in J \tag{1j}$$
$$\sum_{j'\in J, j\neq j'} x_{jj'}^{t'} \le \hat{\ell}_j^{t'-1} + g_j^{t'} - e_j^{t'},\ \forall j \in J \tag{1k}$$
$$\hat{\ell}_j^{t'-1} + g_j^{t'} - e_j^{t'} \le q_j z_j^{t'},\ \forall j \in J \tag{1l}$$
$$\sum_{j\in J} x_{kj}^{t'} \le \hat{\ell}_k^{t'-1},\ \forall k \in K \tag{1m}$$
$$\ell_k^{t'} = \hat{\ell}_k^{t'-1} - \sum_{j\in J} x_{kj}^{t'} + x_k^{t'},\ \forall k \in K \tag{1n}$$
$$z_j^t \in \{0,1\},\ \forall j \in J,\ \forall t \in \mathcal{T}';\ z_k^t \in \{0,1\}, \forall k \in K, \forall t \in \mathcal{T}' \tag{1o}$$
$$\ell_k^t \ge 0,\ \forall k \in K,\ \forall t \in \mathcal{T}';\ \ell_j^t \ge 0,\ \forall j \in J,\ \forall t \in \mathcal{T}';\ u_i^{t'}, e_i^{t'} \ge 0,\ \forall i \in I;\ e_j^{t'}, h_j^{t'}, g_j^{t'} \ge 0,\ \forall j \in J;\ x_k^{t'} \ge 0,\ \forall k \in K \tag{1p}$$
$$y_{ij}^{t'} \ge 0,\ \forall i \in I,\ \forall j \in J;\ x_{kj}^{t'} \ge 0,\ \forall k \in K, \forall j \in J;\ x_{jj'}^{t'} \ge 0,\ \forall j, j' \in J : j \neq j' \tag{1q}$$

The second-stage model

$$\min \sum_{s\in S} p_s \sum_{t\in\{t'+1,\dots,T\}} \left(\sum_{i\in I}\sum_{j\in J} c_{ij}^E y_{ijs}^t + \sum_{j\in J} c_j^{invE} e_{js}^t + \sum_{k\in K} c_k^P x_{ks}^t + \sum_{j\in J}\sum_{j'\in J} c_{jj'}^R x_{jj's}^t + \sum_{j\in J}\sum_{k\in K} c_{kj}^R x_{kjs}^t + \right.$$
$$\left. \sum_{j\in J} c_j^G g_{js}^t + \sum_{j\in J} c_j^H h_{js}^t + \sum_{i\in I} c_t^{PE} u_{is}^t \right) \tag{2a}$$
$$\text{s.t. } e_{is}^{t'} = e_i^{t'},\ \forall i \in I, \forall s \in S \tag{2b}$$
$$e_{js}^{t'} = e_j^{t'},\ \forall j \in J, \forall s \in S \tag{2c}$$

https://doi.org/10.21872/2023IISE_1720

Constraints similar to (1f) – (1n), $\forall t \in \{t'+1, ..., T\}$, $\forall s \in S$ with the first-stage variables replaced by their second-stage copies

The first two terms of the first-stage objective (1a) represent the fixed cost of opening SPs and DCs, respectively. The third term represents the total cost of procuring relief items at DCs. The fourth and fifth terms represent the cost of transporting relief items among SPs and from DCs to SPs. The last two terms represent the inventory holding costs of relief items at DCs and SPs. Constraints (1b) and (1c) ensure that the DCs and SPs stay open until the end of periods once they are decided to be open. Equations (1d), (1l), and (1m) are the inventory and supply capacity constraints of the DCs. Similarly, constraints (1e) and (1k) represent SPs' inventory and supply capacities. Constraints (1n) and (1j) are the balance constraints for the supply and inventory of relief items at DCs and SPs, respectively. Constraint (1f) ensures demand satisfaction, while constraint (1g) represents the number of evacuees at DPs. Constraint (1h) represents the number of evacuees at SPs, and constraint (1i) is the capacity constraint for the number of evacuees at SPs. Objective (2a) is the expected value of second-stage cost from period $t'+1$ until the hurricane's landfall at T. We use the standard Benders decomposition method to solve 2SSP models by treating the first-stage and the second-stage models as the master and subproblems, respectively.

2.1 The rolling horizon approach

A 2SSP model at $t = 1$ can only provide a static decision policy. In this model, the logistics and evacuation decisions $\forall t \in \mathcal{T}$ are made at $t = 1$ that depend on the point forecast at $t = 1$ and the hurricane attributes' possible future realizations. Since a hurricane's attributes and forecasts change over time, a static model fails to adapt to the dynamic evolution of the hurricane's information. The rolling horizon approach naturally generates a dynamic policy in an online fashion where a SLAM is solved at every period, but only the decisions that apply to the current period are implemented. At every period $t' \in \mathcal{T}'$, the SLAM at t' utilizes the actual realization at t' and the most updated forecast available. Let $X_{t'}$ and $Y_{t'}$ be the state and local variables of SLAM at t', and after we solve the SLAM at t', we implement an optimal solution $\hat{X}_{t'}$ and $\hat{Y}_{t'}$ and compute the immediate cost $Z^*_{t'}(\hat{X}_{t'}, \hat{Y}_{t'})$ for the current stage t'. The values of the state variables $\hat{X}_{t'}$ are used as input parameters for the SLAM at $t'+1$ as we roll forward to the next stage. After solving SLAM for all $t' \in \mathcal{T}$ over a sample path, we compute the cumulative cost of the rolling-horizon policy on this sample path, $Z^*_{RH} = \sum_{t' \in \mathcal{T}} Z^*_{t'}(\hat{X}_{t'}, \hat{Y}_{t'})$.

3. Experimental setup and results

In this section, we discuss our experiment setup and some preliminary computational results. We start by discussing how we model the forecast uncertainty. Our stochastic model for the forecast error (FE) is constructed based on data available from the NHC's forecast verification report [11], which is released each year at the end of every hurricane season, including both the track FE and the intensity FE. Our preliminary calculation indicates that the Pearson correlation coefficient of FE between two subsequent periods is greater than 0.7, which motivates us to model the track and intensity FE data using auto-regressive models of order 1 (AR-1). Let ξ^t represent the FE at period t, then according to the AR-1 model: $\xi^t = \rho\xi^{t-1} + \varepsilon^t$, where $\varepsilon^t \overset{\text{iid}}{\sim} N(0, \sigma^2)$, $\forall t \in \mathcal{T} \setminus \{1\}$ and ξ^1 is deterministic. To validate the model assumption, we provide the Q-Q plots of historical intensity and track FE at 24 hours prior to landfall in Figures 1a and 1b. The Q-Q plot supports the normality assumption of intensity FE. In contrast, the non-normal distribution of the track FE is due to heavy tails observed. To address this, we transform the track FE data using a log transformation, and we see that the normality assumption is supported by the Q-Q plot in Figure 1c for the transformed data.

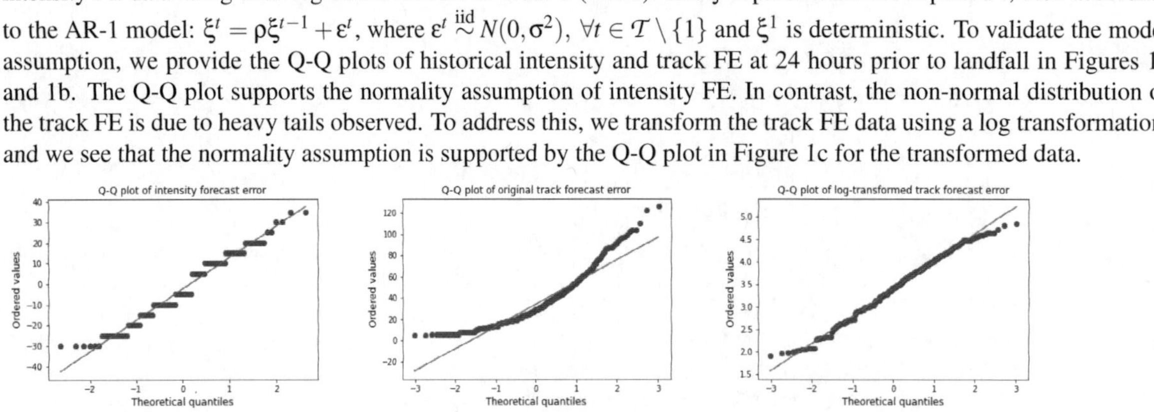

(a) Intensity forecast error (b) Original track forecast error (c) Transformed track forecast error

Figure 1: Q-Q plot of historical hurricane forecast errors at 24 hours prior to the hurricane's landfall

We estimate parameters (ρ, σ) in the AR-1 model using maximum likelihood estimation (MLE), as $(\rho_{mle}, \sigma_{mle})$ based on the NHC's official FE data of the storms from 2017 to 2021 [11]. For a given realization of ξ^{t-1}, ρ_{mle}, and σ^2_{mle}, a

https://doi.org/10.21872/2023IISE_1720

realization of ξ^t can be obtained by sampling $\varepsilon^t \sim N(0, \sigma^2_{mle})$. We thus randomly sample a number of realizations of $\varepsilon^t, \forall t \in \mathcal{T}$ to create an ε-grid for track and intensity errors. To generate a scenario (sample path), for each $t = 1, 2, \ldots, T$, we randomly pick a realization of ε^t from the tth column vector of ε-grid to compute ξ^t.

3.1 Case study

Our case study focuses on Hurricane Florence 2018 in South Carolina (SC), assuming the coastal line to be straight for simplicity. We define the study region as a 200-miles area around the endpoints of the coastal line, encompassing eight coastal SC counties with high hurricane vulnerability as the Demand Points (DPs). We represent the DPs' locations by projecting the respective counties' central latitudes (lats) and longitudes (longs) onto the coastal line. The SC Emergency Management Division (SCEMD) divides the state of SC into four hurricane regions. We select 11 hurricane sub-regions with high vulnerability in the north, central, and south regions as our candidate SPs, with their central locations and aggregated shelter capacities. Additionally, we choose 12 counties in the western hurricane region with their central locations (lats, longs) as our DCs.

We consider a planning horizon of 10 periods, each being 12 hours long, starting at 0-hour (t_1) to 120-hour (t_{11}). At t_1, the hurricane's track and intensity forecast from t_2 until t_{11} is given by NHC's official (point) forecast. We assume that the hurricane makes landfall deterministically at t_{11}. The track FE at every scenario represents the deviation between the realized hurricane's position and the point forecast, along the axis of the coastal line. Figure 2 presents the locations and a set of sample hurricane tracks. We set D_i^{POP}, $\forall i \in I$ to be 5% of the vulnerable population of the respective DPs [2].

We assume that the demand at each DP at any period is a function of the hurricane's location and intensity. To represent the hurricane's track locations, we define the x- and y- axes by the coastal line (which is assumed to be a straight line) and its normal. A positive demand for DP $i \in I$ at a period $t \in \mathcal{T}$ will incur if its location is within certain threshold values in both the x- and y- axes, (x_{max}, y_{max}), from the hurricane's location, i.e., $D_{is}^t = \min\left\{\frac{int_s^t}{int_{\max}}, 1\right\} \times \max\{1 - \frac{dx_i^{t,s}}{x_{max}}, 0\} \times \max\{1 - \frac{dy_i^{t,s}}{y_{max}}, 0\}$ where $int_{\max}$ is the maximum intensity of a hurricane considered, int_s^t is the hurricane's intensity at period t on sample path s, and $dx_i^{t,s}$ and $dy_i^{t,s}$ are the distances between DP i and the hurricane's location at period t on sample path s, respectively.

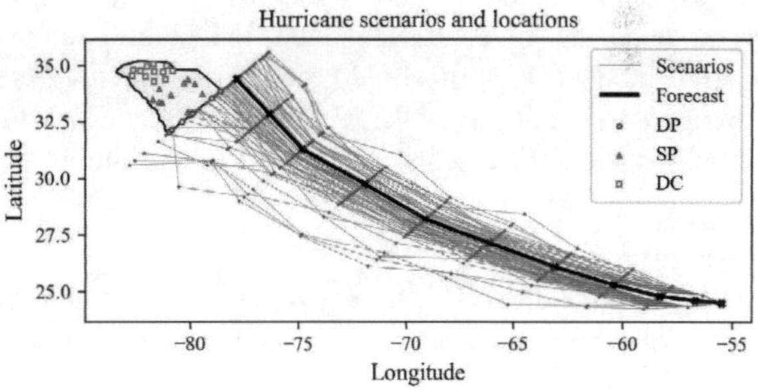

Figure 2: Location of DPs, SPs, DCs and hurricane scenarios

3.2 Out-of-sample evaluation

The out-of-sample (OOS) evaluation for the RH policy is straightforward. To conduct an OOS evaluation on the static 2SSP policy for sample path s^o, we first solve the 2SSP model at $t = 1$. Then for each period $t \in \{2, ..., T\}$, we pick the recourse solution of scenario s closest to s^o among all scenarios used in the 2SSP model at $t = 1$ in terms of the DP demand vector. Using the number of evacuees $\hat{y}^t_{ijs}$ as an upper bound for the number of evacuees on arc (i, j), we solve a deterministic optimization problem to get the values of other local variables.

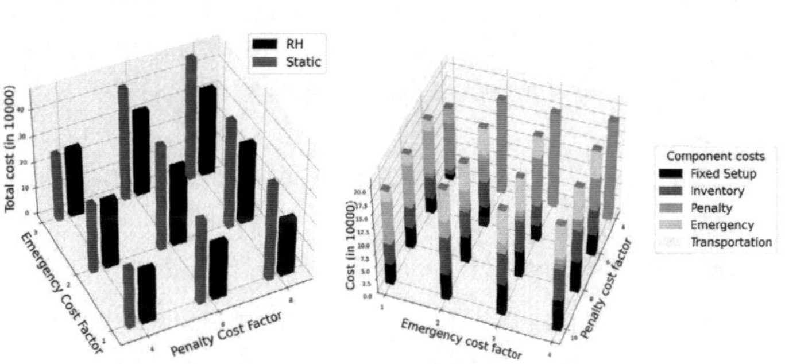

(a) OOS evaluation (b) Sensitivity analysis of the 2SSP model

Figure 3: OOS evaluation and sensitivity analysis on various cost factors.

3.3 Results

We now present numerical results from our preliminary experiments. Firstly, we pick $S = 50$ to be the number of scenarios to use through an in-sample stability test. Furthermore, Figure 3a presents the average OOS cost over 50 replications of both the RH and static policy for different cost factors, G_{fact} and P_{fact}. The RH model exhibits an average total cost that is 13% lower than the static policy, highlighting the benefits of adaptive decision making. In contrast, the static 2SSP model is limited in its ability to evacuate beyond a certain number of evacuees as it must satisfy demand on a sample path with the static solution obtained without the sample-path information. As c^{PE} increases, the static policy results in higher penalty costs and fewer evacuations. Moreover, the rate of change of the cost difference for RH and static 2SSP is less for the rate of change of c^G than c^{PE}. This indicates that the cost of the static 2SSP model on the evacuation side is more significant than on the relief logistics side. Thus, the static policy, when applied to an OOS, results in higher penalty costs and fewer evacuations. In contrast, RH adapts to demand realizations in an online fashion, leading to lower overall costs.

Figure 3b demonstrates the sensitivity analysis of a 2SSP model for different cost factors G_{fact} and P_{fact}, which controls the values of c^G and c^{PE}, respectively. Specifically, the penalty cost dominates among all cost components for smaller values of c^{PE}. As c^{PE} increases, the total penalty cost decreases, and evacuation and relief supplies play a more significant role in fulfilling demand. It is important to note that emergency costs only occur after the SPs are activated to provide relief items to evacuees. In contrast, penalty costs can be incurred independently of all other costs. Furthermore, emergency costs become significant only for small c^G values but become negligible for higher c^G values, as relief logistics are relatively less expensive in the latter case. Thus, decision-makers can utilize this trade-off analysis to select appropriate values of c^{PE} and c^G based on their impact on the overall cost structure.

4. Conclusion

This paper considered the use of dynamically evolving uncertain hurricane attributes in an integrated HRLEP problem. We modeled and solved the integrated HRLEP problem using a *static* 2SSP model and an *online* RH approach. We fit an AR-1 model using historical FE data, from which we created evacuation demand scenarios for the 2SSP model. We demonstrated the benefit of the RH approach over a static 2SSP model with an out-of-sample test. For future research, we will consider the case where the hurricane's landfall time is random, using a multi-stage SP model.

References

[1] Kates, R. ,Colten, C. , Laska, S., and Leatherman, S., 2006, "Reconstruction of New Orleans after Hurricane Katrina: a research perspective," Proceedings of the National Academy of Sciences, 103(40), 14653-14660.

[2] Centers for Disease Control and Prevention (CDC), "Sheltering During Hurricanes," [Online] Available: https://www.cdc.gov/nceh/hsb/disaster/sheltering-during-hurricanes.html [Accessed: Jan. 30, 2023].

[3] South Carolina Emergency Management Division (SCEMD), "South Carolina Hurricane Plan," [Online]. Available: https://www.scemd.org/em-professionals/plans/south-carolina-hurricane-plan. [Accessed: Jan. 23, 2023].

[4] Baker, E., 1991, "Hurricane evacuation behavior," International Journal of Mass Emergencies & Disasters, 9(2), 287-310.

[5] Shapiro, A., 2021, "Tutorial on risk neutral, distributionally robust and risk averse multistage stochastic programming," European Journal of Operational Research, 288(1), 1-13.

[6] Fattahi, M. and Govindan, K., 2022, "Data-driven rolling horizon approach for dynamic design of supply chain distribution networks under disruption and demand uncertainty," Decision Sciences, 53(1), 150-180.

[7] Lodree Jr, E. J., Ballard, K. N., and Song, C. H., 2012, "Pre-positioning hurricane supplies in a commercial supply chain," Socio-Economic Planning Sciences, 46(4), 291-305.

[8] Alem, D., Clark, A., and Moreno, A., 2016, "Stochastic network models for logistics planning in disaster relief," European Journal of Operational Research, 255(1), 187-206.

[9] Regnier, E., and Harr, P. A., 2006, "A dynamic decision model applied to hurricane landfall," Weather and Forecasting, 21(5), 764-780.

[10] Dalal, J., and Üster, H., 2018, "Combining worst case and average case considerations in an integrated emergency response network design problem," Transportation Science, 52(1), 171-188.

[11] "National hurricane center and central pacific hurricane center, 2022, "National Hurricane Center Forecast Verification," [Online]. Available: https://www.nhc.noaa.gov/verification/verify7.shtml. [Accessed: Feb. 4, 2023].

https://doi.org/10.21872/2023IISE_1735

Proceedings of the IISE Annual Conference & Expo 2023
K. Babski-Reeves, B. Eksioglu, D. Hampton, eds.

Evaluating the efficacy of a warning system for a smartphone-based application using the FaceReader software

Abstract ID: 1735

Anika Rimu, Sudipta Sharif, Jaivardhan Sood, Ankit Dubey, Ashutosh Nagargoje and Shuchisnigdha Deb,
The University of Texas at Arlington, Arlington, TX

Abstract

Cyclists are considered vulnerable road users as they are prone to suffering fatal consequences in motor vehicle crashes. While many researchers focused on driver assistance technology, there is little to no research on assisting cyclists using modern technologies. This research tested different warnings for a smartphone-based collision detection and alerting app. The objective is to alert cyclists about an impending collision, ahead of time, without causing further distractions. A bicycle-simulator experiment was designed around two collision scenarios: the *right-hook* collision, a car turning right in front of a cyclist, and the *frontal collision*, a car brakes abruptly in front of a cyclist. The environmental factors involved time of the day, vehicle speed, and traffic volume. Four multimodal signals and the no-signal condition (control), selected from a preliminary study, were tested to determine the efficient warning/s in terms of cyclists' interaction with the app-based alerts and their attention to environmental details. FaceReader software analyzed emotional expressions and provided the *surprise, arousal score,* and *valence of surprise* along with participants' gaze direction (N= 12). Results show that the combination of speech and flashing signs had lower surprise scores and positive valence of surprise compared to the combination of tone and flashing signs. However, participants' gaze was fixated on the traffic environment for tones, but they immediately shifted towards the phone or surroundings for speech alerts. This study addresses a significant design issue regarding road-users' distraction for focusing on their attention to traffic environment while interacting with and reacting to assistive technologies.

Keywords
Bicyclist Assistance System, assistive technology, FaceReader, emotional expressions, distraction.

1. Introduction

Cycling, despite being one of the most beneficial modes of transportation in the world, continues to be a major public health concern with increasing fatalities. In the United States, 938 cyclists were killed in traffic crashes in 2020 [1]. Bicyclists and pedestrians are considered vulnerable road-users (VRUs) as they are not protected by physical structures of motor vehicle [2]. VRUs have seen an alarming rate of crashes in the recent years. Especially, bicyclists have an increase in fatality rate by 2.3% between 2017 and 2018 and 9% between 2019 and 2020 [1], [3], [4]. These statistics indicate the need to assist cyclists on the road with assistive and affordable technology that will guide them avoiding collisions, instead of distracting them. Assisting cyclists on the road comes with challenges as they need to be very attentive to their surroundings. The alert system needs to be efficient in conveying the danger to the cyclist in a very short period of time. An alarm system cannot be overwhelming for the cyclist, as over-alertness can be detrimental to the user [5], [6]. This study is focused on developing a warning system for the smartphone-based Bicyclist Assistance System (BAS) application, that can detect *right-hook collision* and *frontal collision* conditions and alert cyclists ahead of time [7].

In order to determine the ideal and non-distractive warning system, along with users' subjective perceptions and their reaction time to the approaching collision situation, it is important to compare the emotions generated by each warning signal [8]. There is not a lot of research done on using emotional data of cyclists to understand their distraction while using assistive technology. Since they are experiencing unexpected stimuli while cycling, it is anticipated that it would instigate surprise and would cause distraction [9]. Once the user processes the stimuli, the resulting response incorporates the valence of surprise i.e., positive or a negative surprise [10]. Thus, surprise is defined as an interruption to the ongoing thoughts and activities [11]. Valence of surprise is associated with the degree to which the stimulus is perceived i.e., either people become happy with a positive surprise or disappointed with a negative surprise [9], [11]. In a study to determine the usability of different interfaces inside a vehicle, researchers developed a usability index system consisting of effectiveness, efficiency, and satisfaction. The satisfaction index considered valence and arousal

https://doi.org/10.21872/2023IISE_1735

values [8]. They correlated both emotions positively with usability. Their result indicated that the facial data was consistent with their subjective scores, however, their further analysis also indicated that indexes of velocity and valence may not reflect the usability effectively. In the context of this study, arousal will be used to characterize situational awareness i.e., higher arousal would indicate higher awareness. In another study, authors used facial expression data to understand emotion of bicyclists during a downhill mountain bike descent [12]. Consistent with their hypotheses and previous studies, they found the cyclists to exhibit less happiness while riding compared to before or after their rides. However, they also found anger and disgust among the cyclists which they did not anticipate. Both studies indicate the lack of research in using emotional data for road users, especially bicyclists, which leads to unanticipated results. It is important to understand distraction from facial data. For instance, in one study, the authors monitored driver's visual and cognitive distractions by using facial features [13]. They found gaze features and action units i.e., the muscle activity that generates different facial expressions, to play an important role in detecting distraction. Another study used visual behavior of cyclists to detect distraction while interacting with mobile phones in a self-paced vs system-paced tasks in traffic [14]. Results based on gaze data showed cyclists to have more attention to the traffic while performing system-paced tasks as opposed to self-paced tasks. This indicates that having a warning app to assist cyclists might not cause significant distraction.

The purpose of this paper is to identify the ideal waring system for the BAS based on the analysis of the cyclist's facial expression and gaze direction. BAS is a smartphone-based application that can alert cyclists only using smartphone and commercial off-the-shelf speaker for collision detection. This study will determine the ideal warning system under different conditions based on speed, traffic volume, time of the day, and type of warning signal. Arousal, surprise, valence of surprise, anger and disgust will be considered based on previous research. The facial expression will help identify which warning causes the least surprise, positive valence of surprise, higher arousal, and less anger and disgust. The gaze direction will identify which warning signal causes the least distraction. It is hypothesized for this study that warnings consisting of speech will have a positive valence of surprise, however, they will also cause high distraction.

2. Method

2.1 Study Design

The research team developed a bicycle simulator-based study to identify the ideal warning signal under different environmental conditions, such as day vs night, light vs heavy traffic volume, low vs high speed, and *frontal* vs *right-hook* collision. A phone was mounted on the bike that had a dummy app developed to generate the warnings. Each participant had a total of 20 trials in random order. The participants (n=12) had to complete a survey including their demographic information, as well as their likeliness of the warning system. The entire study took about an hour. Their video data for this experiment was collected using a video camera and data was analyzed using FacerReader 9.0 by Nodulus Information Technology. The study was approved by the Institutional Review Board (IRB) at the University of Texas at Arlington (UTA).

2.2 Warning Signal

The warning signals were selected from a preliminary focus group study with stakeholders and researchers [15]. The warning signals identified from the study were multimodal signals with a combination of visual and auditory signals. For visual, red and yellow flashing circle were identified, while for auditory, high pitch tone and a speech saying "Slow Down" were identified (Table 1).

Table 1: Warning signals used in the study

Visual	Audible	
	Tone	Speech
	Four repeated high pitch tone	Slow Down!

2.3 Participants

The participants were recruited using the word of mouth. The majority of the participants were from UTA student community. For the FaceReader analysis, data from the participants who didn't wear glasses and had their face clearly

https://doi.org/10.21872/2023IISE_1735

captured by the camera, and had most data recognized by the FaceReader software were used. A total of 12 participants data were considered for this study.

2.3 Experiment Protocol

Participants were screened according to their frequency of cycling before they were invited to participate. Participants who rode bicycles at least once a week were recruited for this study. Once the participants came to the lab on the day of the experiment, the researchers first went over the experimental process and gave the participants the consent form. When the participant signed the form, they were asked to ride the bicycle for three minutes for exposure to the virtual environment. Figure 2 shows the bicycle setup and the app interface. After they were well accustomed to the simulated environment, participants completed 20 randomized trials with different factors and a combination of red or yellow visual signal and speech or tone audible signal, and no signal warnings. The participants received a five-minute break after the first ten trials. To ensure that participants do not experience any simulation sickness or any other discomfort, they were asked to fill out a Simulation Sickness Questionnaire (SSQ) to assess their fitness to virtual exposure. Participants were asked to complete a demographic survey, as well as a rating survey to express their likeliness of the app. They were also asked to express their opinion verbally for better understanding of their perception.

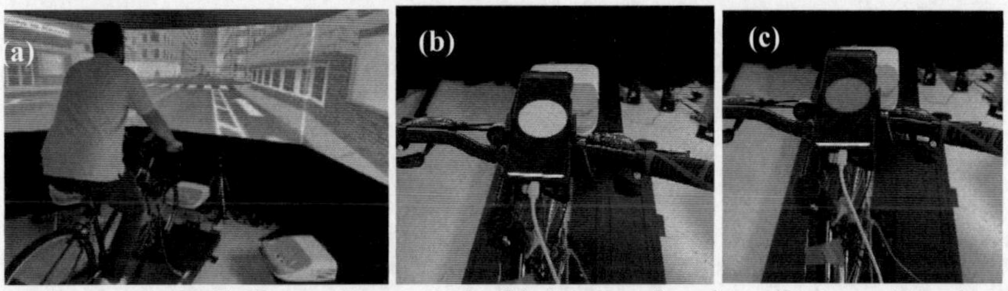

Figure 2: (a) Bicycle setup at UTA Human Factors Lab. (b) App interface for yellow-tone/speech warning. (c) App interface for red-tone/speech warning.

2.3 Data Analysis

The video data of 12 participants were first edited to crop out the surrounding and have more focus on the faces. The brightness and contrast of the videos were also adjusted. The videos were then played on the FaceReader software. The results were analyzed using SPSS and Python. Analysis of Covariance (ANCOVA) was used for statistical analysis.

3. Results and Discussions

3.1. Emotion Data Analysis

For all the emotion variables i.e., arousal, surprise, valence of surprise, and disgust, ANCOVA was performed to analyze the main effect and interaction effects of the factors. Gender was added as a categorical variable in the test. Only *type of warning* had a significant effect with all the emotion variables ($p<0.001$). This is an indicator that the surrounding traffic did not influence how the participants perceived the warning signals. The following sections analyze each emotional variable with respect to each warning signal. There was no significant effect of gender was found.

Arousal with Warning Signals

Participants had a higher average arousal score in the presence of any warning signal compared to when no signal is present *(M*=0.220, *SD*=0.031*)* (Figure 3). Participants had the highest arousal score in the presence of yellow-speech combination warning *(M*=0.424, *SD*=0.039*)*. This result is consistent with another study where researchers compared different warning types, such as verbal (speech), auditory icon and abstract warnings. They found verbal signals to have the highest situational awareness compared to the other two [16]. Overall, all the signals regardless of speech or tone, had a higher arousal than without any warning signal present.

Surprise and valence of surprise with warning signals

The average surprise score was the lowest for red-speech warning (*M*=0.022, *SD*=0.005) (Figure 4). It also had the most positive valence of surprise (*M*=-0.014, *SD*=0.005). The yellow-tone warning had the highest surprise score (*M*=0.050, *SD*=0.020), however, the valence of surprise was more positive than the remaining two signals (*M*=0.024, *SD*=0.003). In previous studies, researchers have found that participants preferred speech-based warning signals to tone-based signals [17], [18]. McKeown [18] compared abstract, environmental, icon-based, and speech-based audible

https://doi.org/10.21872/2023IISE_1735

warnings based on their ability to convey urgency. Participants were divided into four groups for each warning and were asked to determine the urgency of the scenario based on the warning. They found participants to prefer speech-based warnings. They also found speech warning to have consistent level of urgency perceived by the participants. However, they concluded by saying that despite being less pleasant, abstract tone warnings can convey the urgency better in an emergency as opposed to speech warnings. Similarly, Graham [17], found tones, especially auditory icons that can describe the situation without visual warnings, to have the shortest reaction time and most preferred by the participants.

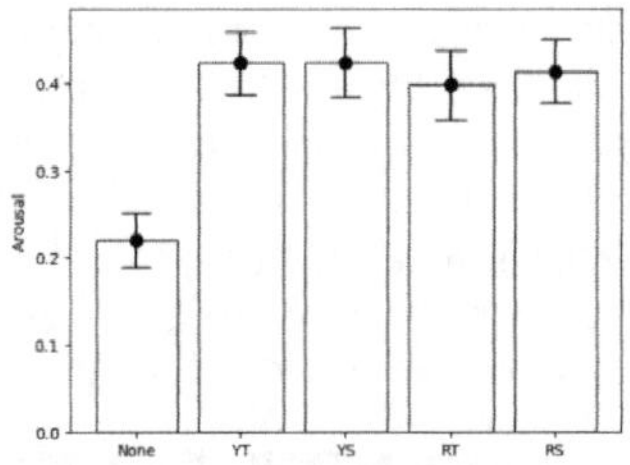

Figure 3: Average arousal score for warning signals.
Note: YT=yellow-tone, YS=yellow-speech, RT=red-tone, RS=red-speech

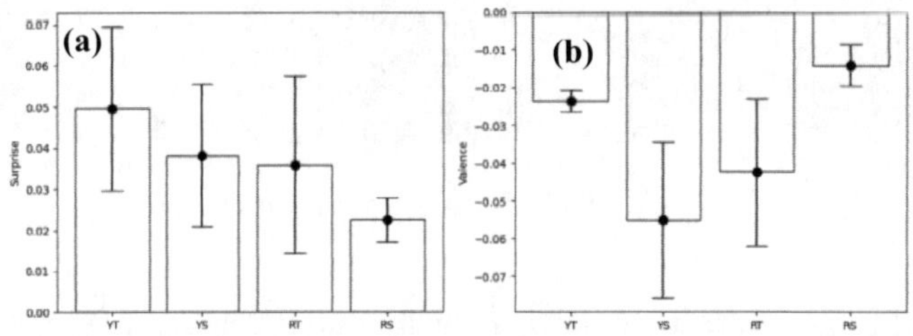

Figure 4: Average surprise (a) and valence of surprise score (b).
Note: YT=yellow-tone, YS=yellow-speech, RT=red-tone, RS=red-speech

Disgust with Warning Signals
In previous studies, disgust and anger had unanticipated effects [12]. As a result, the effect of both emotions was assessed in this study. While anger didn't have any difference across all warning signals, red-tone warning showed the lowest score for disgust (M=0.053, SD=0.010). Yellow-speech on the other hand, had the highest score for disgust (M=0.071, SD=0.0192). Since most participants mentioned during their post-trial verbal interview that they preferred speech warnings over tone warnings, this finding is surprising. Future research should include looking further into disgust emotion by analyzing the action units individually.

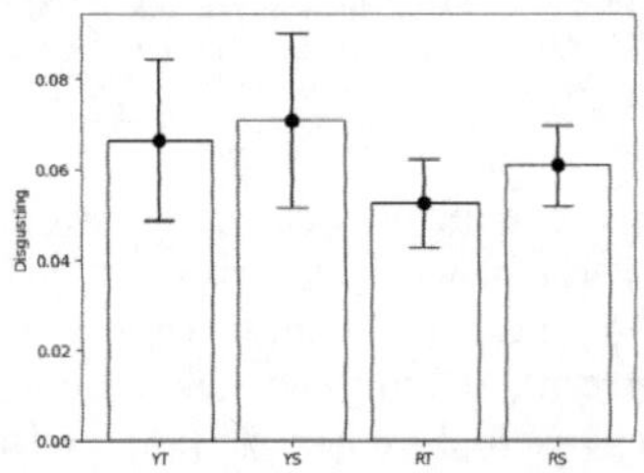

Figure 5: Average disgust score for all warning signals.
Note: YT=yellow-tone, YS=yellow-speech, RT=red-tone, RS=red-speech

3.2. Gaze Data Analysis
Gaze data was analyzed by taking the horizontal and vertical coordinates of each participant for all 20 trials. A heatmap was then used to represent the area where most participants looked at. Figure 6 shows all four warnings for *frontal* (Fig 6a, 6b) and *right-hook* collisions (Fig 6c, 6d). Figure 6a shows that most participants' gaze was focused on the

https://doi.org/10.21872/2023IISE_1735

lower portion of the screen for yellow-speech and red-speech warnings. Figure 6b shows that participants' gaze was focused on the vehicle for red-tone and yellow-tone warning. Figure 6c shows the response to yellow-speech and red-speech warnings for *right-hook* collisions. The gaze, although was focused on the car, some participants also looked at other places on the screen, whereas for red-tone and yellow-tone warning (Fig 6d), their focus was on the car next to them. Several studies have shown that even though speech-based interfaces can prevent visual distraction by providing a clear description of the situation, they are also known to cause cognitive distraction [19]. For the cyclist in this study, cognitive distraction could be occurring from processing the speech warning of BAS. This can increase the mental workload of the cyclist and cause a collision with the vehicle since they are focusing more on the mobile interface. On the road, especially where there is a high amount of ambient noise available, it might even take longer for the cyclist to fully process a speech-based warning. In the post-trial interview, a lot of participants mentioned that they did not notice the yellow visual warning signals at all. This might be more prominent on the roads where there will be natural lights present as well. Based on the emotional and gaze data analysis, as well as post-training interviews, it can be deduced that red-tone warning caused less surprise, disgust and distraction. Therefore, red-tone warning should be incorporated in BAS.

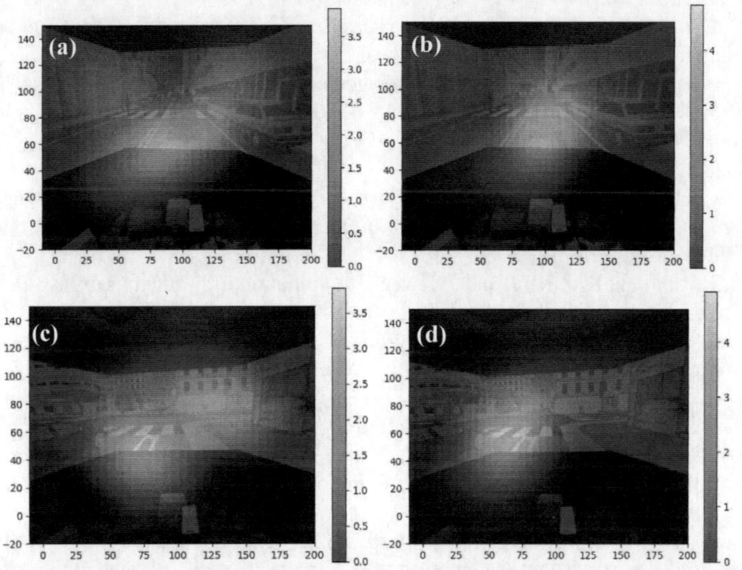

Figure 6: (a) Red-speech and yellow-speech warning for *frontal collision.* (b) Red-tone and yellow-tone warning for *frontal collision.* (c) Red-speech and yellow-speech warning for *right-hook collision.* (d) Red-tone and yellow-tone warning for *right-hook collision*

There are several limitations of this study. The sample size was very low (N=12) due to the visibility requirements of the video and the participants were limited to UTA students, mostly of the same age group. Another limitation is the simulator itself. As it is a controlled and protected environment, participants did not have the fear of collision threats which may have affected their facial expressions and gaze behaviors that they would have on the road. The next step of this study is to run a naturalistic study on the road to determine if the results are consistent with the simulator findings.

Conclusion

This research was focused on identifying the ideal warning signal for a smartphone-based application for cyclists. The Bicyclist Assistance System (BAS) can detect *right-hook collision* and *frontal collision* and alert the cyclist. A bicycle simulator-based study was conducted under different environmental factors. A multimodal signal consisting of red or yellow visual signals, and speech or tone audible signals were tested. 12 participants' video data were collected while riding the bicycle simulator and were analyzed using the FaceReader software to obtain the emotional and gaze data. ANCOVA analysis showed only warning signals to have significant effect with emotional data. Gaze data was analyzed using the average horizontal and vertical coordinates of each participants' gaze. Result showed red-tone warning to be the ideal warning based on surprise, valence of surprise and disgust scores. Gaze data further showed the distraction was lower when using tone signals. Future studies should consider conducting a naturalistic study after incorporating red-tone warning into the BAS app.

Acknowledgements

This research was funded by the by a grant from the Center for Transportation Equity, Decisions and Dollars (CTEDD) funded by U.S. Department of Transportation Research and Innovative Technology Administration (OST-R) and housed at The University of Texas at Arlington.

References

[1] NHTSA, "Traffic Safety Facts - 2020 Data: Bicyclists and Other Cyclists," 2022.

[2] O. L. Ngwu, A. Rimu, S. Deb, J. R. Lummus, C. Kan, and Y. Zhou, "Teenage bicyclists' perceptions toward automated vehicles and their associated traffic infrastructure: A focus group discussion," *Transp Res Part F Traffic Psychol Behav*, vol. 89, pp. 371–384, Aug. 2022, doi: 10.1016/J.TRF.2022.07.006.

[3] NHTSA, "Traffic Safety Facts," 2020. [Online]. Available: https://crashstats.nhtsa.dot.gov/Api/Public/ViewPublication/812850

[4] A. J. Rimu, S. Deb, M. Islam, R. Etminani-Ghasrodashti, and A. Pande, "Roadmap for child-pedestrian training program informed by contextual crash data," *Transp Res Rec*, vol. 2676, no. 11, pp. 250–261, 2022.

[5] H. Smith, C. Howard, and T. Foord, "Alarms management. Priority, floods, tears or gain? Introduction to the 'problem,'" *Measurement and Control*, vol. 36, no. 4, 2003, doi: 10.1177/002029400303600402.

[6] S. Lim, D. Popov, N. Raza, F. Pervaiz, and M. Al-Qadi, "Feasibility Study of Submerged Floating Crossing," 2018.

[7] W. Jin *et al.*, "CycleGuard: A Smartphone-based Assistive Tool for Cyclist Safety Using Acoustic Ranging," *Proc ACM Interact Mob Wearable Ubiquitous Technol*, vol. 5, no. 4, 2021, doi: 10.1145/3494992.

[8] S. H. Han, M. Hwan Yun, K. J. Kim, and J. Kwahk, "Evaluation of product usability: Development and validation of usability dimensions and design elements based on empirical models," *Int J Ind Ergon*, vol. 26, no. 4, 2000, doi: 10.1016/S0169-8141(00)00019-6.

[9] M. K. Noordewier, S. Topolinski, and E. van Dijk, "The Temporal Dynamics of Surprise," *Soc Personal Psychol Compass*, vol. 10, no. 3, 2016, doi: 10.1111/spc3.12242.

[10] M. K. Noordewier and E. van Dijk, "Surprise: unfolding of facial expressions," *Cogn Emot*, vol. 33, no. 5, 2019, doi: 10.1080/02699931.2018.1517730.

[11] G. Horstmann, "Latency and duration of the action interruption in surprise," *Cogn Emot*, vol. 20, no. 2, 2006, doi: 10.1080/02699930500262878.

[12] A. Hetland, E. Kjelstrup, M. Mittner, and J. Vittersø, "The thrill of speedy descents: A pilot study on differences in facially expressed online emotions and retrospective measures of emotions during a downhill mountain-bike descent," *Front Psychol*, vol. 10, no. APR, 2019, doi: 10.3389/fpsyg.2019.00566.

[13] N. Li and C. Busso, "Analysis of facial features of drivers under cognitive and visual distractions," in *Proceedings - IEEE International Conference on Multimedia and Expo*, 2013. doi: 10.1109/ICME.2013.6607575.

[14] C. Ahlstrom, K. Kircher, B. Thorslund, and E. Adell, "Bicyclists' visual strategies when conducting self-paced vs. system-paced smartphone tasks in traffic," *Transp Res Part F Traffic Psychol Behav*, vol. 41, 2016, doi: 10.1016/j.trf.2015.01.010.

[15] A. Rimu, R. Panik, S. Murali, K. Watkins, M. Li, and S. Deb, "Identification of Potential Warning Signals for a Smartphone-based Bicyclist Assistant System.," 2022.

[16] J. Fagerlönn, S. Lindberg, and A. Sirkka, "Combined auditory warnings for driving-related information," in *ACM International Conference Proceeding Series*, 2015, vol. 07-09-October-2015. doi: 10.1145/2814895.2814924.

[17] R. Graham, "Use of auditory icons as emergency warnings: Evaluation within a vehicle collision avoidance application," *Ergonomics*, vol. 42, no. 9, 1999, doi: 10.1080/001401399185108.

[18] D. Mckeown, "Candidates for within-vehicle auditory displays," *Proceedings of the 11th International Conference on Auditory Display (ICAD2005)*, 2005.

[19] C. Brooks and A. Rakotonirainy, "In-vehicle technologies, Advanced Driver Assistance Systems and driver distraction : Research challenges," *Proceedings International Conference on Driver Distraction*, vol. 2005, 2005.

https://doi.org/10.21872/2023IISE_1860

Proceedings of the IISE Annual Conference & Expo 2023
K. Babski-Reeves, B. Eksioglu, D. Hampton, eds.

Planning Distributed Generation Systems Powered with Wind Turbines using a Stochastic Programming Model

Md Fantacher Islam, Clara Novoa, Francis A Mendez
Texas State University
San Marcos, TX 78666

Abstract

This paper presents a non-linear scenario-based stochastic programming model to determine wind turbines' optimal siting and sizing in distributed generation (DG) systems powering twelve houses or industrial facilities. The model's goal is to minimize the expected DG system's annualized installation and operational costs over the system's life cycle. The model is novel because it integrates four practical aspects, wind volatility, realistic loads, system loss-of-load probability and thermal constraints, to design a real-size DG system. The model solves exactly with a relatively low computational time, providing a competitive approach to models in the literature that typically end up solved by heuristic methods or ignoring the stochastic nature of the model parameters. The model is useful to assess if the DG system needs to connect to a centrally located substation or can operate in an islanded mode. The numerical experimentation is performed in six US cities using hourly wind speed collected over thirteen years (2009-2021). The three case studies presented are industrial park, residential houses with substations, and residential houses without substations. The industrial park case results show that in cities with high wind speeds (e.g., averaging 4.6 m/s), the system costs are 7% lower than in cities with low wind speeds (e.g., averaging 2.9 m/s). Costs are highly sensitive to load increases and slightly sensitive to changes in loss-of-load probability. Results for the residential case indicate that the system can be disconnected from the substation, even assuming a low loss-of-load probability (0.001), but the system's cost rises by about 50%.

Keywords
Distributed Generation, Reliability Engineering, Renewable Energy, Stochastic Programming, Wind Turbines

1. Introduction and Related Research

Residential buildings emit 10.9% and industries emit 24.2% of the global greenhouse gases (GHG), respectively [1]. In the US, GHG emission for residential buildings is 20%, and for industries is 32% [2]. Consequently, fossil fuels are losing appeal due to their harmful environmental effects, and the relevance of finding alternative energy sources is growing. For instance, in the US energy market, renewable energy (RE) contributes 13.5% of the overall electricity [3]. However, if focusing on wind generation, one of its main limitations is that the intermittent nature of wind occasionally produces imbalances of power supply and demand. Therefore, it is necessary to consider wind speed variability in the planning of distributed generation (DG) systems using wind turbines (WTs). DG systems provide onsite or decentralized power generation using distributed electricity resources (DERs) to meet all or part of the energy needs [4]. Photovoltaics, WTs, diesel generators, energy storage systems (ESSs), and biogas fuel cells are examples of DERs. Integrating RE and non-RE resources into DG systems is advantageous due to technological, financial, and environmental benefits [5]. DG systems can reduce the number of long-distance transmission lines and power losses.

Siting and sizing for WT have been addressed in the literature. To the best of our knowledge, only [7] considers wind variability through a scenario-based stochastic programming model and solves exactly a DG network smaller than the one proposed in this paper. The study in [8] couples feasible geographical locations for wind farms with the nodes in an electrical grid and solves exactly a deterministic linear programming model to find the optimal sitting and sizing of WTs and maximize the estimated annual wind power produced. In [9], the DG allocation problem looks to minimize power loss and bus voltage deviation of a power network with 33 buses and 94 nodes integrated with (1) photovoltaic arrays and fuel cells and (2) diesel generators and WTs. A swarm optimization technique and fuzzy expert rules find the approximate solution. Authors in [10] investigate the layout design for a real offshore wind farm, assuming a continuous layout approach and proposing a deterministic non-linear programming (NLP) model to maximize wind power at low cost. A mean wind speed of 10.6 m/s is assumed and the model is solved using a genetic algorithm. Integration of WTs in AC distribution networks is studied in [11] using a deterministic NLP that is solved exactly. Newly, the authors use an artificial neural network to incorporate short-term wind power forecasting.

https://doi.org/10.21872/2023IISE_1860

2. Problem Description and Paper Contribution

Figure 1 illustrates the problem researched in this paper. The problem is determining the WTs' siting and sizing of minimum cost in an interconnected DG system that links a substation (node 1) whose size is to be determined to twelve facilities (i.e., nodes 2-13) demanding power. The facilities can be industries in an industrial park or residential houses. Several WTs types can be installed at the facilities, and multiple WTs of the same type can be installed in each node. The WTs types (i.e., capacities) range from 1MW to 3MW. The arrows connecting the nodes are the distribution lines (DL) labeled as DL_2 to DL_{13}. The downward arrows at the nodes represent the expected loads.

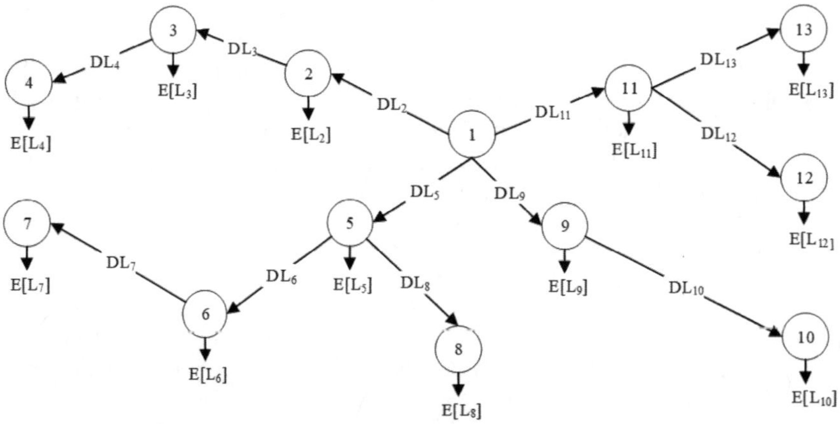

Figure 1: The DG system researched in this paper.

This paper's contribution is on developing a non-linear scenario-based stochastic model to determine the siting and sizing for WTs in a real-size grid-connected DG system considering simultaneously wind volatility, realistic loads (i.e., power demands), system loss-of-load probability and thermal constraints. A model's novelty is that it solves exactly with a relatively low computational time (10 min), providing a competitive approach to previously proposed models that typically end up solved by heuristic methods or ignoring the stochastic nature of the model parameters. The model is also useful to assess the DG system's viability to power the facilities disconnected from the grid. The DG network modeled is the radial distribution system introduced in the study on enhancing grid resilience in [6].

3. Methodology - Stochastic Programming Model

The model and its notation are presented on the next two pages. Equation (1) is the objective function, which minimizes the expected total annual cost of installing and operating the DG system. Constraint (2) is the reliability constraint that ensures that the system power quality is met for a significant percentage of the year (i.e., $100(1-\alpha)$). Given the number of facilities considered in the DG system researched, the system's total power (P) and total load (L) follow a normal distribution using the central limit theorem (CLT). Consequently, the probability density functions for the system's power and load, f_P and f_L, are well-known expressions. The expression for f_P is still a function of the unknown values for the number of WTs of each type to install in each node, x_{ij}. It is encouraging that a non-linear solver can evaluate the constraint at the different x_{ij} solutions visited in the search for the optimal solution. However, the constraint makes the model a non-convex non-linear mixed integer program and consequently challenging to solve.

Thermal constraints (3) apply to the top network nodes, which may supply power to directly connected nodes or to those up to two levels below them. In (3), WTs installed on the nodes will reduce some of the power needs and ensure that the power carried by the DL feeding an upper node and all its lower nodes does not exceed the DL power limits. Accordingly, thermal constraints (4) and (5) apply to nodes with two-layer downward branches and to lower nodes. Thermal constraints (3)-(5) must be satisfied for each wind speed scenario. Constraints (6)-(8) are omitted for brevity. Constraint (6) limits the number of WTs installed at a node. This constraint is practical and may result from space or budget limitations. Equation (7) avoids that WTs be installed at the central node, and constraint (8) requires that only one type of substation be installed. The substation compensates for the deficit of RE when wind generation is low.

https://doi.org/10.21872/2023IISE_1860

Table 1: Sets

Notation	Definition	Units
I	Types of DER (i.e., capacities for the WTs and substations)	MW
R	Types of WTs (i.e., capacities for the WTs)	MW
K	Types of substations (i.e., capacities for the substations)	MW
J	All nodes in the DG system. $J = \{1, 2, \dots 13\}$	
T	Top nodes with distribution lines linking directly to Node 1 and with at least one three-layered downward branch including node t itself, $T = \{2, 5\}$	
C_t	Nodes directly connected down from top node t in set T, $C_2 = \{3\}$, and $C_5 = \{6, 8\}$	
G_t	Nodes connected down two levels from top node t in set T, $G_2 = \{4\}$, and $G_5 = \{7\}$	
F	Nodes with only two-layered downward branches, including node f itself, $F = \{9, 11, 3, 6\}$	
D_f	Nodes connecting down a node f in set F, $D_9 = \{10\}$, $D_{11} = \{12,13\}$, $D_3 = \{4\}$, $D_6 = \{7\}$.	
E	Bottom or terminal nodes, $E = \{4, 7, 8, 10, 12, 13\}$	
S	Wind speed scenarios. One scenario s is a realization of wind speed	m/s

Table 2: Notation for decision variables and parameters

Notation	Definition	Units
Decision variables:		
x_{ij}	Number of DER type i installed at node j. It is an integer decision variable	Item
P	Total power generated in the DG system. P is a random variable with probability density function f_P and a function of x_{ij} and y_s, the wind speed for scenario s. Function arguments dropped to simplify notation. $P \sim N\left(\sum_{i\in I}\sum_{j\in J} E\left[P_{ij}\right]x_{ij}, \sum_{i\in I}\sum_{j\in J}\sigma^2\left(P_{ij}\right)x_{ij}\right)$	MW
Parameters:		
y_s	Wind speed the blades of WTs capture at height h under scenario $s \in S$	m/s
$E[P_{ij}]$	Mean power generated by DER type i at node j. For the WTs, P_{ij} is a random variable that depends on the wind speed that the blades of WTs capture at height h	MW
$\sigma^2(P_{ij})$	Variance of the power output for DER i at node j	MW^2
P_{ijs}	Power output when DER type i is installed at node j and the wind speed scenario is $s \in S$	MW
L	System total load. L is a random variable with probability density function f_L $L \sim N(\sum_{i\in I} E[L_i], \sum_{i\in I}\sigma^2(L_i))$. $E[L_i]$ and $\sigma^2(L_i)$ are node i mean load and variance	MW
$E[L_t], E[L_f]$	Mean peak load at top node t and upper node f, respectively	MW
$E[L_e]$	Mean peak load at lower node e	MW
p_{sj}	Probability associated with wind speed scenario s at node j	
$P_i^{(c)}$	Capacity of DER type i	MW
a_i	Present cost per MW for installing DER type i. Assumed equal at each node j	\$/MW
ϕ	Capital recovery factor, a function of annual interest rate and years to recover an investment	
b_i	Annual operation and maintenance cost per MW for DER type i	\$/MW
c_i	Subsidy or tax penalty for installing DER unit i. A subsidy if a WT is installed, a penalty if a substation is installed.	\$/MW
I_t	Maximum allowed current flow at the distribution line (DL) running from the substation to top nodes t	Amp
I_f	Maximum allowed current flow at the DL reaching from above an upper node f	Amp
I_e	Maximum allowed current flow at the DL reaching lower node e	Amp
V_{DG}	The nominal voltage passing from the substation to the DL	kV
DU	Number of DER types allowed to install per node	Item
α	Loss-of-load probability	

Minimize:

$$g(x) = \sum_{i\in I}\sum_{j\in J}\left(a_i \phi P_i^{(c)}\right)x_{ij} + \sum_{i\in I}\sum_{j\in J}\sum_{s\in S} p_{sj}P_{ijs}(b_i + c_i)\,x_{ij} \tag{1}$$

https://doi.org/10.21872/2023IISE_1860

Subject to:

$$\int_0^{\infty}\left(\int_y^{\infty} f_p(z)dz\right) f_L(y)dy \geq 1-\alpha \tag{2}$$

$$E[L_t] - \sum_{i\in I} P_{its}x_{it} + \sum_{k\in C_t}\left(E[L_k] - \sum_{i\in I} P_{iks}\, x_{ik}\right) + \sum_{g\in G_t}\left(E[L_g] - \sum_{i\in I} P_{igs}\, x_{ig}\right) \leq V_{DG}I_t \quad \forall t\in T, s\in S \tag{3}$$

$$E[L_f] - \sum_{i\in I} P_{ifs}x_{if} + \sum_{d\in D_f}\left(E[L_d] - \sum_{i\in I} P_{ids}\, x_{id}\right) \leq V_{DG}I_f \quad \forall f\in F, s\in S \tag{4}$$

$$E[L_e] - \sum_{i\in I} P_{ies}x_{ie} \leq V_{DG}I_e \quad \forall e\in E, s\in S \tag{5}$$

4. Model Parameters and Results

Table 3: Costs for the DER units

DER Unit	DER Capacity (MW) $P_i^{(c)}$	Total Equipment Cost (\$) $a_iP_i^{(c)}$	Annual O&M Cost (\$/MW) b_i	Annual Penalty Cost (\$/MW) c_i
Industrial and Residential WTs				
1MW	1	1,300,000	42,000	0
2MW	2	2,600,000	45,000	0
3MW	3	3,900,000	48,000	0
Industrial Park Substations				
29 MW	29	9,425,000	22,500	5,000
31 MW	31	10,075,000	18,750	7,500
Residential Substations				
69 kW	0.069	9,425,000	22,500	5,000
115 kW	0.115	10,075,000	18,750	7,500

Table 3 has the values used for the DER's capacities and costs. Figure 2 illustrates the procedure to calculate the probability distribution for the power of each type of WT at each node and city studied. Hourly wind speeds were collected from Weather Underground for 13 consecutive years (2009-2021) in Boston, New York, San Francisco, Dallas, Austin, and Phoenix. Each year's wind speed (WS) data has 8,760 observations per city and is assigned to a node in the network.. This data is tallied into 13 wind speed scenarios defined by the wind speed ranges and midpoints given in Table 4 to estimate the scenario probabilities in each node, p_{sj}. Regarding the power computation, in the first two scenarios WTs can't operate due to low wind, and the power generation is zero. In the operational phase of a WT, the power output is a function of the cube of the wind speed until the wind speed reaches v_r= 12 m/s, the rated power region starts, and the power equals the WT rated power, P_m.

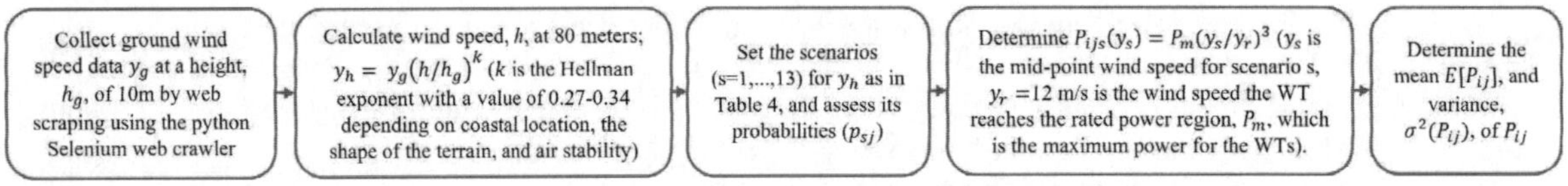

Figure 2: Flow chart used to calculate the probability distribution for the power of each WT type at each node.

Table 5 shows the mean (MW) and variance (MW^2) for the peak load at the nodes for the industrial park and residential case studies. It also displays the maximum current, I_j, flowing in DL_j towards node j. As in [6], the expected peak loads for the facilities in the industrial park are high, ranging from 1.62 MW to 3.47 MW. Peak loads for the residential case are estimated using a database with 17,520 hourly power (MW) demands provided by the New Braunfels Utility (NBU) for all residential houses in this city in 2017 and 2018. Average electricity use (AEU) per house results from

https://doi.org/10.21872/2023IISE_1860

dividing the hourly power demand by the number of residential houses in New Braunfels. The best fit for the AEU per house is a Lognormal distribution with a mean of 0.00596 MW and variance of 0.00145 MW^2. The estimated mean residential peak loads shown in Table 5 were generated with the Arena Input Analyzer. The α values are 0.01, 0.05, and 0.1 for the industrial case and 0.01 and 0.001 for the residential case. The voltage in the DL is 33kV.

Table 4: Predetermined wind speed scenarios (in m/s) used in the computational study.

Scenario	1	2	3	4	5	6	7	8	9	10	11	12	13
WS range	0-1	1-2	2-3	3-4	4-5	5-6	6-7	7-8	8-9	9-10	10-11	11-12	>12
Mid-point WS (y_s)	0	0	2.5	3.5	4.5	5.5	6.5	7.5	8.5	9.5	10.5	11.5	18.5

Table 5: Mean (MW) and variance (MW^2) for the peak load and maximum current (Amp) handled by DL getting into each node.

Nodes	1	2	3	4	5	6	7	8	9	10	11	12	13
Industrial Park													
$E[L_j]$	0	3.44	3.47	2.06	1.8	2.31	2.75	2.54	2.82	2.03	2.51	2.36	1.62
$\sigma^2(L_j)$	0	0.27	0.27	0.1	0.07	0.12	0.17	0.14	0.18	0.09	0.14	0.13	0.06
Residential Case													
$E[L_j]$	0	0.005	0.005	0.004	0.005	0.003	0.01	0.006	0.01	0.003	0.007	0.004	0.005
$\sigma^2(L_j)$	0	2E-6	2E-6	3E-6	3E-6	3E-6	3E-6	3E-6	2E-6	2E-6	1E-6	1E-6	1E-6
I_j	-	500	250	210	600	250	210	210	250	210	450	210	210

4.1 Numerical Results

The model is implemented with Advanced Mathematical Programing Language (AMPL) using the AMPLPY interface for Python and solved by the non-linear solver Knitro. In the industrial park base case, the DG system must meet a total system load that has a mean of 29.70 MW and a variance of 1.73 MW^2 at each city. The model solution shows that the DG system implemented in cities with high wind speeds (e.g., an average of 4.6m/s), such as Boston, has a cost 7% lower than in cities with low wind speeds (e.g., an average of 2.9m/s) such as Phoenix. Boston needs fewer WTs with the least cost (i.e., \$ 21,822,696), whereas Phoenix needs WTs more than any other city, costing \$ 23,523,972. Besides, due to the similar wind speeds, New York, San Francisco, and Dallas need almost the exact number of WTs. In addition, compared to Boston, Austin has mid-wind speeds, and the system requires more WTs, costing \$ 22,306,678. For all cities, the 29 MW substation was chosen.

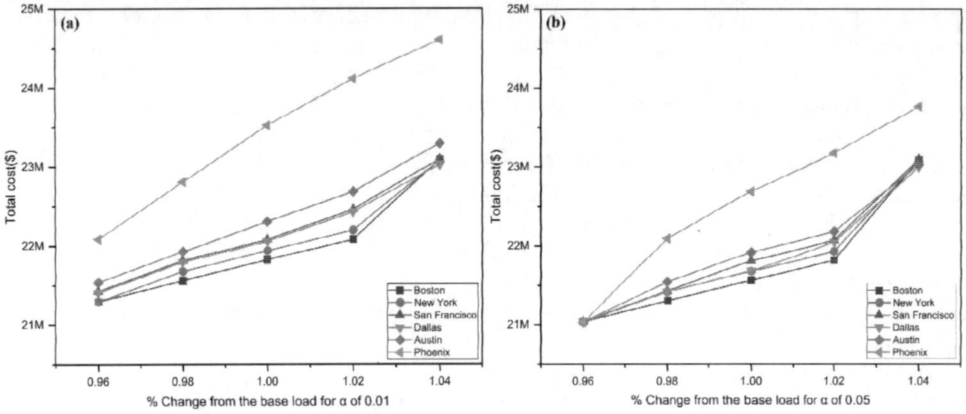

Figure 3: System cost vs. load for industrial park case **(a)** loss-of-load probability, α, of 0.01, and **(b)** α, of 0.05.

Figure 3 shows the effect on the total cost of the DG system if varying the expected loads at the network nodes to become 94% - 106% of the base expected load ($E[L_i]$) in increments of 2% for two different α values. With the increase in load, the system's cost increases in a positive uprising trend due to the need for more WTs, but the size of the substation remains equal. However, if decreasing the load from the assumed base value, the systems require fewer

https://doi.org/10.21872/2023IISE_1860

WTs and consequently incur fewer costs. The difference in annual system costs in Figure 3 for an α of 0.01 vs. 0.05 are practically significant, especially for Phoenix. However, the statistical analysis indicated that the system is more sensitive to changes in load than to changes in α. Of all cities, Phoenix shows the highest cost trend due to low wind speeds, whereas Boston shows a relatively minor cost trend. The trend in system costs for Austin is the second highest. The cost trends for α of 0.1 are similar to the ones presented in Figure 3 for all cities studied.

For the residential case (RC), the DG system needs to meet a total load that has a mean of 0.00553 MW and a variance of 0.000026 MW^2. Like in the industrial park's case, for the RC with substation, Phoenix installs more WTs, at a cost of $ 1,438,822. Phoenix is almost twice as expensive as other cities, costing around $ 538,000. Dallas installs the 69 kW substation and other cities install the 115 kW substation. The RC without substation shows that only WTs can power the DG system. The α used for the base case is 0.001. The absence of a substation and the stricter α are the reasons for most cities doubling the number of WTs vs. the RC with a substation. Consequently, the systems' annual cost also increases for all cities. The model found the feasible solution for all cities only when varying expected loads from the base value by ±10% using α of 0.001 and 0.01. However, the DG system requires 50% more costs if changing α from 0.01 to 0.001. This is because for an α of 0.01 the DG system might not meet the load for almost four days in a year, whereas the stringent value of 0.001 indicates the load will not be fulfilled for less than one day per year.

5. Conclusions

This paper proposed a scenario-based stochastic programming model suitable for determining the siting and sizing of wind turbines in an interconnected DG system comprised of thirteen nodes and a substation. Results show that for the industrial park case, the system implemented in cities with high wind speeds has 7% lower costs. Besides, the annual system cost is more sensitive to load changes than to loss of load probability. For the residential case, the system can operate without connecting to a substation. However, the system's annual cost rises by more than 50%, and the requirement for WTs doubles. Readers can contact the authors for information regarding the number of wind turbines installed in each facility in each case. In addition, the DG systems only running on wind turbines may have satisfaction flaws if loads vary substantially. Future work will integrate energy storage systems and blockchain technology to satisfy the load, trade energy, and increase the system's economic, social, and environmental benefits.

References

[1] H. Ritchie, M. Roser, P. Rosado. "CO_2 and Greenhouse Gas Emissions." Our World In Data [Online]. Available: https://ourworldindata.org/co2-and-other-greenhouse-gas-emissions [Accessed August 2020].

[2] B. Goldstein, D. Gounaridis, J. P. Newell, "The carbon footprint of household energy use in the United States," in *Proceedings of the National Academy of Sciences,* vol. 117, no. 32, July 20, 2020, pp. 19122-19130.

[3] S. Kwon, L. Ntaimo, N. Gautam, "Optimal day-ahead power procurement with renewable energy and demand response," *IEEE Transactions on Power Systems,* vol. 32, no. 5, pp. 3924-3933, September 2017.

[4] M. Begovic, A. Pregelj, A. Rohatgi, D. Novosel, "Impact of renewable distributed generation on power systems," in *Proceedings of the 34th Annual Hawaii International Conference on System Sciences*, Maui, HI, USA, 2001, pp. 654-663.

[5] T. Adefarati, R. Bansal, "Integration of renewable distributed generators into the distribution system: a review," *IET Renewable Power Generation,* vol. 10, no. 7, pp. 873-884, August 2016.

[6] H. Wang, T. Jin, "Prevention and survivability for power distribution resilience: A multi-criteria renewables expansion model," *IEEE Access,* vol. 8, pp. 88422-88433, May 2020.

[7] T. Runsewe, C. Novoa, T. Jin, "Stochastic model for planning distributed wind generation using climate analytics," in *Proceedings of the 2021 Institute of Industrial and Systems Engineers (IISE) Conference*, Orlando. FL, USA, May 18-21, 2019, pp 1755-1760.

[8] H. Cetinay, F. A. Kuipers, A. N. Guven, "Optimal siting and sizing of wind farms," *Renewable Energy,* vol. 101, pp. 51-58, February 2017.

[9] A. El-Fergany, "Optimal allocation of multi-type distributed generators using backtracking search optimization algorithm," *International Journal of Electrical Power & Energy Systems,* vol. 64, pp. 1197-1205, January 2015.

[10] N. Charhouni, M. Sallaou, K. Mansouri, "Realistic wind farm design layout optimization with different wind turbines types," *International Journal of Energy and Environmental Engineering,* vol. 10, no. 3, pp. 307-318, April 2019.

[11] W. Gil-González, O. D. Montoya, L. F. Grisales-Noreña, A-J. Perea-Moreno, and Q. Hernandez-Escobedo, "Optimal placement and sizing of wind generators in AC grids considering reactive power capability and wind speed curves," *Sustainability,* vol. 12, no. 7, pp. 2983, April 2020.

https://doi.org/10.21872/2023IISE_1894

Proceedings of the IISE Annual Conference & Expo 2023
K. Babski-Reeves, B. Eksioglu, D. Hampton, eds.

Classification of Human Enrichment and Refinement in Interactive Optimization

Michelle Zhang, Jennifer Pazour, Sandipan Mishra, John E. Mitchell, Damsara Jayarathne
Rensselaer Polytechnic Institute, Troy, NY

Abstract

Optimization models can recommend the best systematic decisions in the face of exponentially many options, decision trade-offs, systematic interactions and constraints on finite resources, tasks notoriously difficult for human decision makers (DM). Yet, a model – by definition – is only a representation of the system that a DM is interested in optimizing. Human DMs typically have access to important local context, domain knowledge, and goals. Thus, interactive optimization (IO) tools, which are "human-in-the-loop" approaches that iteratively involve the human DM in the optimization process, are critical for decision making. This paper presents a framework for classifying the different approaches in IO to elicit and embed human DM's feedback into a mathematical program. Towards this goal, we build on a previous taxonomy proposed by Meignan et al. in 2015, focusing on problem-oriented interactions. We propose a classification scheme that captures these aspects, where the current model formulation may be incomplete and the human DM provides feedback that can enrich the optimization model and problem instantiation data. Using this scheme, we review the existing literature in this category to find prior work that focuses on eliciting preference information for multiobjective problems. Based on this review, we then present our conclusions and discuss opportunities for creating optimization methods that capture more richly the human side of knowledge.

Keywords
Human-in-the-loop, Interactive Optimization, Math Programming, Operations Research

1. Introduction

Interactive optimization (IO) is a field of research that supports a human decision maker's (DM) active participation in the optimization process by incorporating their feedback in an iterative fashion. This human interaction may occur during any stage of the optimization process. A DM can be asked to update the problem domain with new data, update their preferences, and add or remove constraints (among other interactions). Generally, interactions between a human DM and the optimization system are provided through an interface and interactions continue in an iterative manner until the DM is satisfied with the solution found.

IO is a key tool for addressing several challenges in the integration of advanced optimization methods into decision support tools [1]. First, it is often difficult for human DMs to specify, beforehand, all facets of their real world problem, whether that is applicable constraints, available data, or selection criteria for the problem. Thus, asking the DM pointed questions and to provide updated information can help elicit a richer model formulation. Further, as DMs must understand and trust the solutions generated by the optimization system in order to implement, modify, and justify them, IO approaches also aid in building a DM's understanding of the methodology and the system being optimized. Therefore, IO is an advantageous approach that can help bridge the gap between the information captured in the optimization model and the real world problem and can increase the likelihood for deployment.

Consequently, there has been significant research and prior literature on IO methods. IO approaches range from rudimentary strategies like trial-and-error [2, 3] to complex methods, such as multiobjective (MO) optimization [16] and human-guided search [5]. In 2015, Meignan et al. [1] reviewed the literature in IO and proposed a useful classification of interactive approaches based on a DM's interaction with the optimization system. Specifically they classified IO methods in terms of the purpose of the interaction and the role of the DM, and characteristics of the model, including the type of feedback integration, the preference information lifetime, and the type of optimization procedure. Follow up reviews such as Dudley et al. in 2018 [6] provide a detailed overview on interactive machine learning systems, highlighting user interface designs. Although different from IO, these concepts and associated definitions are closely related. Other subsequent review studies have focused specifically on interactive MO optimization, see [7, 8].

https://doi.org/10.21872/2023IISE_1894

In this work, we develop a classification scheme to emphasize the potential for IO approaches that more richly engage with a human DM. To do so, we use the taxonomy proposed by Meignan et al. 2015 [1] as a guide-post to define certain terms and characteristics in IO. We then build a new classification scheme focusing on the different ways IO methods capture and decode the knowledge from a DM and use this additional information to refine the optimization system, and then apply this scheme to review existing IO methods and to identify promising future research areas.

2. Classification

Our review's scope is limited to IO papers that meet three criteria. First, a human DM must have an interaction with the optimization system. We include papers that use artificial DMs only if the intended purpose of the methodology is for a human DM. Methods designed to interact with nonhuman DMs are out of scope. Second, the paper must have a mathematical programming formulation that describes in mathematical form specific objective function(s), decision variables, and constraints. Meignan et al. 2015 [1] classifies the purpose of DM interaction to be problem-oriented or search-oriented. Third, the paper must have a problem-oriented interaction where the DM has additional knowledge of the problem domain not yet captured by the formulation. We do not consider search-oriented papers as the interaction with the DM is to improve the solution approach efficiency (but assumes the model formulation is fixed) [5].

2.1 Classification Scheme

Our review focuses on IO approaches and the different ways they capture and interpret the knowledge from a DM and use this additional information to refine the optimization system. To do so, we create a classification schema that considers the following characteristics, which we define as a set of questions.

(1) Does the human DM play the role of an adjuster, enricher, or both? (We define these terms based on the concepts introduced by Meignan et al. 2015 [1].)

A DM is an adjuster if they provide feedback on the values of some parameters of the constraints or the objectives. Notably, the human DM does not change the functional form of the optimization model, just provides feedback on the input parameter values, or expanding the elements of proposed sets. Hence, the adjuster will only be refining the model with data (see Question 2 for further classification). A DM is an enricher if they provide feedback that leads to structural changes to the optimization formulation. This is typically in the format of adding or removing some constraints or objectives, or changing or expanding previously defined sets. This is different from the role of an adjuster since the assumption here is that the proposed optimization model may be incomplete and does not capture all the facets that the DM's real problem possesses. Thus, the DM will only be refining the model with domain knowledge (see Question 2). A DM is both if they play the role of an adjuster and an enricher.

(2) What type of information is asked from the human DM to refine the optimization system?

Data is refined when a DM, as an adjuster updates parameter values, such as adjusting the optimization model's objective coefficients or the constraints' input parameter coefficients, or changing or expanding the elements of previously defined sets. Notably, this type of refinement does not change the model formulation's functional form. However by asking the DM to provide refined input data, it could lead to better quality or new data. This could be in the form of different parameter values or redefining sets. For example, if the DM adds a new supplier, this will only be an additional element in a well defined set. It will not lead to an additional constraint in the model. A special type of data is based on DM's preference. The majority of IO papers focus on preference updates, where in interactive MO optimization, the human DM is asked to provide information that the system can use to update weights across the set of multiple objectives or criteria. The system asks the DM for domain information in an attempt to refine some aspects about the functional form of the optimization model. Given human DMs typically have access to important local context, domain knowledge, and goals, this type of refinement uses such new information to enhance the optimization formulation. There may be more restrictions on the optimization problem that needs to be specified or a new decision criteria that was previously not captured in the model. In this case, the DM has information that the optimization system does not have. For instance, the DM may know from previous cases that a specific supplier will be closed on weekends. Hence, the system adds a constraint on availability for this supplier.

(3) What is the level of knowledge the DM is expected to have?

A math programming expert (MPE) is expected to know mathematical formulations and how adding or removing constraints/objectives will affect the model. They have extensive knowledge about the math background of the proposed problem. A decision-making context expert (DMCE) understands the domain and decision being made, but is not expected to be an expert in math programming. Both if the DM has math programming and decision-making knowledge.

https://doi.org/10.21872/2023IISE_1894

(4) What is the type of optimization problem?

We first classify if a model is single objective (SO), or multiobjective (MO). We further classify the model based on the problem structure : discrete (D) vs. continuous (C) , and linear (L) vs. nonlinear (NL) . We use the classification defined in neos Guide `https://neos-guide.org/guide/types/`.

(5) Is the interaction question type asked of the human DM static or adaptive?

A static approach will ask the DM the same type of question in all interactions (albeit the context of the question typically changes over time as the output from the optimization model changes). An adaptive approach asks the DM different types of questions, depending on the response of the user and the nature of information needed for refinement.

(6) How do human DMs provide feedback?

DMs provide feedback by answering targeted questions and we classify approaches in terms of if the human DM provides this feedback by (i) list if they select from a pre-populated list of potential options or more specifically (i.1) a specialized list is yes/no if they are only allowed to respond back with an answer of "yes" or "no", (ii) value if they are asked to enter numerical values, (iii) rank if they are asked to rank information provided from the optimization model, or (iv) other if they are asked to provide feedback in another way (which may be to upload new data).

(7) What approach is used in creating the refinement?

Approaches can be math programming informed, such as using dual variables to recommend refinement actions, preference such as trade-off information, reference point approaches, or classification-based methods (see [9]), or other sophisticated approaches.

2.2 Literature Review using the Classification Scheme

In January 2023, we reviewed papers on Google Scholar that cited Meignan et al. 2015 [1]. We combine the 15 papers we found with the papers that Meignan et al. 2015 [1] reviewed, which incorporate all three of the features identified in Section 2. In Table 1, we classify these 28 papers, using the scheme identified in Section 2.1.

3. Conclusions and Discussion of Promising Research Directions within IO.

An extensive amount of work in IO has been done in the past two decades. Most reviewed papers (21 out of 28) view the DM's role as an adjuster that provides feedback on the values of some parameters used in the constraints or the objectives. The primary focus in problem-domain IO methods has been on MO problems, where the interaction is to ask the DM to provide new preference data among multiple objectives. Existing IO methods are well designed to gain and integrate additional information about a human's preferences among a set of criteria into optimization models.

All reviewed papers in Table 1 take a static approach to asking the DM questions. Static approaches provide useful structure, which is often exploited to create sophisticated, math programming informed refinement approaches that integrate the human's additional knowledge in such a way that preserves or improves some performance metric of interest. Yet, given existing approaches focus on a single type of refinement action, repeatedly asked to the human DM, this also limits the ability of existing IO methods to extract a very specific and narrow piece of knowledge from the human DM. The majority of the existing work assumes the question to ask the DM is known a priori and is not something that the IO method is deciding over time based on the state of the solutions being produced or satisfaction with the solution by the DM. Further, DMs are typically incorporated after a mathematical program has been formulated. Thus, there is a need to include the DM throughout the design and decision process.

A general observation is that the current state of IO focuses on how the optimization formulation can be enhanced by a human, however, the existing approaches tend to under-utilize the human's additional knowledge and goals, and primarily focus on obtaining better estimates of a DM's preferences among a set of criteria. This is valuable information, yet, the human DM has access to a wide range of additional information. Needed are new approaches that can extract a broader range of knowledge from the DM. To do this effectively, these approaches should be adaptive, in that they can dynamically change what type of information to ask from the DM, and math programming informed, in that what additional information asked of the human DM is guided by properties of the optimization problem.

Acknowledgements

This research was supported by the Office of Naval Research under grant number N00014-22-1-2542.

https://doi.org/10.21872/2023IISE_1894

Table 1: Problem-Oriented Interactive Optimization Papers Classified Using the Scheme in Section 2.1.

LITERATURE	Characteristics						
	Role of DM?	**Type of information asked from DM to refine the model?**	**Level of knowledge the DM is expected to have?**	**Type of optimization problem**	**Type of interaction question?**	**How do DMs provide feedback?**	**What approach is used in creating refinement?**
	Adjuster, Enricher, or Both	Data, Preference, Domain	MPE, DMCE, or Both	SO or MO - D vs. C, & L vs. NL	Static, or Adaptive	(i) List, (ii) Value, (iii) Rank, (iv) Other	(i) Math program, (ii) Preference, (iii) Other
Hamel et al. 2012 [10]	Both	Preference	DMCE	SO - D & L	Static	(i) List	(iii) Optimality range
van Vliet et al. 1992 [11]	Both	Data, Preference, and Domain	DMCE	MO - D & L	Static	(i) List, (ii) Value, (iv) Fix or change part of the solution	(i) Math program, (iii) Checks feasibility of changed solutions
Meignan, 2014 [12]	Enricher	Domain	DMCE	MO - D & L	Static	(ii) Value	(iii) Introduces soft constraint
Meignan, 2015 [13]	Enricher	Domain	DMCE	MO - D & L	Static	(ii) Value	(iii) Add two supplementary objectives
Deb and Sundar, 2006 [14]	Adjuster	Preference	DMCE	MO - D, C & L, NL	Static	(ii) Value	(ii) Preference - Reference point
Jaszkiewicz and Słowiński, 1999 [15]	Adjuster	Preference	DMCE	MO - C & L, NL	Static	(ii) Value	(ii) Preference - Reference point
Miettinen and Mäkelä, 2000 [16]	Adjuster	Preference	DMCE	MO - C & NL	Static	(iv) Other	(ii) Preference - Classification based
Greco et al. 2008 [17]	Adjuster	Preference	DMCE	MO - D & L	Static	(i) List, (iv) Other	(ii) Preference - Pareto optimal set
Miettinen et al. 2010 [18]	Adjuster	Preference	DMCE	MO - D & L	Static	(ii) Value, (iii) Rank	(iii) NAUTILUS method
Laukkanen et al. 2012 [19]	Adjuster	Preference	DMCE	MO - D & NL	Static	(ii) Value, (iii) Rank	(iii) Bilevel optimization method
Miettinen, 2007 [20]	Adjuster	Preference	DMCE	MO -D & NL	Static	(i) List	(iii) Scalarization-based method and genetic algorithm
Ruotsalainen et al. 2010 [21]	Adjuster	Preference	DMCE	MO - D & L	Static	(i) List, (iv) Classify objective functions	(ii) Preference
Tveit et al. 2012 [22]	Adjuster	Preference	DMCE	MO - D & L	Static	(iv) Classify objective functions	(iii) Generate regression models for each objective
Ahani et al. 2021 [23]	Adjuster	Data	DMCE	SO - D & L	Static	(iv) Matching process and fine-tune the optimization results	(iii) Crossreferencing
Trachanatzi et al. 2020 [24]	Adjuster	Preference	DMCE	MO - D & L	Static	(iii) Rank	(iii) Preference
He et al. 2019 [25]	Enricher	Domain	DMCE	SO - D & L	Static	(iv) Adding or deleting structural elements	(i) Math program
Ye et al. 2022 [26]	Adjuster	Preference	DMCE	MO - D & L	Static	(i) List	(iii) Adding auxiliary factor values of solutions
Feit et al. 2021 [27]	Both	Both	Both	SO - D & L	Static	(iv) Other	(i) Math program, (ii) Preference
Piemonti et al. 2017 [28]	Adjuster	Preference	DMCE	MO - D & L	Static	(i) List (iii) Rank	(i) Math program
Ruiz et al. 2019 [29]	Adjuster	Preference	DMCE	MO - D & L	Static	(i) List, (ii) Value, (iv) Reference point	(iv) NAUTILUS method
Hu et al. 2021 [30]	Adjuster	Preference	DMCE	MO - D & L	Static	(iii) Rank	(ii) Preference
Jatschka et al. 2021 [31]	Adjuster	Preference	DMCE	SO - D & L	Static	(i) List	(i) Math program
Hanine et al. 2021 [32]	Adjuster	Preference	DMCE	MO - D & L	Static	(i) List	(i) Math program
Jatschka et al. 2019 [33]	Adjuster	Preference	DMCE	SO - D & L	Static	(i) List, (ii) Value	(i) Math program
Jatschka et al. 2019 [34]	Adjuster	Preference	DMCE	SO - D & L	Static	(ii) Value	(i) Math program
Jatschka et al. 2022 [35]	Adjuster	Preference	DMCE	SO - C & L	Static	(i) List, (ii) Value	(i) Math program
Figueiredo et al. 2022 [36]	Adjuster	Data, Preference	DMCE	MO - C & L	Static	(ii) Value	(i) Math program
Meignan 2018 [37]	Enrincher	Domain, Preference	DMCE	SO - D & L	Static	(ii) Value	(iii) Reoptimization

References

[1] Meignan, D., Knust, S., Frayret, J.M., Pesant, G. and Gaud, N., 2015. "A review and taxonomy of interactive optimization methods in operations research," ACM Transactions on Interactive Intelligent Systems, 5(3), 1-43.

[2] Cesta, A., Cortellessa, G., Oddi, A. and Policella, N., 2003. "A CSP-based interactive decision aid for space mission planning," In Congress of the Italian Association for Artificial Intelligence (511-522). Springer, Berlin.

[3] Cesta, A., Cortellessa, G., Denis, M., Donati, A., Fratini, S., Oddi, A., Policella, N., Rabenau, E. and Schulster, J., 2007. "Mexar2: AI solves mission planner problems," IEEE Intelligent Systems, 22(4), 12-19.

[4] Miettinen, K. and Mäkelä, M.M., 2000. "Interactive multiobjective optimization system WWW-NIMBUS on the Internet," Computers and Operations Research, 27(7-8), 709-723.

[5] Klau, G.W., Lesh, N., Marks, J. and Mitzenmacher, M., 2010. "Human-guided search," Journal of Heuristics, 16(3), 289-310.

https://doi.org/10.21872/2023IISE_1894

[6] Dudley, J.J. and Kristensson, P.O., 2018. “A review of user interface design for interactive machine learning,” ACM Transactions on Interactive Intelligent Systems (TiiS), 8(2), 1-37.

[7] Afsar, B., Miettinen, K. and Ruiz, F., 2021. “Assessing the performance of interactive multiobjective optimization methods: a survey,” ACM Computing Surveys (CSUR), 54(4), 1-27.

[8] Xin, B., Chen, L., Chen, J., Ishibuchi, H., Hirota, K. and Liu, B., 2018. “Interactive multiobjective optimization: A review of the state-of-the-art,” IEEE Access, 6, 41256-41279.

[9] Miettinen, K., Ruiz, F. and Wierzbicki, A.P., 2008. “Introduction to multiobjective optimization: interactive approaches,” In Multiobjective optimization (27-57). Springer, Berlin, Heidelberg.

[10] Hamel, S., Gaudreault, J., Quimper, C.G., Bouchard, M. and Marier, P., 2012, October. “Human-machine interaction for real-time linear optimization,” In 2012 IEEE International Conference on Systems, Man, and Cybernetics (SMC) (pp. 673-680). IEEE.

[11] van Vliet, A., Boender, C.G.E. and Rinnooy Kan, A.H., 1992. “Interactive optimization of bulk sugar deliveries,” Interfaces, 22(3), pp.4-14.

[12] Meignan, D., 2014, July. “A heuristic approach to schedule reoptimization in the context of interactive optimization,” In Proceedings of the 2014 Annual Conference on Genetic and Evolutionary Computation (pp. 461-468).

[13] Meignan, D., 2015. “An experimental investigation of reoptimization for shift scheduling,” In Proceedings of the 11th Metaheuristics International Conference (pp. 1-10).

[14] Deb, K. and Sundar, J., 2006, July. “Reference point based multi-objective optimization using evolutionary algorithms,” In Proceedings of the 8th annual conference on Genetic and evolutionary computation (pp. 635-642).

[15] Jaszkiewicz, A. and Słowiński, R., 1999. “The ‘Light Beam Search’approach–an overview of methodology applications,” European Journal of Operational Research (EJOR), 113(2), pp.300-314.

[16] Miettinen, K. and Mäkelä, M.M., 2000. “Interactive multiobjective optimization system WWW-NIMBUS on the Internet,” Computers & Operations Research, 27(7-8), pp.709-723.

[17] Greco, S., Matarazzo, B. and Słowiński, R., 2008. “Dominance-based rough set approach to interactive multiobjective optimization,” Multiobjective optimization: Interactive and evolutionary approaches, pp.121-155.

[18] Miettinen, K., Eskelinen, P., Ruiz, F. and Luque, M., 2010. “NAUTILUS method: An interactive technique in multiobjective optimization based on the nadir point,” EJOR, 206(2), pp.426-434.

[19] Laukkanen, T., Tveit, T.M., Ojalehto, V., Miettinen, K. and Fogelholm, C.J., 2012. “Bilevel heat exchanger network synthesis with an interactive multi-objective optimization method,” Applied Thermal Engineering, 48, pp.301-316.

[20] Miettinen, K., 2007. “Using interactive multiobjective optimization in continuous casting of steel,” Materials and Manufacturing Processes, 22(5), pp.585-593.

[21] Ruotsalainen, H., Miettinen, K. and Palmgren, J.E., 2010. “Interactive multiobjective optimization for 3D HDR brachytherapy applying IND-NIMBUS,” In New Developments in Multiple Objective and Goal Programming (pp. 117-131). Springer Berlin Heidelberg.

[22] Tveit, T.M., Laukkanen, T.P., Ojalehto, V., Miettinen, K. and Fogelholm, C.J., 2012. “Interactive Multi-objective Optimisation of Configurations for an Oxyfuel Power Plant Process for CO 2 Capture,” Chemical Engineering Transactions, 29, pp.433-438.

[23] Ahani, N., Andersson, T., Martinello, A., Teytelboym, A. and Trapp, A.C., 2021. “Placement optimization in refugee resettlement,” Operations Research, 69(5), pp.1468-1486.

[24] Trachanatzi, D., Rigakis, M., Marinaki, M. and Marinakis, Y., 2020. “An interactive preference-guided firefly algorithm for personalized tourist itineraries,” Expert Systems with Applications, 159, p.113563.

https://doi.org/10.21872/2023IISE_1894

[25] He, L., Gilbert, M., Johnson, T. and Pritchard, T., 2019. "Conceptual design of AM components using layout and geometry optimization," Computers & Mathematics with Applications, 78(7), pp.2308-2324.

[26] Ye, Q., Li, F. and Hu, Y., 2022. "Interactive portfolio optimization with cognition-limited human decision making assisted by auxiliary factors."

[27] Feit, A.M., Nancel, M., John, M., Karrenbauer, A., Weir, D. and Oulasvirta, A., 2021. "AZERTY amélioré: computational design on a national scale," Communications of the ACM, 64(2), pp.48-58.

[28] Piemonti, A.D., Babbar-Sebens, M., Mukhopadhyay, S. and Kleinberg, A., 2017. "Interactive genetic algorithm for user-centered design of distributed conservation practices in a watershed: an examination of user preferences in objective space and user behavior," Water Resources Research, 53(5), pp.4303-4326.

[29] Ruiz, A.B., Ruiz, F., Miettinen, K., Delgado-Antequera, L. and Ojalehto, V., 2019. "NAUTILUS Navigator: free search interactive multiobjective optimization without trading-off," Journal of Global Optimization, 74(2), pp.213-231.

[30] Hu, S., Li, D., Jia, J. and Liu, Y., 2021. "A Self-Learning Based Preference Model for Portfolio Optimization," Mathematics, 9(20), p.2621.

[31] Jatschka, T., Raidl, G.R. and Rodemann, T., 2021. "A general cooperative optimization approach for distributing service points in mobility applications," Algorithms, 14(8), p.232.

[32] Hanine, Y., Lamrani Alaoui, Y., Tkiouat, M. and Lahrichi, Y., 2021. "Socially responsible portfolio selection: an interactive intuitionistic fuzzy approach," Mathematics, 9(23), p.3023.

[33] Jatschka, T., Rodemann, T. and Raidl, G.R., 2019. A cooperative optimization approach for distributing service points in mobility applications. In Evolutionary Computation in Combinatorial Optimization: 19th European Conference, Proceedings 19 (pp. 1-16). Springer International Publishing.

[34] Jatschka, T., Rodemann, T. and R. Raidl, G., 2019. Exploiting similar behavior of users in a cooperative optimization approach for distributing service points in mobility applications. In Machine Learning, Optimization, and Data Science: 5th International Conference, Proceedings 5 (pp. 738-750). Springer International Publishing.

[35] Jatschka, T., Rodemann, T. and Raidl, G.R., 2022. A Large Neighborhood Search for a Cooperative Optimization Approach to Distribute Service Points in Mobility Applications. In Metaheuristics and Nature Inspired Computing: 8th International Conference (pp. 3-17). Springer International Publishing.

[36] Figueiredo, M.V., Silva, A.F. and Marins, F.A.S., 2022. Multi-Objective Optimization and Design of Experiments Applied on Orthopedic Assets Distribution Process. Available at SSRN 4178121.

[37] Meignan, D., 2018. A user experiment on interactive reoptimization using iterated local search. Recent Developments in Metaheuristics, pp.399-413.

https://doi.org/10.21872/2023IISE_2145

Order Acceptance and Detailed Scheduling in a Job Shop with Discrete and Batch Processing Machines

Dheeban Kumar Srinivasan Sampathi[a], and Purushothaman Damodaran[a]

[a]*Northern Illinois University, DeKalb, IL, USA*

Abstract

Customization of products to increase customer satisfaction has resulted in a high mix low volume production. Many manufacturers adopt a Make-To-Order (MTO) approach to minimize costs and meet customer deadlines. This study deals with processing customer orders with a set of operations on resources in a job shop environment, with linear precedence constraints, deterministic processing times, due dates, job size, and selling price. Each process step can be performed on one or more machines in a department. Jobs may also recirculate. The objective is to maximize profit and ensure that all accepted orders are completed on time. The job shop includes discrete processing machines and a batch processing machine that can process multiple jobs if its capacity is not exceeded. The research aims to formulate the problem to assist the operations manager. The Mixed Integer Linear Program helps to identify which orders to accept and a detailed schedule for all the accepted orders.

Keywords
Order acceptance, Capacity planning, Job shop, Batch processing machine

1 Introduction

Order acceptance and capacity planning are two important components in a job shop environment. In a job shop, each customer order is unique and requires specialized resources and production processes, making order acceptance and capacity planning particularly challenging. Order acceptance involves evaluating each customer order and determining whether to accept or reject it based on factors such as available capacity, production lead time, and profitability. The goal of order acceptance is to maximize the profitability of the job shop by accepting only orders that can be fulfilled by the due dates using the available resources and other constraints. Capacity planning, on the other hand, involves determining the production capacity of the job shop and allocating resources in the most efficient way to meet customer demand. In a job shop, capacity planning requires a deep understanding of the production process, resource constraints, and the availability of specialized equipment and skilled personnel. Effective order acceptance and capacity planning are critical for ensuring the smooth and efficient operation of a job shop, as well as for maximizing profitability and customer satisfaction. This requires the use of advanced tools and techniques, such as mathematical models, to make informed decisions based on real-time data and market conditions.

In this research, a job shop with both single (or discrete) and batch processing machines are taken into consideration. The use of batch-processing machines can significantly improve the efficiency and profitability of a manufacturing plant by reducing the production time and labor required to produce many identical products.

2 Problem Description

A make-to-order (MTO) environment is a manufacturing process in which products are only produced after a customer has placed an order. In an MTO environment, production is triggered by customer demand rather than being based on pre-determined production schedules. This contrasts with a make-to-stock (MTS) environment, where products are manufactured and stored in anticipation of customer demand. In an MTO environment, customer orders are typically evaluated on a case-by-case basis, considering the resources and capabilities of the manufacturing plant, as well as any specific customer requirements. This allows for a high degree of customization and flexibility in the production process, as products can be tailored to meet the specific needs of each customer. The MTO manufacturing process is

https://doi.org/10.21872/2023IISE_2145

often used for high-value or high-complexity products, where the unique requirements of each customer demand a custom-made solution. This can include products such as machinery, medical devices, and specialized equipment. However, the MTO process can also be more time-consuming and resource-intensive than other manufacturing processes, as production is only initiated once an order has been placed, and each order must be evaluated and planned individually. Careful planning and management of resources are essential in an MTO environment to ensure that customer orders are fulfilled in a timely and efficient manner while maximizing profitability. Examples are engineering tooling, Industrial boilers, and contracting industries.

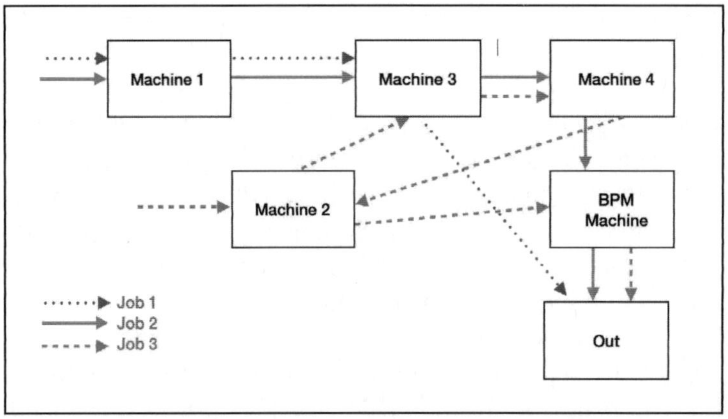

Figure 1. Problem representation

Figure 1 shows a schematic diagram of the problem under study. In this example, Machines 1 through 4 are discrete processing machines and BPM is a batch processing machine. The process routes for three jobs are shown. For example, Job 2 needs to be processed on Machine 2, Machine 3, Machine 4, back to Machine 2 and finally by the BPM.

This research includes both single and Batch Processing Machine (BPM). Each job has a predefined routing step. A job may or may not visit a BPM. If it visits BPM, it is the final operation. Recirculation among the single machines is allowed. Consider a job shop with a set of customer orders. The choice to be made is which customer order to accept and how to plan it to maximize profit and meet the deadlines for the accepted orders. Both decisions should be made at the same time; Otherwise, an order may be approved but delivery would be delayed due to available residual capacity. Order contains a set of activities that must be completed in a linear order with deterministic processing time, due date, and sales price. Late deliveries are not permitted. The goal is to maximize the operational profit, solely considering the sales price and production expenses. The goal is to formulate the MTO problem under study to assist operations managers to make informed decisions. The problem under study can be modeled as a MILP. MILP inherits the block diagonal structure where each subproblem is for a job/order. In the future, it can be decomposed using the Dantzig–Wolfe decomposition process and develop solution approaches to solve large-scale problem instances.

3 Literature Review

Previous research in MTO operations has focused on either order acceptance or capacity planning, but not both problems together with BPM and discrete processing machines. This research addresses the gap in the literature by presenting a MILP formulation that can solve the combined problem. This approach considers the trade-off between order acceptance and production capacity and provides an optimal solution for the problem.

3.1 Order Acceptance

The issue of short-term capacity planning in a make-to-order setting is addressed by Chen et al. (Chen, Mestry, Damodaran, & Wang, 2009), focuses on the capacity planning problem in make-to-order (MTO) enterprises. It uses mathematical and computer modeling to develop a framework for solving the capacity planning problem in MTO enterprises. The framework considers the constraints of available capacity and order acceptance to determine the optimal production schedule. The results of the study show that the proposed framework can be used to effectively manage capacity and improve the efficiency of MTO enterprises. It was concluded that capacity planning is a crucial factor in the success of MTO enterprises and that their framework can provide a useful tool for managing this process.

Tarhan and Oğuz (Tarhan & Oğuz, 2021) worked on generalized order acceptance and scheduling (GOAS) problems in a single-machine environment with release time and sequence-dependent setup times. (Tarhan & Oğuz, 2021) study developed an effective matheuristic for GOAS problems. Rahman et. (Rahman, Sarker, & Essam, 2015) proposed a strategy for solving the Order Acceptance and Scheduling (OAS) problem for a permutation flow shop. In this problem, the orders are dynamic with a predefined processing time. In a flow shop, all orders have the same processing sequence. In a dynamic environment, two decisions must be made. (1) either to accept or reject the order. (2) If accepted, the order scheduling must be made. (Rahman, Sarker, & Essam, 2015)developed a heuristics-based decision for the first task and a Genetic Algorithm (GA) based approach for the second task.

3.2 Solution Approach

Mestry et al. (Mestry, Damodaran, & Chen, 2011), presents a solution approach for the combined problem of order acceptance and capacity planning in make-to-order (MTO) operations. (Mestry, Damodaran, & Chen, 2011) uses a branch and price method to find an optimal solution for the problem, considering the trade-off between order acceptance and production capacity. The results of the study show that the proposed approach can effectively solve the combined problem of order acceptance and capacity planning in MTO operations. They conclude that the branch and price approach provide a practical tool for managing order acceptance and capacity planning in MTO operations and that it can help improve the efficiency of MTO operations.

Tarhan and Oğuz (2021) (Tarhan & Oğuz, 2021) presents a matheuristic approach for solving the generalized order acceptance and scheduling problem. The authors develop a matheuristic algorithm that integrates mathematical programming and heuristics to solve the problem, which involves deciding which orders to accept and scheduling the accepted orders to minimize the total cost. The contribution of this article to the literature on order acceptance and scheduling is significant, as it presents a new approach to solving the problem. The matheuristic algorithm developed in this study can help companies in the manufacturing industry make more effective decisions regarding order acceptance and scheduling, leading to improved efficiency and cost savings.

Rahman et., (2015) (Rahman, Sarker, & Essam, 2015), The authors propose a real-time order acceptance and scheduling approach that combines the use of a heuristic algorithm and a simulation-based optimization method. The heuristic algorithm is used to quickly identify the feasible orders, while the simulation-based optimization method is used to determine the optimal scheduling plan. Herbots et al., (2007) (Herbots, Herroelen, & Leus, 2007), applied stochastic dynamic programming to determine the profit based on order accept/reject in a dynamic order environment. Also, deterministically allocate a single resource for the accepted orders.

3.3 Research Justification

The literature reviewed did not find any research article that has attempted to solve the proposed problem with the stated assumptions. Despite the extensive literature review, there were no studies that combined batch processing machines and discrete machines. The MILP formulation helps to study this complex and real-life problem commonly observed in many manufacturing companies.

4 Formulation

The notations used in the formulation are as follows,

Sets

$Machine$	$\{k \in K\}$
$Jobs$	$\{i \in I\}$
$Operations$	$\{j \in J\}$
$Time$	$\{t \in T\}$
$Batch$	$\{b \in B\}$

Parameters

d_i	Due Date of job *i*
q_i	Size of job *i*
s_i	Selling price of job *i*

https://doi.org/10.21872/2023IISE_2145

c_k	Cost of using machine *k*
p_{ij}	Processing time of operation *j* of job *i*
l_t	Length of each time period *t* (i.e., shift length)
a_{kt}	Availability of machine *k* in time period *t*
$BigM$	Large positive number
MS	Machine Size

Decision Variables

z_i	1 - if job *i* is accepted; 0 otherwise
X_{ijkt}	Hours of operation *j* of job *i* processed on machine *k* in time period *t*
Y_{ijkt}	1 – if operation *j* is of job *i* is processed on machine *m* on time *t*.
β_{ib}	1 – if job *i* is in batch *b;* 0 otherwise
JC_i	Completion time of job *i*
JR_i	Ready time of job *i*
BP_b	Processing time of batch *b*
BR_b	Ready time of batch *b*
BC_b	Completion time of batch *b*
FC_i	Final completion time of job *i*

Maximize

$$\sum_{i\in I} s_i\, z_i - \sum_{i\in I}\sum_{j\in J}\sum_{k\in K}\sum_{t\in T} c_k\, X_{ijkt} - \sum_{k\,=\,Batch}\sum_{b\in B} c_k\, BP_b \tag{1}$$

Subject to

$$\sum_{t\in T} X_{ijkt} \geq p_{ij}\, z_i \qquad \forall\, i \in I, k \in K \tag{2}$$

$$\sum_{i\in I}\sum_{j\in J} X_{ijkt} \leq a_{kt} \qquad \forall\, k \in K\backslash\{|K|\}, t \in T \tag{3}$$

$$\sum_{j\in J}\sum_{k\in K-1} X_{ijkt} \leq l_t \qquad \forall\, i \in I, t \in T \tag{4}$$

$$X_{ijkt} \geq Y_{ijkt} \qquad \forall \in I, j \in J, k \in K\backslash\{|K|\}, t \in T \tag{5}$$

$$X_{ijkt} \leq p_{ij}\, Y_{ijkt} \qquad \forall \in I, j \in J, k \in K\backslash\{|K|\}, t \in T \tag{6}$$

$$\sum_{t'=1}^{t-1} X_{i(j-1)kt'} + X_{i(j-1)kt} \geq p_{ij} \sum_{k'\in K} Y_{ijk't} \qquad \forall \in I, j \in J\backslash\{1\}, k \in K\backslash\{|K|\}, t \in T \tag{7}$$

$$JC_i \geq t\, Y_{ijkt} \qquad \forall\, i \in I, j \in J, k \in K\backslash\{|K|\}, t \in T \tag{8}$$

$$JR_i \geq JC_i \qquad \forall\, i \in I \tag{9}$$

$$\sum_{k\in K} t\, Y_{i|J|kt} \leq d_i\, z_i \qquad \forall\, i \in I, t \in T \tag{10}$$

$$\sum_{b\in B} \beta_{ib} = 1 \times z_i \qquad \forall\, i \in I \tag{11}$$

$$\sum_{i\in I} q_i\, \beta_{ib} \leq MS \qquad \forall\, b \in B \tag{12}$$

$$BP_b \geq \frac{p_{ij}}{8}\, \beta_{ib} \qquad \forall\, b \in B, i \in I, j \in J \tag{13}$$

$$BR_b \geq JC_i + BigM\,(\beta_{ib} - 1) \qquad \forall\, b \in B, i \in I, j \in J \tag{14}$$

$$BC_b \geq BR_b + BP_b \qquad \forall\, b \in B \tag{15}$$

$$BC_b \geq BR_{b-1} + BP_b \qquad \forall\, b \in B\backslash\{1\} \tag{16}$$

$$BC_b \geq BC_{b-1} + BP_b \qquad \forall\, b \in B\backslash\{1\} \tag{17}$$

$$FC_i \geq BC_b + BigM\,(\beta_{ib} - 1) \qquad \forall\, i \in I, b \in B \tag{18}$$

$$FC_i \leq d_i z_i \qquad \forall i \in I, b \in B \qquad (19)$$

$$X_{ijkt}, Y_{ijkt}, \beta_{ib}, JC_i, JR_i, BP_b, BR_b, BC_b, FC_i \geq 0$$
$$z_i, \beta_{ib} \in \{0,1\} \qquad (20)$$

The objective (1) is to maximize the profit by selecting jobs that generate revenue, the second term is the cost associated with using single machines to process those jobs, and the third term is the cost associated with BPM. (2) ensures that if a job is accepted, it is required that every operation of the job has sufficient processing time at least greater than p_{ij} on the machine to which it is assigned. (3) ensure that the machine should not be overused more than the availability in any given time period. (4) ensure that the hours of operation on any machine do not exceed the shift length. (5) ensure that X_{ijkt} should only be positive only if $Y_{ijkt}$is also positive. (6) The hours of operation of each job *i* should be less than or equal to the processing time of operation *j* of job *i* on machine *k* in time period *t*. (7) is the linear precedence constraint. If operation *j* of job *i* is processed on machine *k* in time period *t*, then the previous operation *j-1* of job *i* should be completed before starting operation *j* on machine *k*. (8) Job completion time should be greater than or equal to the time period *t*. (9) the job ready time should be greater than or equal to the job completion time. (10) ensure that the final operation of job *j* in a discrete machine is done before the due date if it is selected.

From equation (11) the constraints for batch processing machine are listed. The following constraints are only for jobs having the last operation in the BPM. (11) Each job should belong to exactly one batch. (12) The sum of job sizes in a batch should not exceed the machine size. (13) The batch processing time should be greater than or equal to the processing time of the operations in the batch, Since the processing time is in hours, the conversion is done by dividing the processing time batch by 8 (assuming 8 hours per time period). (14) The batch-ready time should be greater than or equal to the completion time of the last operation of the previous job. (15) relates to the completion time of a batch. It ensures that the completion time of batch b is greater than or equal to its ready time plus of its processing time. (16) constraint is similar to (15), but it relates to batch *b* and its previous batch *(b-1)*. It ensures that the completion time of batch *b* is greater than or equal to the ready time of batch *(b-1)* plus the processing time of batch *b*. (17) is also similar to constraint (15), but it relates to batch *b* and its previous batch *(b-1)*. It ensures that the completion time of batch *b* is greater than or equal to the completion time of batch *(b-1)* plus the processing time of batch *b*. This constraint applies to batches other than the first batch. (18) ensures that the final completion time of job *i* is greater than or equal to the completion time of batch *b* based on the job in the batch. (19) ensure that the final completion time is less than or equal to the due date of the job if it is selected. (20) ensures that all decision variables are non-negative and $z_i, \beta_{ib}$is a Boolean.

5 Results

An example problem instance was solved using IBM ILOG CPLEX. The data for the problem is summarized in table 1 and 2. The usage of resources is shown in figure 2. In this example two job with multiple operations is shown. The final operation is in the BPM i.e., the machine 3 in this example. The Gantt chart shows the scheduling of job on each machine available. Table 3 shows the result of different job and machine instances ran using CPLEX.

Table 1. Operation details

Jobs	Operation	Machine	Processing time
1	1	1	6
1	2	2	4
1	3	3	4
2	1	2	2
2	2	3	4

Table 2. Job details

Job	Due Date	Size	Selling Price
1	2	1	1000
2	2	4	1000

https://doi.org/10.21872/2023IISE_2145

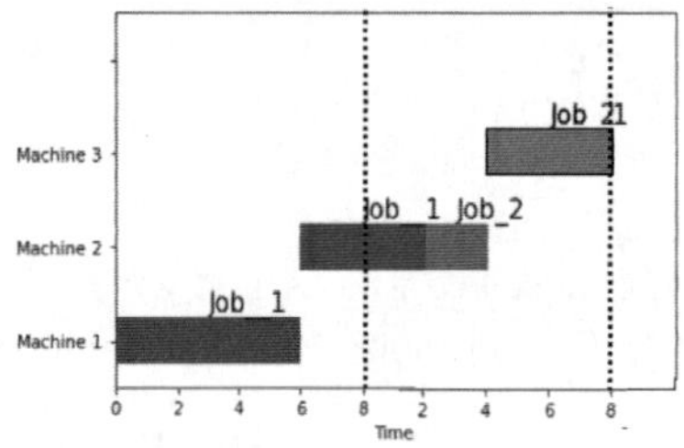

Figure 2. Gantt Chart

Table 3. CPLEX results

Dataset	# Jobs	# Machines	Jobs to select	Run time (seconds)	Gap
1	10	7	[1,2,5,7,8,9,10]	2	0
2	10	7	[1-9]	4.23	0
3	10	7	[2-9]	5.91	0
4	20	10	[2,3,5,7,9,10,11,12,13,15,18,20]	1800	44.48%
5	20	10	[1,2,3,4,6,8,9,11,12,14,15,16]	1800	49.85%
6	20	10	[3,4,6,7,8,9,10,15,16,17,19,10]	1800	46.13%

6 Conclusion and Future scope

In conclusion, it is important to note that the scheduling of a job shop and the acceptance of orders are key factors that might affect a manufacturing company's success or failure. Businesses in the highly competitive market must create efficient order acceptance procedures to satisfy client demand, while the scheduling process must productive and efficient. To maximize the production capacity of the job shop, organizations must use the proper technologies and software systems. Establishing clear channels of communication among team members is also crucial to prevent misunderstandings and delays in the execution of orders.

The job shop order acceptance and scheduling problem can be further expanded to include additional sources such as overtime and outsourcing. By incorporating these sources, businesses can better manage their resources and production capacity to meet customer demands. Furthermore, as the size of the problem increases, the time taken to solve it also increases. Therefore, developing efficient solution strategies to solve large-scale job shop order acceptance and scheduling problems. The use of advanced algorithms such as branch and price algorithms can improve the efficiency of solving large-scale problems. Another possible extension is the application of constraint programming and comparing the solution quality obtained from the different approaches.

7 References

[1] C.-S. Chen, S. Mestry, P. Damodaran and C. Wang, "The capacity planning problem in make-to-order enterprises," *Mathematical and Computer Modelling,* vol. 50, no. 9-10, pp. 1461-1473. doi:10.1016/j.mcm.2009.07.010, 2009.

[2] İ. Tarhan and C. Oğuz, "A matheuristic for the generalized order acceptance and scheduling problem," *European Journal of Operational Research,* vol. 299, no. 1, pp. 87-103. doi:10.1016/j.ejor.2021.08.024, 2021.

[3] H. F. Rahman, R. Sarker and D. Essam, "A real-time order acceptance and scheduling approach for permutation flow shop problems," *European Journal of Operational Research,* vol. 247, no. 2, pp. 488-503. doi:10.1016/j.ejor.2015.06.018, 2015.

[4] S. Mestry, P. Damodaran and C.-S. Chen, "A branch and price solution approach for order acceptance and capacity planning in make-to-order operations," *European Journal of Operational Research,* vol. 211, no. 3, pp. 480-495. doi:10.1016/J.EJOR.2011.01.002, 2011.

[5] J. Herbots, W. Herroelen and R. Leus, "Dynamic order acceptance and capacity planning on a single bottleneck resource," *Naval Research Logistics,* vol. 54, no. 8, pp. 874-889. doi:10.1002/nav.20259, 2007.

https://doi.org/10.21872/2023IISE_2149

Proceedings of the IISE Annual Conference & Expo 2023
K. Babski-Reeves, B. Eksioglu, D. Hampton, eds.

Bid Packaging Decisions for Construction Project Organizations

Abstract ID: 2149

Dr. Panagiotis Mitropoulos and Ms. Kawalpreet Kaur, San Diego State University, San Diego, CA

Abstract

In construction projects, bid packaging involves dividing the project scope into work packages that are assigned to trade contractors. The division of the work determines the scope and complexity of each bid package, as well as the dependencies with other bid packages and consequently the need for coordination between project participants. Improper division of bid packages can create coordination difficulties and breakdowns, with potential delays, added costs, and contractual friction between project participants. The construction literature, however, provides limited information on the topic both descriptive (how managers make the bid packaging decisions) and normative (methods to perform effective bid packaging). The objective of this paper is to increase understanding of the factors affecting bid packaging decisions, and the project implications of bid packaging. The investigation consists of three exploratory case studies on educational building projects. The findings show that the Construction Managers (CM) on these projects had multiple goals, which created trade-offs between the goals. The findings identify three basic configurations for organizing scope elements and discuss their project implications for bid price, and performance risks. The study develops new insights into the complexity, challenges and implications of bid packaging decisions.

Keywords
Bidding, Bid packages, CM, Educational facilities, Project organization

1. Introduction
On every construction project, a new production organization is formed as the work is divided into bid packages and assigned to trade contractors. This division of work establishes the organizational structure of the project and configures the supply chain. Bid packaging is typically performed by the managing contractor, which could be a General Contracting (GC) firm or the Construction Management (CM) firm, during the procurement phase. The CM/GC may award bid packages to specialty contractors, or may self-perform a certain scope of work. The work division determines the dependencies and interfaces among the scope elements. These dependencies create the need for coordination between trade contractors [1]. Poor work division can create coordination difficulties and "breakdowns" in work process, with potential delays and added costs, and contractual friction between project participants. For example, errors or delays in one scope of work may delay subsequent work, or require adjustment by following activities. Thus, bid packaging has significant implications for project performance as it affects the production, coordination and transaction costs of the project [2-5].

2. Problem Description
Despite the importance of bid packaging, the construction engineering literature provides very limited knowledge regarding the factors that drive bid packaging decisions, the strategies and practices that managers use to make these decisions and the implications for project coordination, risks and performance [6]. In addition, practitioners have very limited tools to help them organize the work in bid packages [7]. To address these gaps, this study investigated the bid packaging approach used on three recent construction projects. The study was guided by the following main goals: First, to develop a deeper understanding of the managerial goals and considerations that drive these decisions. Second, to investigate different bid package configurations and examine their project implications.

https://doi.org/10.21872/2023IISE_2149

3. Related research

The way the work scope is divided to different actors greatly influences the need for coordination [1] in order to manage the dependencies between activities [8]. Activity dependencies can be due to workflow, where a task produces an output used by another task or shared resources. Workflow dependencies are of three types: (a) the output of one task must be available at the time it is needed by the other, (b) the output must be of adequate quality, and (c) the output must be available at the right place. Researches recognize that a major part of the problems in construction projects originate at the interfaces of different parties or functions [9] and that tight handoffs and interface issues such as the mismatched parts, system performance, coordination difficulties, and assembly conflicts between trades are critical for project performance [10, 11].

The process of defining the bid packages is critical to the project success [12]. Inappropriate work division leads to complicated interdependencies among work packages which result in the complexity of interfaces in projects and increase the potential for disruptions, rework and delays [5,13]. A mismatch between the work breakdown structure and the organizational structure has a negative impact on project progress [4]. Very few studies have focused on the configuration of the bid packages. The work breakdown structure (WBS) is the typical method used to decompose the project scope into smaller components [7]. Nicolini et al. [14] proposed that another way to address and manage the interfaces is the procurement of work as "clusters." Overall, the construction literature provides very limited information both descriptive (how managers make such decisions) and normative (methods to perform effective bid packaging). In order to develop more effective project organizations, it is necessary to better understand how the bid packaging decisions are made and how they impact the project.

4. Methodology

This research used an exploratory case study method at the project level to investigate the bid packaging issues on three recent construction projects. The projects involved the construction of educational facilities in three community colleges in Southern California. Data were collected through interviews with project personnel and review of project documents. Project documents included project bid packages, contract documents related to the bid packages, documentation of rationale behind the bid packaging decisions, documents and graphs used to visualize bid packaging issues, and other documents such as schedules, etc.

Project 1 was a four story science building. The construction cost was $54.2 million and the schedule was 24 months. The project was design-bid-build with a CM multi-prime approach. The project scope was divided in 35 bid packages. The sources of information of this project were two managers—the Owner's project manager and the CM's Vice-President of Operations. Project 2 was a gymnasium and swimming pool building complex with a construction budget of $47.8 million and a schedule of 36 months. The project delivery system was CM at Risk. The project scope was divided in 27 bid packages. The sources of information of this project were two managers of the CM team: Director of Business Acquisition and Project Engineer. Project 3 was a sports complex with a central utility building. The construction budget of the project was $45.1 million and the schedule was 34 months. The project delivery system was Design-Bid-Build using a "Construction Manager at Risk." The project scope was divided in 22 work packages. The CM's Project Manager, Field Superintendent and project documents were the sources of information on this project.

5. Results

5.1. Management Goals Driving Bid Packaging

The findings indicate that the CM's bid packaging decisions was driven by five primary goals: (1) Satisfy owner's requirements for community benefits, (2) Attract competitive bid price, (3) Attract qualified contractors, (4) Maintain control of the work, and (5) Manage critical dependencies.

(1) **Owner's Requirements for Community Benefits**: The Community College owners in this study established two project requirements aiming at community benefits—participation of local, small and disadvantaged contractors, and a "Community Benefit Agreement (a.k.a. Project Labor Agreement). On all three projects, the owners aimed at attracting local contractors, with emphasis on small and disadvantaged contractors. Participation of small contractors involves performance risks due to limited resources and management skills. To address this risk, the CMs created smaller bid packages of lower risk work items (items that were not critical and did not require special skills). Bid packages targeting small contractors may reduce bidders' interest and result in higher overall bid values.

https://doi.org/10.21872/2023IISE_2149

(2) **Competitive bid pricing**: In order to secure competitive bid pricing, the CM aims at the following: create economies of scale, attract bidders, avoid double markup, and minimize scope exclusions. To create "economies of scale" the CM combines similar work in a large package. Work scope often combined includes electrical and low voltage work, steel and miscellaneous metals, roofing and sheet metal, cladding and glazing and storefront. Larger packages have greater profit potential and can attract more bidders. In building construction, three to six bidders for each BP is considered acceptable bid coverage. To attract bidders it is not enough to create larger bid packages. The BPs need to be consistent with the local contractors' capabilities. Efforts to create larger BPs can sometimes increase the complexity of the packages beyond the capabilities of the contractors' and reduce bidders' interest. On the Sports facilities project, the CM intended to combine carpet and tile in one bid package for "flooring" to facilitate the coordination of transitions and levelness, but finally the scopes were separated because most contractors do not perform both. "Double markup" is created when the lead contractor needs to hire second tier contractors to perform portions of the work. This increases the bid cost (a) both the trade that performs the work and the contractor who hires them add a markup. (b) the lead contractor is responsible for the coordination of the trades which increases their management cost. (c) The complexity of the bid package may reduce the bidders' interest, as contractors may be reluctant to hire and manage a second tier contractor. Scope exclusions are created when the bidders do not have the interest or capability to perform a scope item. Excluded scope may require re-bidding or may be added to one of the other awarded contracts. To prevent exclusions the BPs must be consistent with the potential bidders' capabilities. The CMs in all three cases requested input from local trade contractors regarding their willingness to bid on the scope elements of a bid package.

(3) **Attract qualified contractors**: To increase the chances for a successful project, the CM makes efforts to attract qualified contractors for critical work and specialty items. For the critical project items, such as structural system, mechanical, electrical, etc., where failure or delay would have significant impact on project performance, the CMs in all three cases made efforts to attract larger contractors, as they have greater technical expertise and management capabilities. To attract larger contractors, the CMs formed larger bid packages combining similar work. Specialty items involve specialized equipment and require manufacturer-certified installers in order to provide the required system warranties. The buildings in the case studies required several specialty items such as lab equipment, fume hoods, millwork, pool and pool equipment, wood sports flooring, scoreboard & sports equipment. The CMs placed specialty items in separate bid packages because it gives them direct control over the selection of the contractors. The bid packaging decision of specialty items is also influenced by coordination considerations, as discussed later.

(4) **Control the project**: The CMs want to have control of two issues: selection of the contractors, and control of the work and the schedule. The CMs have greater control when they have direct contracts with each trade as the CM selects the contractors, and can use the contract to establish performance requirements and enforcement mechanisms to direct the contractors. The CM has reduced control over the selection and direction of the 2nd tier contractors as their work is directed by their lead contractor. In this situation, it is important that the CM and the "lead" contractor can work well together. To this end, the CM of the Sports facility includes a clause in the contract that gives the CM the right to replace the contractor's manager if not satisfied with the manager's cooperation.

(5) **Manage critical dependencies**: Work processes that require very close coordination of different trades are better coordinated when the scope elements are placed in the same bid package with one contractor responsible for overall system performance. Close coordination is required for system performance, production sequence, and prevention of defects among trades with extensive interfaces.:

- System performance and warranty: The CM on the Sports facility, placed all the work related to the building enclosure system under one bid package, so that one contractor would be responsible for all issues related to the building enclosure system performance and provide warranty. This included the stone veneer on the exterior wall, which was a specialty item that required manufacturer-certified installers. Typically, this would be placed in a separate bid package, but a review of the technical details of the wall system indicated that it was very important to have close coordination between the stone veneer contractor and the exterior wall trades, to reduce the risk of defects and water penetration.
- Highly coupled production tasks: In the Sports facility project the earthwork operation called for soil nails to stabilize the excavation. The work process required close coordination between excavation and soil nails and shotcrete installers. To facilitate this coordination, the CM combined all these elements as part of the excavation package, and the soil nails contractor was a second tier contractor under the earthwork contractor. On the Science building project, the building concrete BP included the reinforcing steel, and the concrete contractor had the overall responsibility for the work of the two trades.

https://doi.org/10.21872/2023IISE_2149

- "Friction" among trades with extensive interfaces. Contractors of interior systems and finishes work in close proximity and have many interfaces. Problems between trades are very common: problems with transitions (gaps, clearances, alignment, levelness, etc.); errors by an earlier trade create problems to subsequent trades; and downstream trades may damage the work of earlier trades. Such problems increase when the trades work under schedule pressure and congested conditions. Furthermore, repairing a defect often requires rework by several earlier trades. The result is increased production costs (rework and delays) as well as excessive transaction costs, as the CM has to document which trade caused the problem and who needs to be compensated, and issue appropriate contract changes. This can result in disputes, negotiations and claims. To prevent such waste, the CM of the Sports facility project combined the interior work under one multi-trade bid package. The contractor awarded this BP self-performed the majority of the trades. Under this approach the lead contractor was responsible to coordinate the interfaces, the CM did not have to make contract adjustments for issues between the trades, and the trades had a shared interest to avoid causing problems and damages to each other.

5.2. Bid Packaging Trade-offs

The bid packaging objectives identified above support different configurations with regards to the size and the complexity of the bid packages. Some objectives encourage smaller rather than larger BPs. Some objectives "push" towards separating the scope to simpler BPs, while others "pull" towards combining scope into more complex ("multi-trade") packages. Table 1 summarizes some of the important trade-offs identified.

Table 1. Trade offs between bid packaging goals

Objective	Approach	Trade-off
Increase participation of small, local contractors	Separate scope into more and smaller BPs	Reduced bid coverage and higher bids Increased risk of scope gaps Increased performance risks due to lower capabilities. Higher coordination load for CM
Attract bidders Attract larger qualified contractors.	Create larger bid packages	Reduced participation of smaller contractors May require 2nd tier contractors Complex packages may reduce bidders' interest
Attract qualified contractors for specialty work	Separate BPs for specialty items	Increase performance risks due to coordination difficulties. Higher coordination load for CM
Avoid double markup	Separate BPs to avoid second tier contractors	Higher coordination load for the CM Increased performance risks due to coordination problems. Problems with systems warranty and liability
Maintain control of project	Separate BPs for direct contract with trades	Higher coordination load for the CM May increase performance risks due to coordination problems. Problems with systems warranty and liability
Manage critical dependencies.	Combine interdependent work in complex BPs.	Requires 2nd tier contractors May reduce bid coverage Increased markup due to increased risk

5.3. Bid Packaging Configurations and Project Implications

The CM's decisions on how to separate or combine scope elements result in different project configurations. They involve different contractual relationship between the trades who perform the scope elements and have implications for the project organization, the bid price, the coordination, the performance risks and the contractual adjustments when there are changes. The study identified three basic BP configurations—that is, three basic ways in which a set of scope elements can be organized. The configurations are classified in two categories: (1) Fragmented and (2) Combined, which is further divided in (2.A) Combined / Multi-party, and (2.B) Combined / Integrated. Figure 2 illustrates the three configurations.

https://doi.org/10.21872/2023IISE_2149

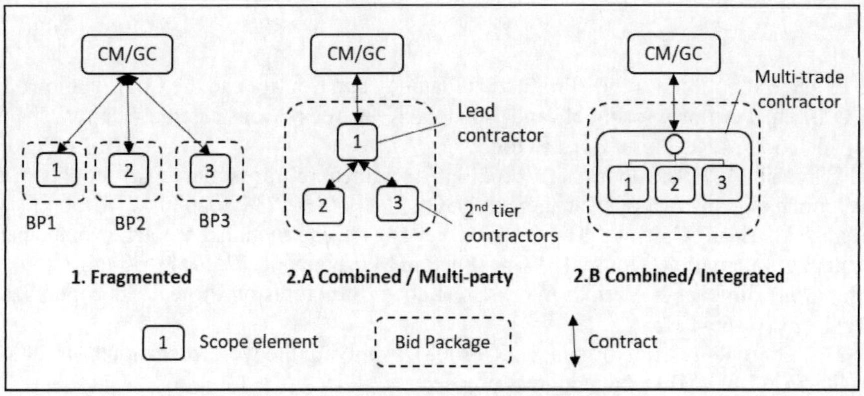

Figure 2. Bid packaging configurations

In the Fragmented configuration, the scope elements are divided into separate BPs and each BP is awarded to a different trade contractor. The division of scope in BPs and the competitive bidding for each package, can result in a lower bid price, however there is higher risk of scope gaps. The CM has high control – selects the trade contractors, and has a direct contract with each; there are no contracts between the trade contractors. The CM is responsible for coordination; for scope elements with extensive dependencies, the coordination load for the CM is very high. Consequently, there is greater risk for coordination problems. When multiple BPs are part of a system, there is no single point of responsibility for system performance. Change management is costly and difficult as several contracts need to be adjusted, generating high transaction costs.

In the Combined configurations, the CM places different scope elements under one bid package and awards it to one lead contractor. In the Multi-party configuration (2.A) most of the work is performed by second-tier contractors who are selected by the lead contractor based not only on bid price, but also on working relations. This configuration allows more collaborative relations between the project participants [15 – 18]. In the Integrated configuration (2.B), one multi-trade contractor performs the majority of the work with their own labor force, possibly with a small number of second-tier contractors.

The Combined configurations have several advantages—it attracts larger contractors with higher capabilities; it reduces the risk for scope gaps and overlaps; the lead contractor provides a single point of responsibility and has strong incentive to minimize problems at the interfaces between the trades in the BP; the coordination within the BP can be more effective due to collaborative relations; and change management is simpler with lower transaction costs. Disadvantages include: (1) a higher mark up and bid price due to increased coordination and effort and increased risk for overall performance, (2) reduced CM control over the selection and direction of the trades; and (3) fewer bidders, as contractors may not want the risk of managing 2nd tier contractors, and the availability of multi-trade contractors may be limited. As a result, CMs may be reluctant to create combined bid packages [19]. In recent years however, there has been increased use of single-point contractors especially for building enclosure systems [20], driven by the high costs of coordination failures. Thus, in addition to the bid price, BP decisions need to carefully consider the coordination requirements between scope elements, the coordination potential of the bid packaging configuration, and the consequences of potential coordination failures.

5.4. Conclusions

This exploratory investigation contributes to a better understanding of the complexity of the bid packaging decisions and makes the following contributions. First, it discovers the multiple goals that drive these decisions. Second, it identifies the trade offs between the goals. Third, the study identified different configurations for organizing dependent scope elements and fourth it increases understanding of the project implications of the different configurations for bid costs, coordination, performance risks and transaction costs. The results indicate that the division of the work scope in bid packages creates project risks and complexities purely due to the work division: coordination load, coordination risks, scope gaps, performance risks, accountability challenges, risks due to misalignment of interests and conflict. The study identifies the need for systematic methodologies to assist managers evaluate BP alternatives for their multiple implications. Such methodologies can make a significant contribution to the design of effective project organizations.

https://doi.org/10.21872/2023IISE_2149

References

[1] Formoso, C. T., and Isatto, E. L. (2009) "Production Planning and Control and the Coordination of Project Supply Chains" In: O'Brien, Formoso, Vrijhoef, and London, K. A (eds) Construction Supply Chain Management Handbook, pp. 3.1–3.25, CRC Press, Boca Raton.

[2] Abdel-Galil E., Ibrahim A. H. & Alborkan A. (2020) "Assessment of transaction costs for construction projects." International Journal of Construction Management, DOI: 10.1080/15623599.2020.1738204

[3] Safa, M., Shahi, A., Haas, C.T. and Hipel, K.W. (2017). "Construction contract management using value packaging systems." International Journal of Construction Management, 17(2017), 50-64.

[4] Globerson, S. (1994). "Impact of Various Work-Breakdown Structures on Project Conceptualization" Intern. J. of Project Mgt., 12 (3): 165-171.

[5] Brotherton, S. A., Fried, R. T. and Norman, E. S. (2008) "Applying the work breakdown structure to the project management lifecycle." In PMI global congress proceedings, 1-15.

[6] Kaur, K., and Mitropoulos, P. (2017) "Work Structuring Principles, Strategies and Tools: A Review of the Literature" In: Proceedings of the 6th CSCE/CRC International Construction Specialty Conference, pp: CON130-1 – 130-9, May 31- June 3, Vancouver, Canada.

[7] Granahan, V. (2013). "PM WBS guide: Program Management Improvement-PIP", Department of Transport and Main Roads, Queensland Government, Australia.

[8] Malone, T. W. and Crowston, K. (1994) "The Interdisciplinary Theory of Coordination" ACM Computing Surveys, 26(1): 87-119.

[9] Vrijhoef, R., and Koskela, L. (2000) "The four roles of supply chain management in construction" European Journal of Purchasing & Supply Management 6 (3): 169-178.

[10] Tommelein, I. and Beeche, G. (2001). "De-coupling cladding installation from other high-rise building trades: A case study." In: Proceedings of the 9th conference of the international group for lean construction, Singapore, 6-8 August.

[11] Chen, Q., Reichard, G., & Beliveau, Y. (2008) Multiperspective Approach to Exploring Comprehensive Cause Factors for Interface Issues" J of Const. Eng. & Mgt., 134(6): 432-441.

[12] Pavitt, T. C., and Gibb, A. G. F. (2003). "Interface management within construction: In particular, building façade." J of Const. Eng. & Mgt., ASCE 129(1): 8–15.

[13] O'Connor, James T., Stephen E. Rusch, and Martin J. Schulz. 1987. "Constructability Concepts for Engineering and Procurement". J. Const. Eng. & Manage. 113 (2): 235-248.

[14] Nicolini, D., Holti, R., Smalley, M. (2001). "Integrating project activities: The theory and Practice of Managing the Supply Chain through Clusters." Construction Management and Economics, 19(1):37-47.

[15] Bygballe, L. E., Jahre, M., and Swärd, A. (2010) "Partnering relationships in construction: A literature review" Journal of Purchasing and Supply Management, 16(4): 239-253.

[16] Anvuur, A.M. and Kumuraswamy, M.M. (2007) "Conceptual model of partnering and alliancing" J. Const. Eng. & Manage, 133 (3), 225-234.

[17] Gadde, L. E., and Dubois, A. (2010) "Partnering in the construction industry—Problems and opportunities" Journal of Purchasing and Supply Management, 16(4): 254-263.

[18] Osipova, E. & Eriksson P. E. (2011) "How procurement options influence risk management in construction projects" Construction Management and Economics, 29:11, 1149-1158, DOI: 10.1080/01446193.2011.639379

[19] Kaur, K. Bid Packaging Decisions in Construction Projects. Master's Thesis, San Diego State University, San Diego, CA, USA, 2018. Available online: https://www.researchgate.net/profile/Kawalpreet-Kaur-2/publication/344306448_BID_PACKAGING_DECISIONS_IN_CONSTRUCTION_PROJECTS/links/5f65478da6fdcc00862d11b3/BID-PACKAGING-DECISIONS-INCONSTRUCTION-PROJECTS.pdf

[20] Shields, D. (2017) "Coordinating Collaboration: Benefits of collaborating with a single-source building enclosure contractor" Walls and Ceilings Magazine, November 2, 2017, https://www.wconline.com/articles/91131-coordinating-collaboration-benefits-of-collaborating-with-a-single-source-building-enclosure-contractor accessed February 2023.

https://doi.org/10.21872/2023IISE_2204

Proceedings of the IISE Annual Conference & Expo 2023
K. Babski-Reeves, B. Eksioglu, D. Hampton, eds.

A Simulation Testbed and Reservation-Based Asynchronous AMR Fleet Management Protocol in Storage Process of Retail Warehouse

Abstract ID: 1234

Zekun Liu and Prof. Seokcheon Lee, Purdue University, West Lafayette, IN

Abstract

Autonomous Mobile Robot (AMR) has been publicly recognized as the upgrade for the comprehensively adopted material handling system Automated Guided Vehicle (AGV). Compared with AGV, the improvements of AMRs in movement flexibility, information connectivity, and resilience to uncertainty provide both advantages and challenges for AMR's fleet management. However, at the current stage of AMR fleet management research, there are gaps between the conceptual framework of AMR intralogistics and its industrial application – the subproblems of AMR Fleet Management are seldomly discussed from an integrated perspective, and the assumptions for motion synchronization and coordination among the fleet of AMRs are unrealistic. Furthermore, most research targets low vehicle-density application scenarios, so particularly challenging applications like Warehouse Intralogistics are not properly addressed. To fill these gaps, we developed a kinetically enabled Simulated-Reality simulator with Unity™ to address the specific application of AMR material handling in the reception-to-storage process of retail warehousing systems. In addition, an asynchronous implementation of the pathfinding algorithm and a cluster-based Task Allocation protocol named PNSC was developed and tested in the simulation testbed. With simulation on randomly generated problem sets, we found the proposed fleet management protocol significantly outperformed the commonly adopted protocols regarding hourly throughputs and congestion management.

Keywords
AMR, Fleet Management, Intralogistics, Parallelism, Unity Simulation

1. Introduction

Warehouse automation has seen various technologies applied, with the link between loading stations and storage units drawing the most research attention. Conveyor belts have traditionally been used to handle shipping items, but the emergence of flexible transportation in warehouses has led to a shift towards vehicle-based transportation, such as automated guided vehicles (AGVs), due to their superior failure response and capacity modification capabilities [1], [2]. The AMR system offers several advantages over its predecessor, including the ability to operate independently from the fixed guiding rails, which grants increased flexibility and maneuverability for resolving congestion and deadlocks. Additionally, the control hierarchy of the AMR system differs from that of the AGV, as a centralized controller is no longer required as each AMR is capable of making pathfinding decisions and determining task assignments independently [3]. Furthermore, the AMR system is considered to be more resilient than AGV systems as the breakdown of a single vehicle can be compensated for by the flexible path-planning of other AMRs, whereas an AGV system's entire segment of guiding rail can be paralyzed by a single breakdown [1].

However, these benefits are accompanied by corresponding challenges. The higher motion flexibility of the AMR system leads to an increase in the computation complexity of the pathfinding and motion planning stages [4]. Additionally, the ability to evade motion like timeout protocols and replanning for detours can result in interruptions and conflicts with other planned paths. Furthermore, while the decentralized and interconnected architecture improves robustness and resilience to changes in resources and partial breakdowns, coordination and collaboration between vehicles must be re-evaluated and addressed [5]. Hence, the development of more sophisticated and efficient fleet management protocols is required.

Retail warehouses, whose primary function is to move materials, pose significant challenges for AMR intralogistics. Our research focuses on the reception-to-storage process, in which a fleet of AMRs transport items from the reception buffer area to the PD station connecting to the designated storage unit. The primary challenge comes from navigating

https://doi.org/10.21872/2023IISE_2204

vehicles through heavy traffic, as modern retail warehouses often serve a wide range of customers that orders from the same truck can be highly mixed, and usable floor space is typically minimized to prioritize storage area which leads to a high vehicle density.

The remainder of this paper is structured as follows. A succinct literature review is presented in Section 2. Section 3 proposes a framework for a simulated-reality simulation testbed for AMR systems in a warehouse environment. In Section 4, the fleet management protocol of AMRs is developed, including a modified Bounded Multiagent A* based motion coordination protocol and a cluster-based task allocation algorithm. The proposed and benchmark protocols are simulated in Section 5. Finally, the conclusion and contributions of this research are summarized in Section 6.

2. Literature Review

Fleet management protocols are usually decomposed into three subproblems: Task Allocation, Pathfinding, and Conflict Resolution. Despite the abundance research on individual topics, gaps remain to hinder the practice of AMR intralogistics.

Task Allocation, involving the sequential assignment of transportation tasks to resources, is a fundamental aspect of Multi-Robot Task Allocation (MRTA) problems. Research in this area typically falls into two categories: deliberative and reactive schemes. Deliberative schemes schedule all tasks initially and often extend well-known problems such as the Vehicle Routing Problem (VRP) or Pickup and Delivery Problem (PDP), using Mixed Integer Linear Programming (MILP) models based on task locations [4], [6], [7]. However, given the NP-hard nature of VRP and PDP, obtaining optimal solutions in a reasonable time frame is challenging. As a result, heuristic and meta-heuristic methods are employed to speed up convergence or reduce model sizes [4], [6], [8]–[12]. These approaches usually assume deterministic travel times, despite the inherent limitations in accuracy. Conversely, reactive schemes adapt to tasks that dynamically arrive [13]–[15], but may produce suboptimal objective scores in static scenarios because they prioritize short-term gains over long-term effects.

Pathfinding problem evaluates the feasibility, plans the steps of travel, and proposes a traveling time estimation for transportation tasks. Common algorithms for this problem are the famous A* algorithm [16] and its variants [17]–[19]. These algorithms are robust for finding paths in an environment with static obstacles but struggle to incorporate information on dynamic objects. Consequently, only the feasibility at the zero-traffic condition is validated, and potential conflicts and deadlocks are not considered. Besides, distance-oriented algorithms like A* and its variants may fail to consider the volume of traffic in the planning phase, leading to negligence of congestion issues.

The Conflict Resolution problem in Fleet Management Protocols involves determining optimal evading strategy when vehicles encounter conflicts. Relevant research primarily focuses on resolving conflicts in a decentralized manner [20]–[22]; however, local resolution strategies can lead to inaccurate time estimation in pathfinding. Some research combined conflict resolution with pathfinding to prevent conflicts through high-quality path planning, resulting in the Multiagent Pathfinding (MAPF) problem [9], [21], [23]–[28], where a time dimension is usually added to the road network, and conflicts are excluded by ensuring no more than one agent is assigned to a node at an instance. However, there are still gaps between research and dynamic applications of these MAPF algorithms – many algorihtms rely on the assumption of agents' perfect motion synchronization or constant speed of the agents [12], which is not always realistic for a cyber-physical system like AMR systems [28]; and [21], [25], [26], [28] plan paths for all agents in the environment at every decision instances, leading to enormous computation time, making the plan infeasible and inefficient in dynamic and sequential arrival of transportation tasks.

3. Design of Simulated-Reality Simulator

To incorporate the non-linearity introduced by the heterogeneous motion steps and imperfect motion synchronization, we are proposing a Simulated-Reality simulator with the Unity™ engine [29] to incorporate kinematic features.

3.1. Design of the Environment

We aim to create a simulation testbed for intralogistics study in retail warehouses by constructing a model of a warehouse environment that mimics real-life operations. The model will include a reception buffer area for temporarily holding packages, multiple Pickup/Delivery Stations (PD Stations) connecting to storage units, and AMRs for transporting packages from the reception buffer area to the correct PD Station. Packages will be generated in

https://doi.org/10.21872/2023IISE_2204

queues of a fixed number of pallets to mimic cargo trucks unloading packages into the reception buffer area. The simulation also includes an invisible path network baked into the floor space for future pathfinding needs.

3.2. Design of AMR Agents

AMR agents execute assigned tasks by maintaining a series of variables, such as a sequence of tasks, a sequence of nodes to traverse, and the agent's configuration (location, orientation angle). They also continuously monitor kinematic features, such as current speed and direction of movements. Hardware specifications, such as acceleration rate, turning radius, and loading operations time, are used as hard-coded parameters to influence the behaviors of the agents.

During task execution, the AMR navigates through intermediate location nodes, adjusting its orientation towards each node with restricted angular speed and moving forward with accumulated speed ($v + a \cdot dt$) when the node is in front of the robot, or moving backward if the node is behind the robot. When the direction of movement alters, the speed will be reset to zero. Arrival at a node is determined by reaching a predefined distance δx from the node, and the process is repeated until task completion.

4. Fleet Management Protocols

In this section, we develop a comprehensive fleet management protocol using the MAPF with motion coordination and Task Allocation paradigm. We begin with selecting a proven MAPF algorithm as the solver and devise a motion coordination protocol for integration of the execution phase, and we propose a cluster-based task selection algorithm.

4.1. Multiagent Pathfinding Solver

The Bounded Multiagent A* (BMAA*) algorithm, an enhanced version of A* with a time dimension for each path network node, has demonstrated strong online planning performance with minimal impact on existing plans, which grants it with better scalability comparing to other popular MAPF algorithms. Unlike A*, BMAA* only considers idle nodes during exploration, excluding conflict nodes from path solutions. Once a conflict-free path is identified, node usage is updated, and occupied nodes become obstacles. However, applying BMAA* in dynamic environments with varied motion steps and asynchronous execution can compromise conflict-free conditions. Thus, we developed an asynchronous BMAA* variant for comparison with the synchronous version.

4.2. Asynchronous BMAA*

Figure 1: The flow chart of the Asynchronous BMAA* protocol with partial path validation.

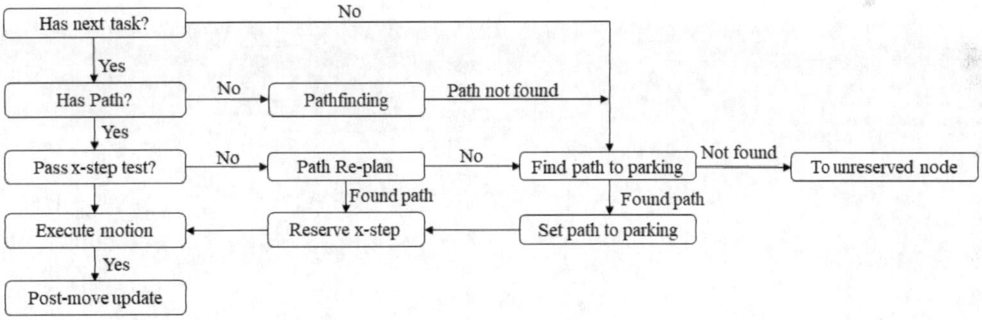

The Asynchronous BMAA* protocol employs a Plan-Check-Do-Update loop (Figure 1). Initially, the AMR obtains a conflict-free path plan from BMAA* when the motion step queue is empty. It then checks for contradictions in the next x-step by detecting other AMRs on the same node at the same time. If no contradiction exists, the AMR executes the motion step; otherwise, it initiates replanning. Upon completing the motion step, the AMR updates the motion queue and releases the previous node's occupancy. In crowded environments, if conflict-free paths are unavailable, the AMR is directed to the nearest parking spot or, if unreachable, attempts to reserve and move to neighboring unreserved nodes, ensuring smooth operation.

4.3. Parallelized Next Step Clustering Task Allocation (PNSC)

The proposed Parallelized Next-Step Clustering (PNSC) protocol is a cluster-based single-task selection mechanism that aims to minimize pickup distance and fulfill as many tasks as possible. It addresses criticisms of traditional single-task selection algorithms by incorporating parallelism [30], [31], which considers global information and assigns both

https://doi.org/10.21872/2023IISE_2204

idle and busy resources to unassigned tasks. Additionally, the protocol addresses congestion issues by implementing work zoning, which assigns AMRs to different clusters of tasks to reduce interference between AMR movements. Even though path crossover is inevitable as task destinations are predesignated, the PNSC encourages AMR to select from local cluster to minimize potential conflict opportunities.

$$\underset{t\in C}{\operatorname{argmin}}\left(\sum_{\forall r\in R}\left(\sum_{\forall t'\in r}\left(\text{Distance}(t',r)\right)^2\right)^2\right) \tag{1}$$

$$\underset{t\in C}{\operatorname{argmin}}\left(\gamma\cdot\text{Distance}(r,t)+(1-\gamma)N\left(\sum_{\forall r'\in R}\left(\sum_{\forall t'\in r}\left(\text{Distance}(t',r')\right)^2\right)^2\right)\right) \tag{2}$$

In PNSC, tasks are selected by AMRs using their current locations or destination location of current-fulfilling tasks. A fitness function is defined to measure the quality of the cluster by summing the distances between tasks and closest AMRs, with augmented convexity to ensure even task distribution and prevent long pickup distances from being compensated by short ones. Furthermore, we incorporated a tunable myopic coefficient γ in the PNSC algorithm to balance the trade-off between reducing instant pickup distances and considering future global pickup distances (Equation 2). A normalization function $N(x)=\frac{|R|}{|T|}x^{\frac{1}{4}}$ is required to help myopic coefficient preserve its impact on the final objective score, as a higher power is used in Equation (1). In the task selection phase, AMRs iterate through all tasks in current cluster, evaluate the scores of potential next-step clusters, and select the task that minimizes the instant pickup distance and global next-step pickup distance.

5. Simulation Case Study

(a)

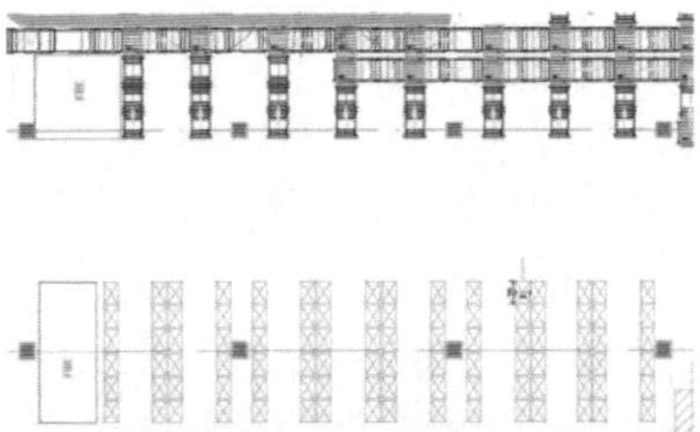

(b)

(c)

Vehicle Specs of Blueswords ForkLiftII-F	
Width	0.8 m
Length	1.7 m
Acceleration	0.5 m/s^2
Top Speed	2 m/s
Turning Speed	1.2 m/s
Turning Radius	1.4 m
Navigation Accuracy	± 5 mm
Loading Time	15 sec

Figure 2: **(a)** The layout of the reception-storage section of a retail warehouse sponsored by Bluesword Intelligence Technology; **(b)** Blueswords ForkLift II – F; **(c)** Vehicle specs of Blueswords ForkLift II -F used in the simulation case study; **(d)** A snapshot of 10 AMRs operating in the simulation environment with the proposed protocol.

We evaluate the performance of our proposed protocols for reception-storage processes using AMRs in a retail warehouse layout (Figure 2-a) provided by Blueswords Intelligent Technology, a Chinese warehouse solution provider. The 32m by 8m area features 14 buffer queues and 8 PD stations, currently employing 12 AGVs with an hourly throughput of 257 pallets. The ForkLift II-F vehicles (Figure 2-b, specs in Figure 2-c) are LIDAR and visual-SLAM equipped unmanned forklifts operating along dynamic virtual guidance rails, technically AMRs but deployed as AGVs due to lacking efficient fleet management protocols. Tasks are generated in batches of 6, simulating package unloading from cargo trucks and queuing at reception buffers. We assume an ample package supply and a new batch arrival within 30 seconds after fulfilling the previous queue, enabling evaluation of the performance of protocols under high-demand conditions.

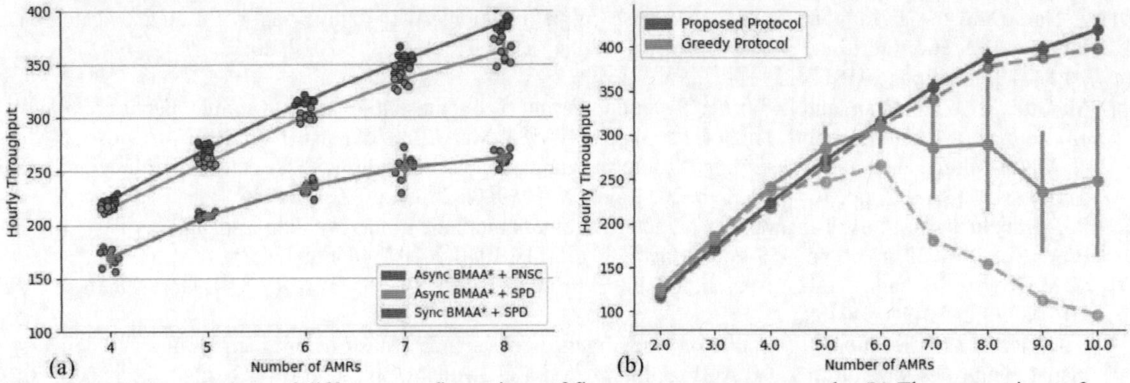

Figure 3: **(a)** the comparison of different configurations of fleet management protocols; **(b)** The comparison of our proposed protocol with greedy protocol, and the dashed line refers to the lower bound of 95% CI of hour throughput

In the simulation, we assessed three protocol configurations: Synchronous BMAA* with SPD rule, Asynchronous BMAA* with SPD rule, and Asynchronous BMAA* with PNSC task allocation. We compared these with a greedy protocol (grid-free path planning, Shortest Pickup Distance rule, and enumerated conflict resolution). Our findings demonstrated a notable increase in maximum throughput capacity and a minor decrease in efficiency under less crowded conditions; while the greedy protocol's grid-free feature failed to manage excessive traffic and led to numerous conflicts, peaking throughput with 6-8 AMRs in the environment.

6. Conclusion & Contribution

In this study, we propose a Simulated-Reality model with vital kinetic features to evaluate fleet management protocols in retail warehouse storage. We developed an Asynchronous BMAA* motion coordination protocol for feasible AMR intralogistics applications and a PNSC task allocation protocol to address congestion in confined scenarios by considering instant pickup distances and minimizing global distances. The proposed protocol configuration demonstrated effectiveness in simulations, achieving high hourly throughput in crowded environments.

7. Reference

[1] S. Adinandra, J. Caarls, D. Kostić, J. Verriet, and H. Nijmeijer, "Flexible Transportation in Warehouses," in Automation in Warehouse Development, R. Hamberg and J. Verriet, Eds., London: Springer, 2012, pp. 191–207. doi: 10.1007/978-0-85729-968-0_14.

[2] R. Hamberg and J. Verriet, "The Falcon Project: Model-Based Design of Automated Warehouses," in Automation in Warehouse Development, R. Hamberg and J. Verriet, Eds., London: Springer, 2012, pp. 3–13. doi: 10.1007/978-0-85729-968-0_1.

[3] K. Furmans and K. Gue, "A Framework for Modeling Material Handling with Decentralized Control," Prog. Mater. Handl. Res., Jan. 2018, [Online]. Available: https://digitalcommons.georgiasouthern.edu/pmhr_2018/23

[4] S. Jun, S. Lee, and Y. Yih, "Pickup and delivery problem with recharging for material handling systems utilising autonomous mobile robots," Eur. J. Oper. Res., Aug. 2020, doi: 10.1016/j.ejor.2020.07.049.

[5] G. Fragapane, R. de Koster, F. Sgarbossa, and J. O. Strandhagen, "Planning and control of autonomous mobile robots for intralogistics: Literature review and research agenda," Eur. J. Oper. Res., vol. 294, no. 2, pp. 405–426, Oct. 2021, doi: 10.1016/j.ejor.2021.01.019.

[6] T. I. Faiz, C. Vogiatzis, and Md. Noor-E-Alam, "A column generation algorithm for vehicle scheduling and routing problems," Comput. Ind. Eng., vol. 130, pp. 222–236, Apr. 2019, doi: 10.1016/j.cie.2019.02.032.

[7] J. Lin, W. Zhou, and O. Wolfson, "Electric Vehicle Routing Problem," Transp. Res. Procedia, vol. 12, pp. 508–521, Jan. 2016, doi: 10.1016/j.trpro.2016.02.007.

[8] S. Mitrović-Minić, R. Krishnamurti, and G. Laporte, "Double-horizon based heuristics for the dynamic pickup and delivery problem with time windows," Transp. Res. Part B Methodol., vol. 38, no. 8, pp. 669–685, Sep. 2004, doi: 10.1016/j.trb.2003.09.001.

[9] K. Jose and D. K. Pratihar, "Task allocation and collision-free path planning of centralized multi-robots system for industrial plant inspection using heuristic methods," Robot. Auton. Syst., vol. 80, pp. 34–42, Jun. 2016, doi: 10.1016/j.robot.2016.02.003.

[10] Z. Han, D. Wang, F. Liu, and Z. Zhao, "Multi-AGV path planning with double-path constraints by using an improved genetic algorithm," PLOS ONE, vol. 12, no. 7, p. e0181747, Jul. 2017, doi: 10.1371/journal.pone.0181747.

[11] M. Otte, M. J. Kuhlman, and D. Sofge, "Auctions for multi-robot task allocation in communication limited environments," Auton. Robots, vol. 44, no. 3, pp. 547–584, Mar. 2020, doi: 10.1007/s10514-019-09828-5.

[12] N. Matin-Moghaddam and J. A. Sefair, "Route assignment and scheduling with trajectory coordination," IISE Trans., vol. 53, no. 2, pp. 164–181, Feb. 2021, doi: 10.1080/24725854.2020.1774096.

[13] P. J. Egbelu and J. M. A. Tanchoco, "Characterization of automatic guided vehicle dispatching rules," Int. J. Prod. Res., vol. 22, no. 3, pp. 359–374, May 1984, doi: 10.1080/00207548408942459.

[14] C. M. Klei and J. Kim, "AGV dispatching," Int. J. Prod. Res., vol. 34, no. 1, pp. 95–110, Jan. 1996, doi: 10.1080/00207549608904893.

[15] Y.-C. Ho and S.-H. Chien, "A simulation study on the performance of task-determination rules and delivery-dispatching rules for multiple-load AGVs," Int. J. Prod. Res., vol. 44, no. 20, pp. 4193–4222, Oct. 2006, doi: 10.1080/00207540500442401.

[16] P. E. Hart, N. J. Nilsson, and B. Raphael, "A Formal Basis for the Heuristic Determination of Minimum Cost Paths," IEEE Trans. Syst. Sci. Cybern., vol. 4, no. 2, pp. 100–107, Jul. 1968, doi: 10.1109/TSSC.1968.300136.

[17] S. Koenig and M. Likhachev, "D*Lite," Aaai/iaai, vol. 15, pp. 476–483, 2002.

[18] A. Bolu and Ö. Korçak, "Path Planning for Multiple Mobile Robots in Smart Warehouse," in 2019 7th International Conference on Control, Mechatronics and Automation (ICCMA), Nov. 2019, pp. 144–150. doi: 10.1109/ICCMA46720.2019.8988635.

[19] L. Cohen, T. Uras, S. Jahangiri, A. Arunasalam, S. Koenig, and T. K. S. Kumar, "The FastMap Algorithm for Shortest Path Computations," ArXiv170602792 Cs, Dec. 2017, Accessed: Feb. 24, 2021. [Online]. Available: http://arxiv.org/abs/1706.02792

[20] K. Zheng, D. Tang, W. Gu, and M. Dai, "Distributed control of multi-AGV system based on regional control model," Prod. Eng. 74, Mar. 2013.

[21] L. Juntao, D. Tingting, L. Yuanyuan, and H. Yan, "Study on Robot Path Collision Avoidance Planning Based on the Improved Ant Colony Algorithm," in 2016 8th International Conference on Intelligent Human-Machine Systems and Cybernetics (IHMSC), Aug. 2016, pp. 540–544. doi: 10.1109/IHMSC.2016.215.

[22] P. Prabhu and A. R. Chowdhury, "Feasibility Study of Multi Autonomous Mobile Robots (AMRs) Motion Planning in Smart Warehouse Environment," in 2021 18th International Conference on Ubiquitous Robots (UR), Jul. 2021, pp. 380–385. doi: 10.1109/UR52253.2021.9494634.

[23] K. Dresner and P. Stone, "A multiagent approach to autonomous intersection management," J. Artif. Intell. Res., vol. 31, pp. 591–656, 2008, doi: 10.1613/jair.2502.

[24] D. Sigurdson, V. Bulitko, W. Yeoh, C. Hernández, and S. Koenig, "Multi-Agent Pathfinding with Real-Time Heuristic Search," in 2018 IEEE Conference on Computational Intelligence and Games (CIG), Aug. 2018, pp. 1–8. doi: 10.1109/CIG.2018.8490436.

[25] G. Sharon, R. Stern, A. Felner, and N. R. Sturtevant, "Conflict-based search for optimal multi-agent pathfinding," Artif. Intell., vol. 219, pp. 40–66, Feb. 2015, doi: 10.1016/j.artint.2014.11.006.

[26] J. Yu and S. M. LaValle, "Optimal Multirobot Path Planning on Graphs: Complete Algorithms and Effective Heuristics," IEEE Trans. Robot., vol. 32, no. 5, pp. 1163–1177, Oct. 2016, doi: 10.1109/TRO.2016.2593448.

[27] N. V. Kumar and C. S. Kumar, "Development of collision free path planning algorithm for warehouse mobile robot," Procedia Comput. Sci., vol. 133, pp. 456–463, Jan. 2018, doi: 10.1016/j.procs.2018.07.056.

[28] W. Hönig, S. Kiesel, A. Tinka, J. W. Durham, and N. Ayanian, "Persistent and Robust Execution of MAPF Schedules in Warehouses," IEEE Robot. Autom. Lett., vol. 4, no. 2, pp. 1125–1131, Apr. 2019, doi: 10.1109/LRA.2019.2894217.

[29] Unity Technologies, "Unity - Game Engine," Accessed: Mar 22, 2021. [Online]. Available: https://unity.com/

[30] J. C.-Y. Lee and S. Lee, "Ambulance dispatching driven by parallelism, myopia, and no-reservation," IIE Annu. Conf. Proc., pp. 1–6, 2011.

[31] S. Lee, "Role of Parallelism in Ambulance Dispatching," IEEE Trans. Syst. Man Cybern. Syst., vol. 44, no. 8, pp. 1113–1122, Aug. 2014, doi: 10.1109/TSMC.2013.2296280.

https://doi.org/10.21872/2023IISE_2547

Proceedings of the IISE Annual Conference & Expo 2023
K. Babski-Reeves, B. Eksioglu, D. Hampton, eds.

Optimal launch strategy for modular products under resource constraint and module sequencing complexity

Adewole Adegbola and Venkat Allada

Engineering Management and Systems Engineering Department
Missouri University of Science and Technology
Rolla, MO 65401, USA

Abstract

This study seeks to address the resource constraint and module development sequencing issues associated with the non-preemptive product development (PD) process. We specifically consider the development of multiple product variants from a common product family where the product variants may share one or more common modules, and each product variant emerges only after the completion of its associated module(s). Furthermore, the modular configuration of each product variant is known a priori, and each module is defined by a specific set of attributes. Given the constraint of PD resource and the stringency of modular configuration for product modules, the goal of this study is to develop an approach for determining the development sequence which maximizes the net revenue associated with the PD process upon product release. The approach will utilize parameters relevant to the conventional PD process and we proceed to develop a mathematical model to define the problem. We conclude by demonstrating the applicability of the model using a realistic case example and discuss the insights and managerial implications of the study.

Keywords

Non-preemptive PD process, product variant, modular product family, module development sequencing.

1. Introduction

The non-preemptive PD process can be described as a situation whereby products are developed without interrupting the PD process. The non-preemptive PD process is applicable in cases where a firm seeks to minimize the manufacturing setup time and setup cost, or in cases where product quality can be compromised if the PD process is interrupted (Adegbola and Allada) [1]. In spite of these many advantages, one issue that manufacturing firms have to contend with in adopting the non-preemptive PD process is the issue of determining the best strategy for realizing multiple products variants at their expected time to market, particularly when there is a limit to the development personnel resource available within a particular period of time. This challenge is common because some firms adopt the functional organization structure where personnel are drafted to other ongoing projects. In addition, some of the personnel are sub-contracted personnel that only render services on a temporary basis. Another reason why this problem is of major concern is that product variants are typically developed from modules, where the modules have varying development attributes such as development cost, development duration, and the modules may share dependent, independent or coupled relationships among themselves. This problem as described suggests the existence of many alternatives to develop the product variants, depending on the possibility of each alternative to satisfy the development personnel and module configuration stringencies. With each feasible alternative yielding different levels of revenue and product launch timing, the problem becomes that of determining the best strategy which yields the maximum revenue for the product variants upon market launch, while taking into account the module sequencing complexities and development personnel constraint throughout the PD planning horizon.

2. Problem Description

*"Given the set of modules **M** = {M1, M2, M3, ……., Mn}, where the relationship between the modules can be independent, dependent or coupled; the set of product variants **P** = {P1, P2, P3, ………, Pn}, where products may share one or more common modules; **T** available periods for realizing and launching the product variants contained in set P; and the available development personnel per period A_T, determine the sequence for completing the development of the products contained in the set P which maximizes net revenue upon product launch".*

3. Related Research

Launching a product at the right time is essential to the success of the product because it provides a competitive business advantage which is necessary for sustainability; however, the realization of this goal comes with numerous issues which researchers have invested efforts to address. Cui et. al [2] investigated the reason for product post-

https://doi.org/10.21872/2023IISE_2547

market launch failure and stressed that the unfavorable situation frequently occurs because market uncertainties after initial launch are typically not taken into consideration. With regard to finding optimal timing of new product preannouncement (NPP) and launch (NPL) in a competitive market, Su and Rao [3] employed a game-theoretic approach in conducting numerical studies to examine the effects of NPP on demand, market share, competitor's reaction and profit margin, asserting that anticipating competitor's timing choice is the most significant factor to be considered under NPP. Yao and Xin [4] stressed the need to focus on marketing strategies as an external factor that determines the success of new product launch across different market segments.

Furthermore, several other issues affecting product launch timing are encountered as a result of the complexities associated with PD processes. Vickery et al. [5] investigated the nature of the relationship between modular product development and timeliness and deduced that product modularity supported by product platforming and manufacturing flexibility had positive effects on launch speed. In order to reduce manufacturing lead times for make-to-order manufacturing firms, a robust *CRISP-DM* incorporated data-analytics tool was proposed by Sun et al. [6] to aid product lifecycle management. Bolanos and Barbalho [7] improved on the predictive accuracy of PD cycle time from a manufacturing lead time viewpoint by extending the quantification of product complexity to include interconnections between components, number of product functions and customization of the final system. Maoqi et. al. [8] proposed a geometric programming model to determine optimal design rates which minimizes PD lead time. In a view to minimize lead time, Ogura et al. [9] focused on applying convex optimization methods to PD projects having varying time dependencies.

Launching multiple products under resource-constrained product development conditions, is another product launch related problem which has been investigated by researchers. Ogura et al. [9]; Zhao et al. [10]; Hutchison-Krupat & Kavadias [11] addressed budget allocation issues associated with the PD process. Dash et. al [12] also investigated the effect of available product development teams on product portfolio planning and launch timing. This work considers the impact of resource constraint and module sequencing complexities on the launch timing of modular products for the non-preemptive PD process.

4. Methodology

Modular Products

Modular products are products that are built form parts that can be interchanged so as to aid repair and customization of products Gershenson et al. [13]. In many cases of modular product development, some relationships may exist between the modules required for the emergence of a product during product architecting stage. These modular relationships were described in Yassine [14]. In this work, we consider the dependent (sequential), independent (parallel) and coupled (interdependent) modular relationships. These relationships are shown in Table 1.

Table 1: Modular Relationships

Relationship	Dependent	Independent	Coupled
Representation	(A) ⟶ (B)	(A) (B)	(A) ⟷ (B)
Description	Here, module A must be developed before module B can be developed.	Here, the development of modules A and B can be carried out separately.	This implies that modules A and B are to be developed simultaneously.

Product Realization

Product realization can be defined as a collection of all the processes a firm undergoes, and the efforts invested by the firm towards achieving a physical product. In this study, the scope of product realization is confined to the emergence of products from the development of their associated modules, and the hinderance to product realization is viewed from the development personnel and module sequencing perspective. To illustrate the concept of product realization, assuming the set of modules S = {M1, M4, M9} where M1 and M4 are independent modules and M9 is dependent on M4. Product variant P1 comprises of the set P1 = {M1, M9} and P2 comprises of the set {M1}. P1 and P2 can be realized in the development time horizon (T) through any of the following development sequences shown in figure 1.

https://doi.org/10.21872/2023IISE_2547

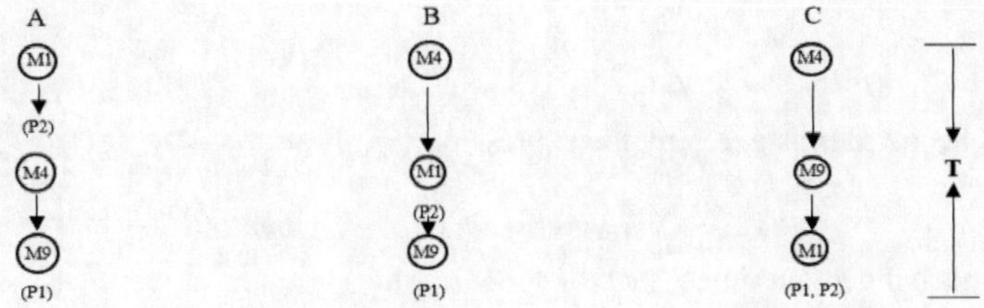

Figure 1: Product development sequence and product realization

From figure 1, assuming the firm wants an early launch for P2, strategy A will be selected. However, if the firm falls short of personnel at the early stages of the process, M1 cannot be developed, and this renders strategy A "infeasible".

Model Notations:

P = product family containing the set of product variants to be launched, where P = {P_1, P_2, P_3, ………, P_n}
M = set of modules required to realize the product variants contained in set P, where M = {M_1, M_2, M_3, ………., M_n}
T = total set of periods in the development/launch horizon.
t = index indicating periods in the development/launch horizon, where t = {1, 2, 3, ………, T}
A_T = available development personnel per period in the development/launch horizon
D_{PT} = development personnel required per period for the module development.
M_{DT} = duration for developing module M
M_{DC} = development cost for module M
M_{IC} = module integration cost for product variant P
C_0 = initial cumulative module development cost before development of the next module
C_{DC} = cumulative development cost starting from t = 1 to the present period in the development/launch horizon
C_{DT} = cumulative module development time starting from t = 1 to the present period in the development horizon
t_l = ideal launch period for product variant P
t_r = period in which product variant P emerged
τ = tardiness in the launch of product variant P defined by:

$$\tau = \max\left\{0\,;\ \sum_{t=1}^{t_r} t - \sum_{t=1}^{t_l} t\right.$$

R_I = revenue associated with the ideal launch of product variant P
R_P = revenue loss per period if product variant P is launched after the ideal launch period.
R_A = revenue associated with the actual launch of product P defined by:

$$R_A = \begin{cases} [R_I - (\tau * R_P)]; & if\ t_r\ occurs\ after\ t_l \\ R_I; & otherwise \end{cases}$$

R_0 = initial cumulative revenue before the emergence of the next product variant.
C_R = cumulative revenue starting from t = 1 to the present period in the development horizon
N_R = net revenue realized for any development/launch sequence
i = index indicating the development/launch sequence, where i = {1, 2, 3, ………, n}
$\hat{S}_i$ = a variable connoting the module development/launch sequence defined by:

$$\hat{S}_i = \begin{cases} 1; & if\ \hat{S}_i\ is\ the\ optimal\ sequence\ for\ the\ product\ variants\ contained\ in\ set\ P \\ 0; & if\ \hat{S}_i\ is\ a\ feasible\ sequence\ for\ all\ product\ variants\ contained\ in\ set\ P\ but\ not\ optimal \\ -; & if\ \hat{S}_i\ is\ an\ infeasible\ sequence\ based\ on\ module\ sequencing\ and\ resource\ availability \end{cases}$$

Model Formulation:

The objective of the problem is to find the optimal sequence which maximizes the net revenue (N_R). Considering the entire horizon in any sequence, the objective function in terms of development cost (C_{DC}) and revenue (C_R) equals

$$Maximize\quad N_R = \hat{S}_i\left(\sum_{t=1}^{T}(R_A) - (M_{DC} + M_{IC})\right) \qquad (1)$$

Since the optimal launch sequence is of interest, then

$$\sum_{i=1}^{n}\hat{S}_i = 1 \qquad (2)$$

Given that the development process is carried out without interruption, then

$$\sum_{M=M_1}^{M_n} M_{DT} = T \quad \forall \ \hat{S}_i \qquad (3)$$

The development personnel constraint is given by the expression:

$$\sum D_{PT} \leq A_T \quad \forall \ t \qquad (4)$$

Assuming M_μ = set of all independent modules, such that: $M_\mu \subset M$
M_α = any independent module with completion time t_c such that: $M_\alpha \in M_\mu$ and $t_c \in T$
M_β = set of dependent and/or coupled modules preceded by M_α, with members $\{M_\omega, M_\psi, \ldots, M_\Omega\}$ and development start times $\{t_\omega, t_\psi, \ldots, t_\Omega\}$ respectively, such that $M_\beta \subset M$ and $t_\omega, t_\psi, \ldots, t_\Omega \in T$. By definition,

$$\sum_{t=1}^{t_c} t < \sum_{t=1}^{t_\omega} t, \ \sum_{t=1}^{t_\psi} t, \ldots \sum_{t=1}^{t_\Omega} t \quad \forall \ M_\alpha \in M_\mu; \ M_\omega, M_\psi, \ldots, M_\Omega \in M_\beta \qquad (5)$$

Likewise, assuming M_ρ = any set of coupled modules with members $\{M_\lambda, M_\emptyset, \ldots, M_\sigma\}$ and development start times $\{t_\lambda, t_\emptyset, \ldots, t_\sigma\}$ respectively, such that: $M_\rho \in M$; and $t_\lambda, t_\emptyset, \ldots, t_\sigma \in T$. Again, by definition,

$$\sum_{t=1}^{t_\lambda} t = \sum_{t=1}^{t_\emptyset} t = \cdots = \sum_{t=1}^{t_\sigma} t \quad \forall \ M_\lambda, M_\emptyset, \ldots, M_\sigma \in M_\rho \ ; M_\rho \in M \qquad (6)$$

Equations 1 to 6 represent the model that describes the resource constrained non-preemptive PD process with module sequencing complexity. Equations 5 and 6 specifically cater for the module sequencing complexities.

Numerical Example

We assume that the marketing team of a firm has identified market opportunities for 6 product variants and the manufacturing team determines the development attributes as follows:

Product Modules: Set of modules M = {M1, M2, M3, M4, M5, M6} with the following relationships:
M1 and M2 are independent; M5 is dependent on M1; M3 and M4 are coupled; and M6 is dependent on M3, M4
Product Variants: Set of product variants P = {P1, P2, P3, P4, P5, P6}, with the following module requirements:
P1 = {M1}; P2 = {M1, M3, M6}; P3 = {M2, M4}; P4 = {M1, M2}; P5 = {M3, M5}; P6 = {M4, M6}
Development/Launch Horizon: T = 40 months

Table 2: Module Development Data

Module	M_{DT} (months)	M_{DC} ($)	D_{PT}
M1	10	400,000	10
M2	10	300,000	10
M3|M4	10	500,000	11
M5	5	150,000	6
M6	5	200,000	7

Table 3: Module Development/Product Launch Data

Product	Integration	M_{IC} ($)	t_l	R_I ($)	R_P ($)
P1	--	--	10	1,000,000	20,000
P2	M1, M3, M6	150,000	20	1,600,000	30,000
P3	M2, M4	100,000	30	1,200,000	25,000
P4	M1, M2	120,000	35	1,500,000	30,000
P5	M3, M5	70,000	40	800,000	15,000
P6	M4, M6	80,000	20	1,000,000	10,000

Table 4: Development Personnel Data

Period Interval (months)	Available Personnel (A_T)
$0 < t \leq 10$	13
$10 < t \leq 20$	11
$20 < t \leq 25$	8
$25 < t \leq 30$	12
$30 < t \leq 35$	10
$35 < t \leq 40$	9

https://doi.org/10.21872/2023IISE_2547

5. Results

The result for all feasible alternatives in terms of cumulative development costs (CDC), cumulative revenue (CR) at each stage of the development process and net revenue (NR) for each alternative is summarized in Table 5.

Table 5: Product realization/launch results

$\hat{S}_1$	$\hat{S}_2$	$\hat{S}_3$	$\hat{S}_4$	$\hat{S}_5$	$\hat{S}_6$	$\hat{S}_7$	$\hat{S}_8$
M1 {P1}	**M1 {P1}**	**M1 {P1}**	**M3|4 { }**	**M3|4 { }**	**M3|4 { }**	**M2 { }**	**M2 { }**
C_{DT} = 10 C_0 = 0 M_{DC} = 400,000 M_{IC} = 0 C_{DC} = 400,000 R_0 = 0 R_A = 1,000,000 C_R = 1,000,000	C_{DT} = 10 C_0 = 0 M_{DC} = 400,000 M_{IC} = 0 C_{DC} = 400,000 R_0 = 0 R_A = 1,000,000 C_R = 1,000,000	C_{DT} = 10 C_0 = 0 M_{DC} = 400,000 M_{IC} = 0 C_{DC} = 400,000 R_0 = 0 R_A = 1,000,000 C_R = 1,000,000	C_{DT} = 10 C_0 = 0 M_{DC} = 500,000 M_{IC} = 0 C_{DC} = 500,000 R_0 = 0 R_A = 0 C_R = 0	C_{DT} = 10 C_0 = 0 M_{DC} = 500,000 M_{IC} = 0 C_{DC} = 500,000 R_0 = 0 R_A = 0 C_R = 0	C_{DT} = 10 C_0 = 0 M_{DC} = 500,000 M_{IC} = 0 C_{DC} = 500,000 R_0 = 0 R_A = 0 C_R = 0	C_{DT} = 10 C_0 = 0 M_{DC} = 300,000 M_{IC} = 0 C_{DC} = 300,000 R_0 = 0 R_A = 0 C_R = 0	C_{DT} = 10 C_0 = 0 M_{DC} = 300,000 M_{IC} = 0 C_{DC} = 300,000 R_0 = 0 R_A = 0 C_R = 0
M2 {P4}	**M3|4 { }**	**M3|4 { }**	**M1 {P1}**	**M1 {P1}**	**M2 {P3}**	**M1 {P1, P4}**	**M3|4 {P3}**
C_{DT} = 20 C_0 = 400,000 M_{DC} = 300,000 M_{IC} = 120,000 C_{DC} = 820,000 R_0 = 1,000,000 R_A = 1,500,000 C_R = 2,500,000	C_{DT} = 20 C_0 = 400,000 M_{DC} = 500,000 M_{IC} = 0 C_{DC} = 900,000 R_0 = 1,000,000 R_A = 0 C_R = 1,000,000	C_{DT} = 20 C_0 = 400,000 M_{DC} = 500,000 M_{IC} = 0 C_{DC} = 900,000 R_0 = 1,000,000 R_A = 0 C_R = 1,000,000	C_{DT} = 20 C_0 = 500,000 M_{DC} = 400,000 M_{IC} = 0 C_{DC} = 900,000 R_0 = 0 R_A = 800,000 C_R = 800,000	C_{DT} = 20 C_0 = 500,000 M_{DC} = 400,000 M_{IC} = 0 C_{DC} = 900,000 R_0 = 0 R_A = 800,000 C_R = 800,000	C_{DT} = 20 C_0 = 500,000 M_{DC} = 300,000 M_{IC} = 100,000 C_{DC} = 900,000 R_0 = 0 R_A = 1,200,000 C_R = 1,200,000	C_{DT} = 20 C_0 = 300,000 M_{DC} = 400,000 M_{IC} = 120,000 C_{DC} = 820,000 R_0 = 0 R_A = 2,300,000 C_R = 2,300,000	C_{DT} = 20 C_0 = 300,000 M_{DC} = 500,000 M_{IC} = 100,000 C_{DC} = 900,000 R_0 = 0 R_A = 1,200,000 C_R = 1,200,000
M5 { }	**M5 {P5}**	**M6 {P2, P6}**	**M6 {P2, P6}**	**M5 {P5}**	**M6 {P6}**	**M5 {Ø}**	**M6 {P6}**
C_{DT} = 25 C_0 = 820,000 M_{DC} = 150,000 M_{IC} = 0 C_{DC} = 970,000 R_0 = 2,500,000 R_A = 0 C_R = 2,500,000	C_{DT} = 25 C_0 = 900,000 M_{DC} = 150,000 M_{IC} = 70,000 C_{DC} =1,120,000 R_0 = 1,000,000 R_A = 800,000 C_R = 1,800,000	C_{DT} = 25 C_0 = 900,000 M_{DC} = 200,000 M_{IC} = 230,000 C_{DC} =1,330,000 R_0 = 1,000,000 R_A = 2,400,000 C_R = 3,400,000	C_{DT} = 25 C_0 = 900,000 M_{DC} = 200,000 M_{IC} = 230,000 C_{DC} =1,330,000 R_0 = 800,000 R_A = 2,400,000 C_R = 3,200,000	C_{DT} = 25 C_0 = 900,000 M_{DC} = 150,000 M_{IC} = 70,000 C_{DC} =1,120,000 R_0 = 800,000 R_A = 800,000 C_R = 1,600,000	C_{DT} = 25 C_0 = 900,000 M_{DC} = 200,000 M_{IC} = 80,000 C_{DC} =1,180,000 R_0 = 1,200,000 R_A = 950,000 C_R = 2,150,000	C_{DT} = 25 C_0 = 820,000 M_{DC} = 150,000 M_{IC} = 0 C_{DC} = 970,000 R_0 = 2,300,000 R_A = 0 C_R = 2,300,000	C_{DT} = 25 C_0 = 900,000 M_{DC} = 200,000 M_{IC} = 80,000 C_{DC} =1,180,000 R_0 = 1,200,000 R_A = 950,000 C_R = 2,150,000
M3|4 {P3, P5}	**M2 {P3, P4}**	**M2 {P3, P4}**	**M2 {P3, P4}**	**M2 {P3, P4}**	**M1 {P1, P2, P4}**	**M3|4 {P5, P3}**	**M1 {P1, P2, P4}**
C_{DT} = 35 C_0 = 970,000 M_{DC} = 500,000 M_{IC} = 170,000 C_{DC} =1,640,000 R_0 = 2,500,000 R_A = 1,875,000 C_R = 4,375,000	C_{DT} = 35 C_0 = 1,120,000 M_{DC} = 300,000 M_{IC} = 220,000 C_{DC}=1,640,000 R_0 = 1,800,000 R_A = 2,575,000 C_R = 4,375,000	C_{DT} = 35 C_0 = 1,330,000 M_{DC} = 300,000 M_{IC} = 220,000 C_{DC} =1,850,000 R_0 = 3,400,000 R_A = 2,575,000 C_R = 5,975,000	C_{DT} = 35 C_0 = 1,330,000 M_{DC} = 300,000 M_{IC} = 220,000 C_{DC}=1,850,000 R_0 = 3,200,000 R_A = 2,575,000 C_R = 5,775,000	C_{DT} = 35 C_0 = 1,120,000 M_{DC} = 300,000 M_{IC} = 220,000 C_{DC} =1,640,000 R_0 = 1,600,000 R_A = 2,575,000 C_R = 4,175,000	C_{DT} = 35 C_0 = 0 M_{DC} = 400,000 M_{IC} = 270,000 C_{DC}=1,850,000 R_0 = 2,150,000 R_A = 3,150,000 C_R = 5,300,000	C_{DT} = 35 C_0 = 970,000 M_{DC} = 500,000 M_{IC} = 170,000 C_{DC}=1,640,000 R_0 = 2,300,000 R_A = 1,875,000 C_R = 4,175,000	C_{DT} = 35 C_0 = 1,180,000 M_{DC} = 400,000 M_{IC} = 270,000 C_{DC}=1,850,000 R_0 = 2,150,000 R_A = 3,150,000 C_R = 5,300,000
M6 {P2, P6}	**M6 {P2, P6}**	**M5 {P5}**	**M5 {P5}**	**M6 {P2, P6}**	**M5 {P5}**	**M6 {P2, P6}**	**M5 {P5}**
C_{DT} = 40 C_0 = 1,640,000 M_{DC} = 200,000 M_{IC} = 230,000 C_{DC} = 2,070,000 R_0 = 4,375,000 R_A = 1,800,000 C_R = 6,175,000	C_{DT} = 40 C_0 = 1,640,000 M_{DC} = 200,000 M_{IC} = 230,000 C_{DC}=2,070,000 R_0 = 4,375,000 R_A = 1,800,000 C_R = 6,175,000	C_{DT} = 40 C_0 = 1,850,000 M_{DC} = 150,000 M_{IC} = 70,000 C_{DC}=2,070,000 R_0 = 5,975,000 R_A = 800,000 C_R = 6,775,000	C_{DT} = 40 C_0 = 1,850,000 M_{DC} = 150,000 M_{IC} =70,000 C_{DC}=2,070,000 R_0 = 5,775,000 R_A = 800,000 C_R = 6,575,000	C_{DT} = 40 C_0 = 1,640,000 M_{DC} = 200,000 M_{IC} = 230,000 C_{DC}=2,070,000 R_0 = 4,175,000 R_A = 1,800,000 C_R = 5,975,000	C_{DT} = 40 C_0 = 0 M_{DC} = 150,000 M_{IC} = 70,000 C_{DC}=2,070,000 R_0 = 5,300,000 R_A = 800,000 C_R = 6,100,000	C_{DT} = 40 C_0 = 1,640,000 M_{DC} = 200,000 M_{IC} = 230,000 C_{DC}=2,070,000 R_0 = 4,175,000 R_A = 1,800,000 C_R = 5,975,000	C_{DT} = 40 C_0 = 1,850,000 M_{DC} = 150,000 M_{IC} = 70,000 C_{DC}=2,070,000 R_0 = 5,300,000 R_A = 800,000 C_R = 6,100,000
N_R = 4,105,000	**N_R = 4,105,000**	**N_R = 4,705,000**	**N_R = 4,505,000**	**N_R = 3,905,000**	**N_R = 4,030,000**	**N_R = 3,905,000**	**N_R = 4,030,000**

Table 5 shows how the firm accumulates different levels of development cost and revenue at different time periods in the development horizon depending on the strategy adopted. This also explains the reason behind the emergence of product variants at different periods in the horizon as each sequence represents a unique sequence for developing the set of modules. In addition, each alternative yields a value for net revenue which represents the strength of the alternative in meeting the firm's objective of revenue maximization. $\hat{S}_3$ guarantees the highest level of revenue.

Although the alternatives are unique, the net revenue shown in { $\hat{S}_1$, $\hat{S}_2$}; {$\hat{S}_5$, $\hat{S}_7$}; { $\hat{S}_6$, $\hat{S}_8$} suggests that two or more sequences may yield the same net revenue. This is because the market performance of any product depends on the period the product variant emerged relative to the proposed market launch. Using $\hat{S}_1$ and $\hat{S}_2$ as example, P1 is realized at $t_r = 10$ which is the same as in $\hat{S}_2$. Thus, there were no delays and no revenue losses in both cases. P4 was realized at $t_r = 20$ and $t_r = 35$ in $\hat{S}_1$ and $\hat{S}_2$ respectively. Even though P4 emerged earlier in $\hat{S}_1$, there are no revenue losses in both cases since $t_l = 35$ for P4. P5 has a product launch period of 40. Since P5 emerged at $t_r = 35$ and $t_r = 25$ in $\hat{S}_1$ and $\hat{S}_2$ respectively, again, no revenue losses are incurred by the firm. Lastly, product variants P2, P3 and P6 were realized at the same periods (after their launch periods) in both $\hat{S}_1$ and $\hat{S}_2$ meaning the revenue losses incurred at the same in both cases. With no difference in revenue across both alternatives, the net revenue will be the same value for both

https://doi.org/10.21872/2023IISE_2547

strategies. The observation on {$\hat{S}_1$, $\hat{S}_2$}; {$\hat{S}_5$, $\hat{S}_7$}; and {$\hat{S}_6$, $\hat{S}_8$} indicates the possibility of having another alternative yielding the same revenue as $\hat{S}_3$. However, such an alternative if it exists, must have been ruled out as "infeasible" by way of not satisfying the development personnel and module sequencing stringencies. Furthermore, although the development personnel requirement suggests that modules M5 and M6 could have been developed simultaneously from t = 1 to t = 5 which could have resulted in early launch of their associated products, the sequencing complexity issues associated with the PD process rendered the simultaneous development of both modules infeasible since M5 and M6 are dependent modules. This insight further corroborates the influence of module sequencing complexities on launch timing for the non-preemptive PD process.

6. Conclusion

This paper investigates the resource constraint and module sequencing issues associated with the non-preemptive modular product development process. The methodology adopted was illustrated using a realistic example yielding different alternative strategies for the PD process and the results show that the nature of the problem has a significant effect on the selection of product launch strategy, product launch timing and revenue generation. In conclusion, the findings of this study will be beneficial to firms and product development managers by empowering them with a decision-making framework for evaluating various product launch strategies for the non-preemptive modular product development in a resource-constrained environment.

Future Work: Although this work has covered a large problem domain in new product development, this study only considered the development personnel aspect of the resource requirements for PD projects which was addressed collectively. In many real-life cases of PD projects, development personnel usually exist in different teams. Furthermore, the module substitution strategy was introduced in Adegbola and Allada [15] as a means of achieving flexibility in launch timing and the technology selection process of the product portfolio planning (PPP) problem. However, the capabilities of the module substitution concept in reducing the PD cycle time is yet to be explored. Hence, future works will focus on adapting the module substitution strategy to address issues relating to product portfolio selection and PD cycle time minimization for multi-constrained PD projects.

References

[1] Adegbola, A., & Allada, V., (2021) Product launch decision making for non-preemptive product development process. *6th North American International Conference on Industrial Engineering and Operations Management (IEOM).*

[2] Cui, A., Zhao, M., & Ravichandran, T., (2011). Market uncertainty and dynamic new product launch strategies: A system dynamics model, *IEEE Transactions on Engineering Management*. 58. 530-550. 10.1109/TEM.2010.2100822.

[3] Su, M., & Rao, V., (2011). Timing decisions of new product preannouncement and launch with competition. *International Journal of Production Economics*. 129. 51-64. 10.1016/j.ijpe.2010.09.001.

[4] Yao, Z., & Xin, G., (2017). A fuzzy optimization method to select marketing strategies for new products based on similar cases. *Journal of Intelligent & Fuzzy Systems*. 32. 2679-2695. 10.3233/JIFS-16723.

[5] Vickery, S.K., Bolumole, Y., Castel, M.J., & Calantone, R. (2015). The effects of product modularity on launch speed. *International Journal of Production Research*, 53, 5369 - 5381.

[6] Sun, K., Li, Y., & Roy, U., (2017) A PLM-based data analytics approach for improving product development lead time in an engineer-to-order manufacturing firm, *Mathematical Modelling of Engineering Problems*, Vol. 4, No. 2, pp. 69-74.

[7] Bolanos, R.D.S., & Barbalho, S.C.M., (2021). Exploring product complexity and prototype lead-times to predict new product development cycle-times. *International Journal of Production Economics*. 235. 108077. 10.1016/j.ijpe.2021.108077.

[8] Maoqi, L., Zheng, L., & Changchun, L., (2020). Faster or fewer iterations? A strategic perspective of a sequential product development project. *IISE Transactions*. 10.1080/24725854.2020.1830207.

[9] Ogura, M., Harada, J., Kishida, M., & Yassine, A., (2019) Resource optimization of product development projects with time-varying dependency structure. *Res Eng Design* 30, 435–452.

[10] Zhao, C., Ogura, M., Kishida, M., & Yassine, Ali., (2021). Optimal resource allocation for dynamic product development process via convex optimization. *Res Eng Design*, 32, pp. 71-90

[11] Hutchison-Krupat, J., & Kavadias, S. (2015)., Strategic resource allocation: Top-down, bottom-up, and the value of strategic buckets. *Management Science*, 61(2), 391–412.

[12] Dash, B., Gajanand, M.S., & Narendran, T.T., (2018) A model for planning the product portfolio and launch timings under resource constraints, *International Journal of Production Research*, 56:15, 10.1080/00207543.2017.1394588

[13] Gershenson, J. K., Prasad, G. J. and Zhang, Y. (2003). Product modularity: Definitions and benefits. *Journal of Engineering Design*, Vol. 14, No. 3, pp. 295-313

[14] Yassine, A., (2004) An introduction to modeling and analyzing complex product development processes using the design structure matrix (DSM) method, *Quaderni di Management*, Vol. 51, No. 9, pp.1–17.

[15] Adegbola, A., & Allada, V., (2022) A revenue-based decision-making approach for evaluating modular product release plans under resource constraints. *Int. J. of Industrial and Systems Engineering* – (Paper Accepted).

https://doi.org/10.21872/2023IISE_2655

Proceedings of the IISE Annual Conference & Expo 2023
K. Babski-Reeves, B. Eksioglu, D. Hampton, eds.

A Multivariate Regression Approach to Modeling Homelessness Trends Using Local and National Indicators

Rumeysa Dagsoy, Ali E. Abbas
Industrial and Systems Engineering Department, University of Southern California
Los Angeles, California 90089
dagsoy@usc.edu aliabbas@usc.edu

Abstract

Homelessness is a growing problem that impacts the national economy and increases the social divide. Furthermore, homelessness is influenced by many economic and social factors, including national and local indicators such as GDP, crime index, and unemployment rate. This paper focuses on the relationship between homelessness and factors associated with the economy and crime rates on both local and national levels. Using homeless data from HUD Exchange and AHAR reports for ten years; we explore the factors that improve the prediction of homelessness for six states (California, New York, Washington, Texas, Florida, and DC). Two regression models were created: Model 1 included local-level indicators, while Model 2 explored the impact of adding three more U.S. macro-level indicators to the regression model. The results showed that Model 2 provided better predictions of the total number of homeless people than Model 1. We comment on the efficacy of homelessness prediction using local and national indicators. Results show that a holistic systems approach is needed when addressing homelessness trends.

Keywords

Homelessness, prediction, model development, macro-level indicators, local factors

1. Introduction

A total of one and a half million Americans experience homelessness each year [1], which has remained consistent since 2007. Homelessness is caused by a range of factors that vary from mental illness and substance abuse to social issues, such as a lack of affordable housing [2]. The U.S. federal government has allocated more than 5 billion dollars to housing programs for Fiscal Year 2021, as reported by the National Alliance to End Homelessness [3]. People experiencing homelessness have difficulty coping with medical and psychological issues [4]. Two theories have been proposed to explain the factors contributing to homelessness [5]: the theory of systematic causes and the personal inadequacy theory. The former attribute homelessness to factors such as mental illness, unemployment, and lack of adequate housing. The latter suggests that personal shortcomings, such as a lack of social relationships, alcohol and substance abuse, leisurely lifestyles, laziness, and avoidance of responsibility, contribute to homelessness. This paper takes a systematic approach to examine the underlying factors that contribute to homelessness. Using multivariate regression analysis, we investigate the relationship between GDP, unemployment rate, and crime rate, which are known to affect homelessness [6]. To provide a comprehensive understanding, we analyze data sets from various U.S. government institutions collected between 2010 and 2020, spanning a decade.

2. Problem Statement

This study aims to investigate the relationship between homelessness and various economic and social factors, such as GDP, unemployment rate, and crime index, for six selected states in the U.S. Previous research on homelessness has primarily focused on local factors such as the effects of the COVID-19 pandemics [7], temperature [8], and political inclination [9]. This paper also considers macro-level U.S. indicators, specifically GDP, unemployment rate, and crime index, to better understand their impact on predicting homelessness.

We constructed two models to develop regression models that accurately predict the number of homeless individuals for the last three years. The first model (Model 1) incorporated local factors (GDP, unemployment rate, crime index) and evaluated its effectiveness. We then incorporated the same three indicators at the U.S. level into the model to examine the impact of macro-level data on the model's prediction accuracy for the six states. This led to the second model (Model 2), which underwent multiple regression analysis for estimating the number of homeless individuals. We compared the accuracy of Model 2 to that of Model 1 to determine whether including the three additional factors improved the precision of the model's predictions.

https://doi.org/10.21872/2023IISE_2655

3. Background and Related Works

Homelessness in the United States is a complex social issue [10] influenced by local factors such as GDP, unemployment, and crime rates. The causes are multifaceted, and this study aims to examine them to inform effective policies and provide social justice. Economic instability, including unemployment, poverty, and unaffordable housing, plays a significant role in homelessness [11, 12]. Although a negative correlation exists between GDP and homelessness [13], other factors like high housing costs and low wages can still cause instability [14]. Further research is needed to address root causes, including mental illness, substance abuse, and domestic violence [15], and reduce crime rates in affected communities. Economic factors can contribute to a higher risk of homelessness when combined with other factors, such as mental illness, substance abuse, and domestic violence. Furthermore, the crime rate is another societal factor related to homelessness. High crime rates in an area can make it difficult for individuals experiencing homelessness to feel safe and find stable housing. A study [16] found that individuals experiencing homelessness had a lower likelihood of exiting homelessness in areas with high crime rates. Several studies have highlighted the impact of economic factors on both homelessness and crime rates. For example, it is found that a positive correlation exists between unemployment rates and crime rates [17]. In addition, rising income inequality has been linked to increased levels of homelessness [18].

The impact of national factors on the homelessness rate can vary from state to state. Policies to reduce the unemployment rate in some states may increase the unemployment rate in other states. Therefore, it is crucial to examine homelessness using both local and national indicators. The presence of a homeless population due to various economic factors, including unemployment, is an undeniable reality [19]. It is crucial to consider these economic impacts when investigating homelessness issues. Examining the cost of homelessness not only reveals its potential cost-effectiveness but also suggests that solutions to homelessness may lead to cost savings [20]. It helps us understand the financial implications of homelessness, revealing whether addressing this issue may be cost-effective.

4. Methodology

4.1. Dataset

Table 1. Raw data for the total number of homeless

Year	CA	NY	WA	TX	FL	DC
2010	123,480	65,606	22,878	35,121	57,551	6,539
2011	125,128	63,445	20,439	36,911	56,687	6,546
2012	120,098	69,566	20,504	34,052	55,170	6,954
2013	118,552	77,430	17,760	29,615	47,862	6,865
2014	113,952	80,590	18,442	28,495	41,542	7,748
2015	115,738	88,250	19,419	23,678	35,900	7,298
2016	118,142	86,352	20,827	23,122	33,559	8,350
2017	131,532	89,503	21,112	23,548	32,190	7,473
2018	129,972	91,897	22,304	25,310	31,030	6,904
2019	151,278	92,091	21,577	25,848	28,328	6,521
2020	161,548	91,271	22,923	27,229	27,487	6,380

We used datasets containing the total number of homeless individuals from the Housing and Urban Development for six U.S. states between 2010 and 2020 (https://www.hudexchange.info/). Table 1 represents the total number of individuals experiencing homelessness during those years.
GDP data for each state were acquired from the U.S. Bureau of Economic Analysis (BEA) at https://www.bea.gov/.
Unemployment rates for each state were sourced from the U.S. Bureau of Labor Statistics (BLS) at https://www.bls.gov/.

https://doi.org/10.21872/2023IISE_2655

Crime rates for each state were obtained from The Disaster Center at https://www.disastercenter.com/.
U.S. macro-level data (national GDP, national unemployment rate, and U.S. crime rate) was obtained from the website (https://www.macrotrends.net/countries/USA/united-states/).

4.2. Multiple Regression Analysis

A multiple regression analysis was conducted to estimate the impact of three local factors on homelessness rates for each state. The model was built using data from 2010 through 2017, and its predictive ability was tested using the years 2018, 2019, and 2020. Model 1 focused on these state-specific factors. The specific factors used in Model 1 were the unemployment rate, GDP, and crime index at the local level metrics for six states. Model 2 included U.S. level metrics of GDP, unemployment rate, and crime rate in addition to the local-level metrics for each state. By incorporating both local and national data, the analysis aimed at the effects of national indicators on local predictions.

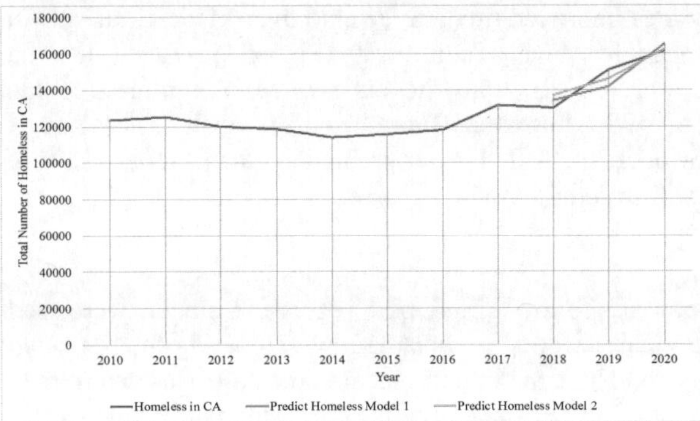

Figure 1. Measured and predicted homeless in CA

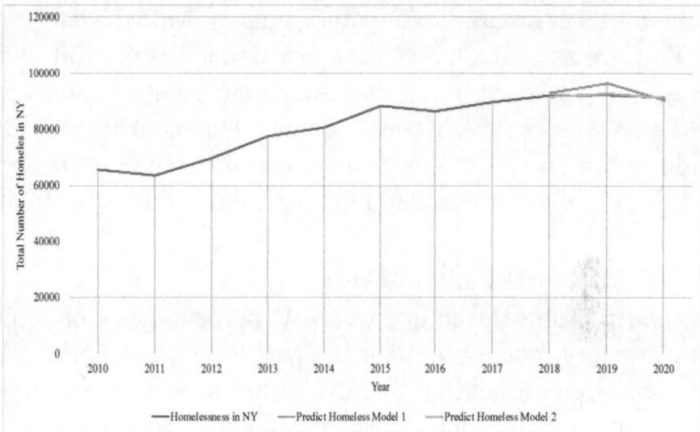

Figure 2. Measured and predicted homeless in NY

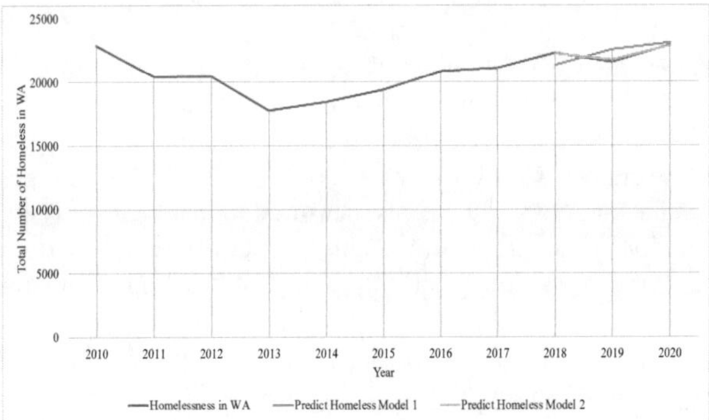

Figure 3. Measured and predicted homeless in WA

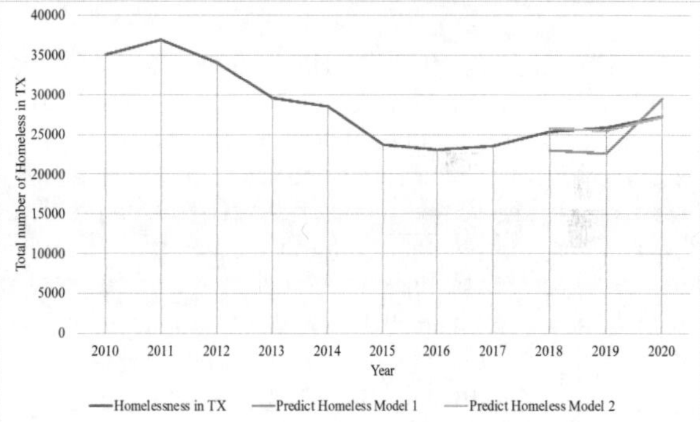

Figure 4. Measured and predicted homeless in TX

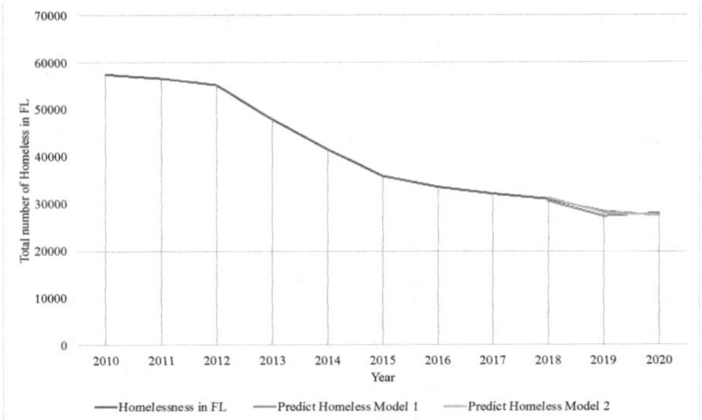

Figure 5. Measured and predicted homeless in FL

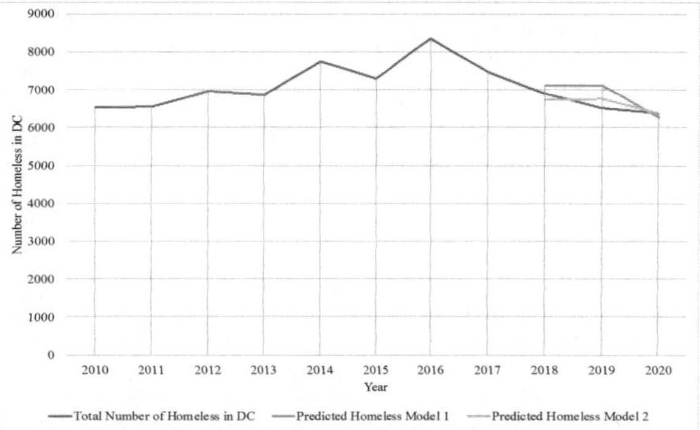

Figure 6. Measured and predicted homeless in DC

https://doi.org/10.21872/2023IISE_2655

Figure 1-6 shows the differences between the measured total number of homeless for 2010-2020 and prediction results according to Model 1 and Model 2 for 2018-2020. Model 1 demonstrated a high level of predictive accuracy for homelessness over the past three years using regional metrics such as GDP, unemployment rate, and crime index; however, by incorporating three additional U.S. metrics, Model 2 significantly improved predictions, resulting in values that closely matched actual homeless numbers over the same period. The results indicate that including U.S. indicators led to enhanced predictions for homelessness.

Figure 1 shows the homelessness data in California and the estimation data from the application of the two models. Figure 2 represents that the overall number of homeless individuals in New York State exhibits an upward trend. The predictions become remarkably accurate when both local and national variables are incorporated into Model 2. In Figure 3-5, Model 2 provided better prediction values than Model 1 in Washington, Texas, and Florida, respectively by estimating almost the same homeless number with real-life data. Although the convergence to the actual homeless data is less compared to the other figures, Model 2 still provides improved predictive value over Model 1, as shown in Figure 6 as well. Our results showed that Texas exhibited the highest level of conformity (Figure 4), while California exhibited the least conformity among the six states analyzed (Figure 1). In 2020, the total number of homeless people in Texas was 27,229. Model 1 predicted the number to be 29,533.95, while Model 2 predicted 27,155.20. These results indicate that both models made reasonable predictions, but Model 2 provided better predictions than Model 1. These findings indicate that including macro-level data can enhance the model's predictive power.

5. Results and Discussion

To clarify Table 2, the improvement percentages were computed as follows. First, both regression models were used to estimate homelessness for the past three years. Next, the square error was calculated by comparing the predicted data for each model to the actual homelessness data for that year. Then, the improvement percentage was determined by subtracting the square error of Model 2 from the square error of Model 1 and dividing the result by the square error of Model 1. The average of the improvement percentages for the last three years was taken. This provides a measure of the percentage change between the two models.

Results indicate that Model 2, which incorporates U.S. indicators such as GDP, unemployment rate, and crime index, provides a more accurate prediction of homelessness compared to Model 1, which only considers state-level metrics. This suggests that including macro-level data can significantly enhance the model's predictive capability. Across all years and models, the square error has improved, except for the state of California, as shown in Table 2. The square errors column shows the square error between the predicted and actual homeless values measured in that year for both models. The improvement column defines the percent reduction in square error. A higher percentage reduction indicates greater efficiency of Model 2. According to Table 2, the most significant improvement in average square error for three years was in TX, with a reduction of 97.93%.

Table 2. Comparing square error for two models in each state for the last three years

State	Year	Square Error for Model 1	Square Error for Model 2	Average Improvement
CA	2018	17099800.28	51077660.04	11.11%
	2019	91043104.65	27898664.54	
	2020	14643569.24	582965.96	
NY	2018	887815.38	245363.59	-89.68%
	2019	17881317.37	307943.85	
	2020	807851.5	13000.91	
WA	2018	887170.58	5818.2	-97.26%
	2019	971922.14	38179.04	
	2020	43982.44	1596.64	

https://doi.org/10.21872/2023IISE_2655

TX	2018	5303094.7	251923.8	-97.93%
	2019	10260496.8	138082.24	
	2020	5312810.13	5446.61	
FL	2018	160430.02	31749.34	-91.56%
	2019	863022.16	31931.3	
	2020	254452.33	4654.77	
DC	2018	40194.22	26662.22	-72.43%
	2019	366595.19	58604.37	
	2020	12179.26	47.92	

Both local and national factors together did not capture as well by the model in California and contributed to the lower improvement in the model's predictive power. There may be other factors that are unique to the state, such as the high cost of living, limited affordable housing, and a large population of people experiencing mental health or addiction issues.

6. Conclusions

This study is a first in a sequence of studies to explore the effects of national indicators on local homelessness trends. This model used a multivariate regression with local and national indicators. Overall, the results indicate that local-level indicators, including GDP, crime rate index, and unemployment, contribute to predicting homelessness. Additionally, incorporating macro-level data from the U.S. significantly improved the ability to predict the number of homeless individuals in each state.

In our analysis, we addressed potential interactions between factors by conducting multicollinearity tests and examining variance multiple correlation values. This approach, along with sensitivity analyses, ensured that our findings were robust and not influenced by spurious correlations or confounding factors. We acknowledge that macro-level indicators can impact local-level indicators, even if they represent the same factors. To account for this, we included both local and national variables in our models, allowing us to assess their individual and combined effects on homelessness rates. This approach helps us better understand the interplay between these factors and their influence on the issue. The study's findings have implications for policymakers, highlighting the need to consider state-level and macro-level data when developing policies and interventions to address homelessness. By incorporating various factors into their decision-making, policymakers can make more informed and effective decisions considering the issue's complexity.

Further research and analysis could be conducted to investigate and better understand the reasons for these differences. One limitation of this study is that not all factors potentially affecting homelessness could be included in the analysis due to data availability and accessibility. Future research can aim to have a more comprehensive and diverse set of factors to explore their impact on homelessness. We will use a probabilistic relevance diagram approach in the future. One limitation of this study is that not all factors potentially affecting homelessness could be included in the analysis due to data availability and accessibility. Future research will also use a probabilistic relevance diagram approach to account for dependencies among the indicators.

References

[1] The Department of Housing and Urban Development, "HUD Exchange," Available: https://www.hudexchange.info/. [Accessed December 10, 2022].

[2] Substance Abuse and Mental Health Services Administration, "SAMSHA" Available: https://www.samhsa.gov/data/. [Accessed December 12, 2022].

[3] "National Alliance to End Homelessness," Available: https://endhomelessness.org/legislation/appropriations-fy21-selected-homelessness-and-housing-programs-funding-chart/. [Accessed November 16, 2022].

https://doi.org/10.21872/2023IISE_2655

[4] V. Tischler, A. Rademeyer, and P. Vostanis, "Mothers Experiencing Homelessness: Mental Health, Support and Social Care Needs," *Health and Social Care in the Community*, vol. 15, no. 3, pp. 246–253 (8 pages), 2007, doi: 10.1111/j.1365-2524.2006.00678.x.

[5] M. V. Ry, "Homelessness Families. Causes, Effects, and Recommendations," NY, USA: Garland Publishing, 1993.

[6] G. Chard, T. Faulkner, and A. Chugg, "Exploring Occupation and Its Meaning Among Homeless Men," *The British Journal of Occupational Therapy*, vol. 72, no. 3, pp. 116–124 (9 pages), 2009, doi: 10.1177/030802260907200306.

[7] C. Parsell, A. Clarke, and E. Kuskoff, "Understanding Responses to Homelessness during COVID-19: An Examination of Australia," *Housing Studies*, vol. 38, no. 1, pp. 8–21 (15 pages), 2022, doi: 10.1080/02673037.2020.1829564.

[8] B. Ramin and T. Svoboda, "Health of the Homeless and Climate Change," *Journal of Urban Health*, vol. 86, no. 4, pp. 654–664 (11 pages), 2009, doi: 10.1007/s11524-009-9354-7.

[9] J. Aulette and A. Aulette, "Police Harassment of the Homeless: The Political Purpose of the Criminalization of Homelessness," *Humanity and Society*, vol. 11, no. 2, pp. 244–256 (12 pages), 1987, doi: 10.1177/016059768701100207.

[10] J. Turnbull, W. Muckle, and C. Masters, "Homelessness and Health," *Canadian Medical Association Journal (CMAJ)*, vol. 177, no. 9, pp. 1065–1066 (2 pages), 2007, doi: 10.1503/cmaj.071294.

[11] D. Culhane, D. Treglia, K. Steif, R. Kuhn, and T. Byrne, "Estimated Emergency and Observational/Quarantine Capacity Need for the US Homeless Population Related to COVID-19 Exposure by County" *Projected Hospitalizations, Intensive Care Units and Mortality*, 2020.

[12] B. O'Flaherty, "Homelessness Research: A Guide for Economists (and Friends)," *Journal of Housing Economics*, vol. 44, pp. 1–25 (25 pages), 2019, doi: 10.1016/j.jhe.2019.01.003.

[13] B. O'Flaherty, "Making Room: The Economics of Homelessness," *Cambridge, Mass: Harvard University Press*, 1996.

[14] M. Shinn, A. L. Greer, J. Bainbridge, J. Kwon, and S. Zuiderveen, "Efficient Targeting of Homelessness Prevention Services for Families," *American Journal of Public Health*, vol. 103, no. 2, pp. S324–S330 (6 pages), December 2013, doi: 10.2105/AJPH.2013.301468.

[15] J. Tsai, A. S. Mares, and R. A. Rosenheck, "Does Housing Chronically Homeless Adults Lead to Social Integration?," *Psychiatric Services*, vol. 63, no. 5, pp. 427–434 (7 pages), 2012, doi: 10.1176/appi.ps.201100047.

[16] C. J. Tompsett and P. A. Toro, "Predicting Overt and Covert Antisocial Behaviors: Parents, Peers, and Homelessness," Journal of Community Psychology., vol. 38, no. 4, pp. 469–485, 2010, doi: 10.1002/jcop.20375.

[17] J. Tsai, C. Y. S. Lee, J. Shen, S. M. Southwick, and R. H. Pietrzak, "Public Exposure and Attitudes About Homelessness," *Journal of Community Psychology*, vol. 47, no. 1, pp. 76–92 (17 pages), 2019, doi: 10.1002/jcop.22100.

[18] M. Desmond, "Unaffordable America: Poverty, Housing, and Eviction: American Journal of Sociology," *The Affordable Housing Reader*, pp. 389-395 (6 pages), Routledge, 2022.

[19] P. Flatau and K. Zaretzky, "The Economic Evaluation of Homelessness Programmes," *European Journal of Homelessness*, vol. 2, pp. 305-320 (16 pages), December 2008.

[20] J. Scott, "Homelessness and Mental Illness," *British Journal of Psychiatry*, vol. 162, no. 3, pp. 314–324 (11 pages), 1993, doi: 10.1192/bjp.162.3.314.

https://doi.org/10.21872/2023IISE_2694

Proceedings of the IISE Annual Conference & Expo 2023
K. Babski-Reeves, B. Eksioglu, D. Hampton, eds.

Simulation-based Modeling and Analysis for the AGV Sorting System of Central Fill Pharmacy Systems

Hicham El Baz, Yong Wang, Yu Jin, and Sangwon Yoon

Binghamton University, Binghamton, New York, USA

Abstract

Recently, the Central Fill Pharmacy (CFP) market expanded with an increased number of medication packages being delivered to customers worldwide. Many typical parcel sorting systems rely on long conveyors, which require large spaces, rigid locations, and high costs. In comparison, the sorting systems equipped with Automated Guided Vehicles (AGVs), which can perform in tight spaces while providing high flexibility with lower costs, have increasingly been utilized in distribution centers and manufacturing systems. To improve the design and operation of such AGV sorting systems, this work focuses on studying the effect of the number of AGVs on the system performance by applying simulation modeling and conducting the sensitivity analysis. Numerical experiments are conducted on the simulation model of an AGV sorting system with two layers (five induction operators and 405 sorting destinations in each layer). The simulation model is established using FlexSim, where six experiments were carried out considering 120, 140, 160, 180, 200, and 220 AGVs evenly distributed among operators. The experimental results show that the number of AGVs has a significant effect on the system throughput. More specifically, the system throughput increases as the number of AGVs increases initially. However, after a certain point, the increase in the number of AGVs will cause a decrease in system throughput, and this is due to congestion, collision, and lack of space. The outcome of this research provides useful insights for AGV sorting system application in CFP systems.

Keywords

Automated guided vehicle, Central fill pharmacy, Performance estimation and simulation

1. Introduction

Due to the growing aging population and the development of chronic conditions that require regular renewed prescriptions, the consumption of prescription medication has rapidly increased in recent years, from 2.1 billion to 3.9 billion between 1994 and 2009 [1]. Automation is now being used not only in small-medium enterprises (SMEs) [2-4] but also in pharmacies to keep up with this development. Thanks to automation, pharmacies can handle greater tasks in less time and with fewer staff members, reducing drug errors, improving productivity, and enhancing inventory management [5]. Recently, CFP has become a solution to process large volume prescription demand in large-scale mail-order pharmacies [6]. The CFP consists of several workstations, including sorting, collation, packing, and auto-filling, all connected by a conveyor system that moves prescriptions as needed until they are finally ready for delivery to the consumer. In warehouses or distribution facilities, sorting systems organize products according to their final destinations, preparing unpredictable flows of goods into shipping-ready orders. An increase in package volumes has been partly driven by commercial clients who demand time-sensitive shipping and trends toward smaller orders [7]. Companies that handle parcels require affordable solutions with high performance [8]. Several options are available to accommodate rising order demand, improved delivery accuracy, and more frequent shipments. AGVs are logistics transport vehicles with excellent safety standards and strong capabilities that can operate entirely unattended. The deployment of AGVs in the CFP can significantly alter the logistics sector by optimizing the entire logistics process [9].

https://doi.org/10.21872/2023IISE_2694

2. Problem Description

Despite the growing usage of AGV sorting systems in CFPs, new techniques for monitoring the throughput of these systems are still necessary. Throughput, defined as the rate at which items are sorted and moved through the system, is a crucial performance indicator for assessing AGV sorting systems in CFPs. However, present approaches for measuring AGV sorting system throughput are unable to effectively evaluate the performance of these systems in real-world settings. To propose a technique for correctly assessing the throughput of AGV sorting systems in CFPs, taking into consideration the particular problems and limitations of these systems. The technique should appropriately assess the performance of AGV sorting systems in terms of their capacity to sort and move objects quickly and efficiently. This research will be confined to analyzing the throughput of the AGV sorting system in CFPs. Simulations will be used to design the approach. The anticipated objective of this research is a technique for precisely assessing the throughput of AGV sorting systems in CFPs that yields significant information into the functioning of these systems. The technique should assist in identifying areas for improvement in AGV sorting systems in CFPs and give suggestions for enhancing the performance of these systems.

3. Related Research

In the order processing and shipping sector of fulfillment centers, items or packages are typically sorted using automated conveyor sorters [10]. The induction station (IS) is used to supply items to be sorted, either manually or mechanically, into the sorter. Once inside, the items are identified and sorted into multiple departure channels before being collected for packaging. Various sorting systems, such as tilt-tray, bomb bay, and cross belt sorters, are available based on the packages to be sorted and the desired throughput capacity. This research focuses on evaluating the performance of automated sorting systems. Bozer and Sharp used simulation to assess the impact of the number of output lanes on the throughput of the warehouse sorting system for different sorting machine designs [11]. Bozer, Quiroz, and Sharp investigated how the number of station lanes and the duration before the next phase is fed into the system affected the performance of a closed-loop conveyor sortation system [12]. They found that, in systems with low output lane obstruction, the next accessible technique outperforms the constant priority method in terms of throughput. Fedtke and Boysen [13] used simulation to investigate how the system throughput of the closed-loop sorting system in the package fulfillment center changed with varying numbers of detection points, pallet counts, related flow directions, and parallel circulating conveyors. According to the study, increasing the number of detecting points would significantly boost throughput, while increasing the number of pallets would have no effect. Additionally, adding parallel conveyors would increase output. Bozer and White formulated a two-server closed queueing model for a mini-load AS/RS sorting system with two ISs and one operator [14]. They enhanced this model in 1996 by giving each operator access to multiple ISs and lanes [15]. The models estimate the system's anticipated throughput and the operator's anticipated use. Foley and Frazelle calculated the predicted throughput limits for a mini-load AS/RS sorting system with two ISs [16-18].

4. Methodology

The A* (A-star) method is a well-known graph search algorithm used in AGV sorting systems to determine the most efficient route for the AGV to travel while carrying objects between locations. The algorithm uses a heuristic function to estimate the cost of reaching the destination from a particular node and calculates the optimal route based on this information. In AGV sorting systems, the A* algorithm is typically used to calculate the ideal route for the AGV between various sites, such as the IS and the destination station. The algorithm takes into account several variables, including the AGV's current position, the target station's location, and any obstacles or constraints in the environment. Numerous discrete event simulation languages and software packages are available, and many successful simulation applications have been documented. FlexSim is an object-oriented discrete event simulation environment that allows for the creation of simulation models to visualize and manage the dynamic-flow processes of a system. It includes a comprehensive set of model-creation tools [19]. FlexSim is commonly used to analyze and optimize AGV sorting systems, allowing users to develop virtual models and test various setups and scenarios. The program also incorporates

https://doi.org/10.21872/2023IISE_2694

the A* algorithm into the model of the AGV sorting system, enabling users to evaluate the system's performance and optimize its efficiency.

An automated material handling system, such as an AGV sorting system or conveyor system, has a dedicated space called an induction station where new packages are fed into the system. Packages are identified and measured at the induction station (IS) in preparation for additional processing before being sorted to their respective places inside the facility. The ISs main objective is to guarantee precise and proficient package processing, routing, and handling in an automated system by the AGV sorting system

4.1. Assignment Rules

In an AGV sorting system, packages can be assigned to an IS in two ways: randomly or through dedicated assignment. When packages are randomly assigned, they are allocated to ISs in an arbitrary manner. In contrast, a dedicated assignment rule involves earmarking specific system resources, such as AGVs, for particular tasks. This method can improve system productivity by minimizing the time required for resources to move between activities and allowing them to focus exclusively on the assigned task. For instance, in an AGV sorting system, a dedicated assignment rule could be used to allocate certain AGVs to specific ISs. These AGVs would be responsible for collecting and conveying packages from their assigned IS and would not be used for other tasks. By utilizing a dedicated assignment algorithm, the AGV sorting system can enhance performance and reduce the risk of congestion or bottlenecks at the ISs.

4.2. Simulation Model Assumptions

This study has developed a simulation model of an AGV sorting system, and a top view of the layout is shown in Figure 1. The simulation model consists of 2 layers, with each layer having 405 destination totes and 5 ISs. Each IS consists of 2 operators, one for the bottom layer and the other for the top layer. In an IS, a worker can immediately place a package on a robot with the barcode facing upward for scanning, and a robot can release an item into a drop-off point in a short period [20]. Robotic sorting systems usually handle items lighter than 33 pounds. Table 1 shows the AGV parameters used in the simulation. We make the following assumptions in this study:

a) The package arrival rate is relatively high to keep loading stations active and determine the maximum throughput capacity of a system with a specific number of AGVs and ISs.
b) The simulation runs for 7 hours and 20 minutes (shift time).
c) The AGV follows dedicated assignment rules when returning to an IS.
d) AGV battery charging time is not considered in the simulation.

Table 1: System Parameters Used in Simulation

Speed (m/s)	Acceleration (m/s^2)	Stops for a Turn (s)	Loading Time (s)	Unloading Time (s)
2	1.5	1	2	2

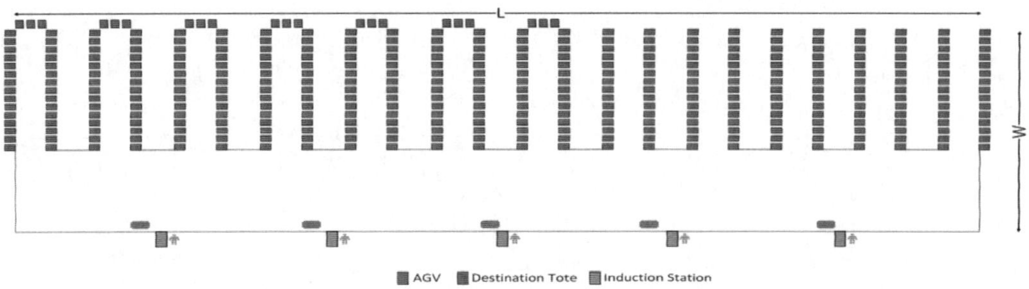

Figure 1: Top view of the two-layer sortation system.

https://doi.org/10.21872/2023IISE_2694

5. Analysis and Inferences

In order to assess the effect of increasing the number of AGVs allocated to each IS, the sortation system's performance was evaluated using throughput and AGV utilization metrics. The simulation model was run for a single shift lasting 7 hours and 20 minutes with a 1-hour warm-up period.

5.1. Throughput and Utilization

Utilization and throughput are two important performance indicators for AGV sorting systems. Throughput measures the number of packages sorted and moved by the AGV system within a given time period, and it is also a metric of the system's efficiency and capacity. A high throughput indicates that the AGV sorting system can handle and transport more parcels in a certain time period. Utilization, on the other hand, represents the percentage of time that the AGV system is actively sorting and carrying items. It takes into account travel times, loading and unloading times, and periods of inactivity or being blocked. High utilization demonstrates that the AGV sorting system is being used effectively and efficiently, with all available resources being utilized to their maximum capacity.

Figures 2 and 3 illustrate the analysis of throughput and utilization in this study. As shown in Figure 2, the system throughput is significantly affected by the number of AGVs. It increases with the number of AGVs and reaches a maximum of 69,202 packages for one shift with 200 AGVs. However, beyond a certain threshold, adding more AGVs would reduce system throughput due to congestion, collisions, and lack of space. With 220 AGVs, the system throughput drops to 68,682 packages. Figure 3 shows that the overall utilization of AGVs starts at 86.06% utilized and 13.94% being blocked with 120 AGVs, and decreases as more AGVs are added, reaching 58.18% being utilized with 220 AGVs. Meanwhile, the proportion of blocked AGVs increases to 41.82% with 220 AGVs.

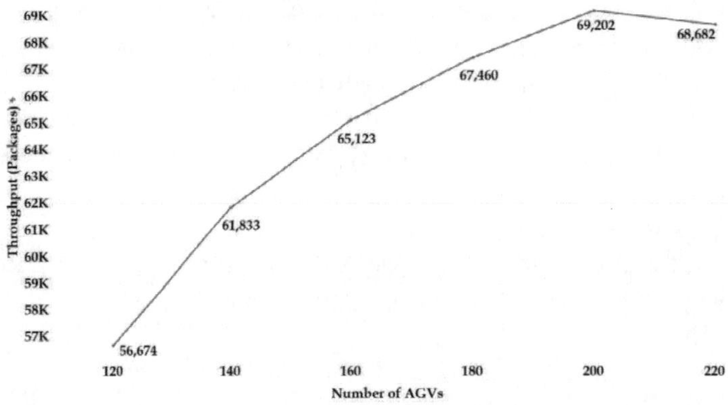

Figure 2: Throughput of the sortation system with regard to different number of AGVs.

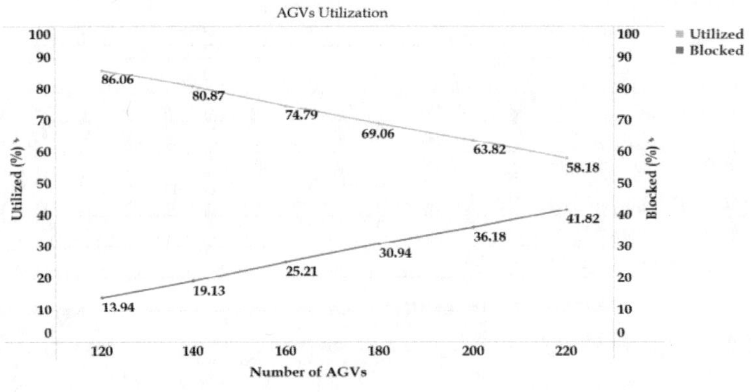

Figure 3: Utilization of the sortation system with regard to different number of AGVs.

https://doi.org/10.21872/2023IISE_2694
Proceedings of the IISE Annual Conference & Expo 2023
K. Babski-Reeves, B. Eksioglu, D. Hampton, eds.

5.2. Travel Time

The cycle time refers to the time elapsed between the AGV's start of loading a package and its return to the IS for sorting another package. In this study, we assessed the travel time for different numbers of AGVs, and the corresponding box plots are displayed in Figure 4. As shown in the figure, when there are 120 AGVs, the average travel time for AGVs is 56.5 seconds, whereas when the number of AGVs increases to 220, the average travel time becomes 86.69 seconds. This indicates an increase of 30 seconds in the average travel time after adding 100 AGVs.

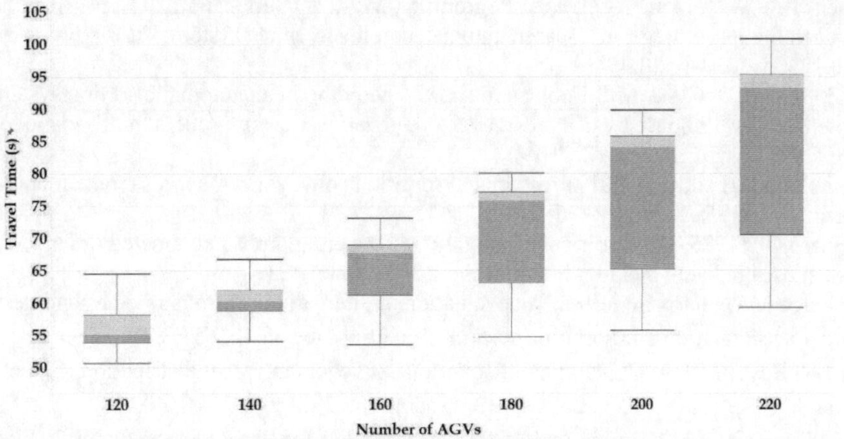

Figure 4: Travel time of the sortation system with regard to different number of AGVs.

6. Conclusions

AGV sorting throughput, utilization, and travel time are critical performance indicators for AGV sorting systems in CFPs. Monitoring and managing these variables is essential to ensure that the AGV sorting system is operating at maximum efficiency. High throughput, utilization, and short travel time can result in improved operational efficiency, increased production, and reduced expenses. Conversely, low throughput, utilization, and high travel time may indicate system inefficiency and contribute to congestion and bottlenecks. To enhance the efficiency of the AGV sorting system, a comprehensive study of the system is necessary to identify potential areas for improvement. This can be achieved by considering the following suggestions:

a) Conduct a thorough examination of the system to identify areas for improvement, such as modifying routing algorithms, optimizing assignment rules, and enhancing the system's overall architecture.

b) Consider using advanced simulation tools to create a virtual version of the system, enabling a detailed examination of the system and the opportunity to test different scenarios and methods.

c) Invest in technological upgrades and maintenance, such as improved sensors and automation tools, to enhance the system's performance and reduce downtime. Regular maintenance of the system can also help prevent issues and ensure consistent performance.

References

[1] A. Jenkins and S. F. Eckel, "Analyzing methods for improved management of workflow in an outpatient pharmacy setting," *American Journal of Health-System Pharmacy*, vol. 69, no. 11, pp. 966–971, 2012.
[2] A.-R. S. Shehata, M. Heshmat, M. A. El-Sharief, and M. G. El-Sebaie, "Identifying sources of variations in cement bags company: a case study," in *2019 IEEE 6th International Conference on Industrial Engineering and Applications (ICIEA)*, 2019, pp. 154–158.
[3] A. S. Shehata, M. Heshmat, and M. A. El-Sharief, "Reduction of variation and control parameters optimising in a cement-bags company," *Int. J. Process Manag. Benchmarking*, vol. 12, no. 3, pp. 321–347, 2022.
[4] A.-R. S. Shehata, M. Heshmat, M. A. El-Sharief, and M. G. El-Sebaie, "Implementation of statistical process control techniques to reduce the defective ratio: a case study," in *Proceedings of the International Conference on Industrial Engineering and Operations Management*, 2018, pp. 666–675.

https://doi.org/10.21872/2023IISE_2694
Proceedings of the IISE Annual Conference & Expo 2023
K. Babski-Reeves, B. Eksioglu, D. Hampton, eds.

[5] S. Oswald and R. Caldwell, "Dispensing error rate after implementation of an automated Pharmacy Carousel System," *American Journal of Health-System Pharmacy*, vol. 64, no. 13, pp. 1427–1431, 2007.
[6] N. Cao, A. Marcus, L. Altarawneh, and S. Kwon, "Priority-based replenishment policy for robotic dispensing in central fill pharmacy systems: A simulation-based study," Health Care Management Science, 2023.
[7] U. Clausen, D. Diekmann, J. Baudach, J. Kaffka, and M. Poting, "Improving parcel transshipment operations - impact of different objective functions in a combined simulation and optimization approach," *2015 Winter Simulation Conference (WSC)*, 2015.
[8] K. J. C. Fransen, J. A. W. M. van Eekelen, A. Pogromsky, M. A. A. Boon, and I. J. B. F. Adan, "A dynamic path planning approach for dense, large, grid-based automated guided vehicle systems," *Computers & Operations Research*, vol. 123, p. 105046, 2020.
[9] Y. Liu, S. Ji, Z. Su, and D. Guo, "Multi-objective AGV scheduling in an automatic sorting system of an unmanned (intelligent) warehouse by using two adaptive genetic algorithms and a multi-adaptive genetic algorithm," *PLOS ONE*, vol. 14, no. 12, 2019.
[10] Boysen N, Briskorn D, Fedtke S (2019) Automated sortation conveyors: A survey from an operational research perspective. Eur. J. Oper. Res. 276(3):796–815.
[11] Bozer YA, Sharp GP (1985) An empirical evaluation of a general purpose automated order accumulation and sortation system used in batch picking. Material Flow 2(2-3):111–131.
[12] Bozer YA, Quiroz MA, Sharp GP (1988) An evaluation of alternative control strategies and design issure for automated order accumulation and sortation systems. Material Flow 4(4):265–282.
[13] S. Fedtke and N. Boysen, "Layout planning of sortation conveyors in parcel distribution centers," Transportation Science, vol. 51, Nov. 2014, pp. 3-18, doi: 10.1287/trsc.2014.0540.
[14] Bozer, Y.A., White, J.A. 1990. Design and Performance Models for End-Of-Aisle Order Picking Systems. Management Science, 36(7), 852-866.
[15] Bozer, Y.A., White, J.A. 1996. A Generalized Design and Performance Analysis Model for EndOf-Aisle Order-Picking Systems. IIE Transactions, 28(4), 271-280.
[16] Foley, R.D., Frazelle, E.H. 1991. Analytical Results for Miniload Throughput and the Distribution of Dual Command Travel Time. IIE Transactions, 23(3), 273-281.
[17] Foley, R.D., Frazelle, E.H., and Park, B.C. 2002. Throughput Bounds for Miniload Automated Storage/Retrieval Systems. IIE Transactions, 34(10), 915-920.
[18] Foley, R.D., Hackman, S., and Park, B.C. 2004. Back-Of-The-Envelope Miniload Throughput Bounds and Approximations. IIE Transactions, 36(3), 279-285.
[19] Abu-Taieh, E. M. O. and EL Sheikh, A. A. R. 2007. Commercial simulation packages: a comparative study. International Journal of Simulation Systems, Science and Technology, Special Issue on: Models and Applications in Engineering and Technology, 8, pp 66-76.
[20] Nordgren, W. B. 2003. Flexsim simulation environment. In: Proceedings of Winter Simulation Conference, Phoenix, Arizona, USA. IEEE, 197-200 Vol. 1.

https://doi.org/10.21872/2023IISE_2751

Proceedings of the IISE Annual Conference & Expo 2023
K. Babski-Reeves, B. Eksioglu, D. Hampton, eds.

Individualized Consideration for Team Collaboration

Rochelle Cook, PhD, ThinkWriteScholar, LLC, Greensboro, NC

Abstract

Globalization continues to drive the need and relevance for organizations to deploy projects which are accomplished largely through the work of internal teams. While Team and Project Management Literature tends to focus on performance outcomes (quality, cost, time), intrinsic metrics such as collaboration are gaining more attention as a strong influencer of project success. Collaboration between team members, or the intentional act of thinking and acting together to achieve a common goal, is driven by the psychosocial needs of the team members assigned to perform the work. However, the disproportionate and essential contributions expected from the cross-functional team tend to make collaboration challenging. While some studies explain the impact of structured approaches on collaboration, scant research describes how the Project Lead role can influence socially salient outcomes such as collaboration with and through project team members. Recent study findings will be compared to the tenets of individualized consideration, using one of the constructs of transformational leadership theory, to understand how the project lead's behaviors can narrow the gap between the project's needs and the team members' interpersonal needs. Opportunities for future research will be identified.

Keywords
Project lead, collaboration, cross-functional team, transformational leadership theory, individualized consideration

1. Introduction
As projects remain a significant contributor to organizations' strategic goals, metrics tend to reflect what project management literature coins the "iron triangle": quality, cost, and time [1]. Despite projects being prioritized as part of a larger business strategy, many fail [2]. Moreover, project objectives are broken down into tasks that internal team members must perform to achieve success. As a result, factors that are thought and felt are being explored as well. One of those factors is collaboration.

Collaboration has evolved over the past two decades from a transactional requirement between Customer and Supplier to an international standard [3]. Studies have described collaboration as a means to achieve cooperation between organizations in efforts such as mergers and acquisitions and construction projects [3]. Collaboration enables learning, knowledge sharing, and consensus building [1] which impacts project outcomes. Collaboration is also applied within organizations through cross-functional teams.

1.1 Problem Statement
Cross-functional teams represent different departments and functional areas and are typically assigned a project lead. A structured process is applied to transition the project from ideas/plans to tangible, implemented, and sustained outcomes [4]. Individuals are typically delegated to contribute as an internal team member by an authoritative stakeholder such as the project Sponsor [2]. The project lead is formally assigned to achieve the project objectives through the team's efforts. However, mere assignment to a project team does not guarantee collaboration between team members. In addition to applying technical tools and techniques, the project lead should leverage the social and knowledge capital that each team member brings to the project in a way that reflects a need to think and act together.

1.2 Purpose and Scope

https://doi.org/10.21872/2023IISE_2751

This paper suggests opportunities for the project lead to initiate behaviors that facilitate collaboration in cross-functional team environments. This will be accomplished through exploring individualized consideration from the perspective of the project lead. For this work, the terms *internal team* and *cross-functional team members,* or individuals from different departments and/or functional areas assigned to perform project work, will be used interchangeably.

1.3 Method

This paper, using one of the constructs of transformational leadership theory (individualized consideration), explores how the Project Leader can influence interpersonal collaboration between team members. Moreover, the benefits and challenges of collaboration as a precursor to performance-based outcomes will be compared to the purpose of individualized consideration from the lens of the Project lead. Lastly, recommended leader behaviors to forge collaboration will be shared.

2. Collaboration

Collaboration, or "the commitment to working together with two or more parties to create value by striving to achieve shared competitive goals and operational benefits through a spirit of mutual trust and openness," [1] has evolved from a preferred posture for project operations to an expectation that is evaluated as a contributor to project success. Collaboration is reflected throughout the project and demonstrated in team behaviors such as learning, information sharing, and consensus building.

2.1 Drivers of Collaboration

Studies suggest that projects should have a degree of standardization or a set way of operating alongside the ability to navigate complexities [4]. Collaboration should be part of that standardization to enable effectiveness during the temporary, unique nature of projects. Recent studies also identify multiple factors that have a significant influence on collaboration. This paper will explore two of them: relationships and coordination.

Relationships imply how two or more individuals are familiar with or connected to each other [1]. Establishing relationships between cross-functional team members is challenging due to the temporal nature of projects or the proximity between the team members. Also, the degree to which the project tasks impact the functional groups tends to drive voluntary interpersonal relationships among team members. This is critical because individuals need to share knowledge and information that results in a collective contribution to the project's success.

Coordination reflects the degree to which the required tasks, talents, and processes are used over time to accomplish project objectives. Studies suggest that collaboration increases as the degree of coordination increases [1]. In addition, structured practices such as scope and risk planning promote coordination while allowing the team to address emerging events. This is important because internal teams perform routine and non-routine tasks within a certain time frame.

2.2 Collaboration Challenges

With the intentional efforts to collaborate comes challenges. Project expectations are translated into taskwork disproportionately delegated to cross-functional team members [3]. Taskwork evolves through techniques such as progressive elaboration, which can make collaboration difficult if what it takes to accomplish the work is unclear. Also, depending on the organizational culture, individuals may not have a working relationship with some team members. This is significant because the degree of relationships is a significant driver of emotional responses such as trust [1].

Moreover, recent studies suggest that unforeseen events in a project expose factors such as the degree of relationships and coordination in a way that requires more than a structured or contractual approach to drive project outcomes [4]. A willingness to think, understand, and work together is essential to avoid emotional responses that can derail a project. As a result, the following questions are posed: How can the project lead's behavior be leveraged to initiate collaboration? How can the project lead optimize key factors of the collaboration process throughout the life of the project?

3. Transformational Leadership Theory

Transformational leadership is a behavioral-based process where actions initiated by the leader result in emotional responses from the recipients of that behavior. This theory is significant because it suggests the interpersonal

https://doi.org/10.21872/2023IISE_2751

exchange between the leader and the cross functional team members can result in outcomes that seemed challenging or impossible before that interaction occurred. For example, the project leader who initiates behaviors that transforms a team member's self-interest into the need to learn, share information, and make decisions can impact psychosocial factors such as team collaboration [5]. Four constructs shape this theory. For this paper, we will explore one of them: individualized consideration.

3.1. Individualized Consideration

Individualized consideration aligns the project work with the team member's needs in a way that influences a higher degree of interest and desire to contribute. This is salient in a team environment because roles and expectations differ among functions but are necessary for project success [1]. Behaviors that drive collective enthusiasm instead of divisiveness and group think occur because of the project lead's interaction [4].

For instance, at the beginning of the planning phase of an improvement project, the problem is usually clear, but how to address it requires team-based techniques. When project lead uses silent brainstorming to get ideas, each person can participate, which enables consensus building when it is time to take that same data and decide on the next steps. Moreover, information that only one team member may be aware of is now available to the entire team.

3.2 Role of the Project Lead

There are opportunities in the life of a project for individualized consideration to be used to activate collaborative team outcomes such as consensus building, learning, and knowledge sharing.

One of the project lead's roles is to manage the intersection of tasks among the team members. Imagine a key milestone or event that advances the project forward is to get the test plan approved. The project lead should make sure that there is sufficient participation among the contributors to the test plan. Therefore, collaborative practices such as sharing the roles and responsibilities matrix before constructing the test plan will help distinguish who is responsible for drafting the plan versus who needs to be consulted or informed [1]. As a result, the risk of team members feeling excluded or disregarded is minimized. Also, time to draft the plan is optimized because the team is aware of the expectations before the work begins.

Another reality in projectized organizations is interpersonal conflict. Different thoughts or perspectives about different aspects of the project is expected, but how it is managed can either reinforce or threaten collaboration [1]. The project lead should ensure that the conflicts are addressed while coordinating the tasks required to meet the project objectives.

For example, let's say that a process improvement project team is planning a validation pilot. One of the team members dominates the conversation and insists that the team adopts a checklist used in a previous pilot to expedite planning. The lead notices that the other team members disengage. The project lead can address this team dynamics by leveraging tools that the team is already aligned on and use it to influence participation.

For instance, the flow chart for the process being piloted can be used to clarify the scope, so that the checklist being suggested can be verified as something the team should use. Moreover, the success factors documented in the project charter can also be leveraged to make sure the team is focused on measuring factors that enable the team to identify the right performance metrics. Initiating behaviors through tools and techniques that the team is already aligned on provides a basis to learn how one task performed by a functional area impacts another later in the process.

The project lead should also remind the team how their roles impact project outcomes. For instance, while the operations function is responsible for doing the work, the logistics function ensures the product is delivered on time and quality is responsible for measuring and documenting the results. The project lead should prioritize transforming the overbearing behavior of one team member into learning and knowledge sharing through these tools and techniques. This should drive the team to arrive at a consensus about how the pilot is to be executed so that the agreed upon success factors are met.

https://doi.org/10.21872/2023IISE_2751

4. Comparative Analysis

This paper aimed to identify ways collaborative behaviors can emerge through the team due to the project lead initiating individualized consideration. Through observing the current team relationship dynamics and effective use of team and project management techniques, the project lead can interject behaviors that promote learning, knowledge sharing, and consensus building. The table below summarizes recommended team lead behaviors to narrow the gap between collaboration challenges and achieving desired collaborative outcomes.

To influence learning, the project lead should make sure the team members understand the relationship between the project requirements and their respective roles at the beginning of the planning phase. Behaviors that shift thoughts of self-interest to a collective understanding of why individuals were delegated to participate drives collaboration among team members.

Sharing information tends to advance the team from learning/discovery to the identification of risks, issues, and constraints that must be addressed. The project lead should insert behaviors that enable team members to not only recognize how they can contribute but express how one function's contribution impacts another. When the project lead adds tools such as Failure Mode and Effects Analysis (FMEA), risks are documented and prioritized which enable the team to identify actions /tasks that will mitigate the chances that risks turn into issues.

Collective decision making helps the team align on the approach to accomplish the project goals. This requires the project lead to make sure there are effective ways to achieve consensus while keeping the project on track. Conflicts tend to arise when there are differences about how to execute a change. These conflicts can be addressed through initiating objective techniques to assist in decision making. Techniques such as cost benefit analysis, force field analysis, and multi-voting bridge the gap between diversity of thought and validated data. When the project lead facilitates interactions that transform multiple perspectives into a single decision, the team members tend to respond with understanding, alignment, and support. Therefore, the following table offers some recommended leader behaviors that can facilitate collaboration.

Table 1: Project Lead Behaviors to achieve collaboration outcomes (learning, knowledge sharing, consensus)

Project Phase	***Learning***	***Knowledge Sharing***	***Consensus Building***
Initiating		Develop responsibility matrix prior to first team meeting	
Planning	Publish agenda prior to status meetings. Share responsibility matrix. Align tasks with function and project expectations. Share communication plan	Have ad hoc work sessions to facilitate risk planning. Centralize information and make it accessible to team members. Build timeline with the team as they identify what it takes to achieve the objectives	Enable individual input using consensus building tools and techniques. Weigh pros and cons with team when multiple options are presented.
Executing		Centralize information and make it accessible to team members	
Monitoring	Close /disband workgroup/functions once they fulfil their project responsibilities	Centralize information and make it accessible to team members	
Closing	Share success factors and actual project outcomes	Enable team members to document their project experience	

https://doi.org/10.21872/2023IISE_2751

When one considers the impact that collaboration has on project outcomes when initiated by the project lead, it is necessary to identify propositions to research in the future.

4.1 Propositions

Proposition 1:
Project leads who initiate collaborative behaviors tend to increase collaboration among team members significantly.

The planning phase tends to occupy most of the life of the project. This phase involves decomposing the milestones into tasks that the team must complete to achieve the success factors of the project. Behaviors that make room for thinking and deciding together while navigating differences enables learning, knowledge sharing, and consensus building. Also, the project lead's engagement tends to decrease as the project goes from planning to execution.

Therefore, Proposition 2:
Project teams who optimize collaboration during the planning phase tend to achieve positive project outcomes compared to those who do not.

5. Implications

This paper will benefit team environments that seek to develop project leads into thought leaders who can effectively influence the team members who perform the work required to accomplish the project objectives. In doing so, an intrinsic metric is prioritized as a driver of tangible outcomes. This may further impact research in the testing and validation of collaboration as a measure of project success.

References

[1] T. J. Bond-Barnard, L. Fletcher, H. Steyn, "Linking Trust and Collaboration in Project Teams to Project Management Success," *International Journal of Managing Projects in Business*, vol. 11, no. 2, pp. 432-457, 2018, doi:10.1108/IJMPB-06-2017-0068

[2] Project Management Institute, *A Guide to the Project Management Body of Knowledge*, 6th ed. Project Management Institute, 2017.

[3] M. Chakkol, K. Selviaridis, M. Finn, "The Governance of Collaboration in Complex Projects," *International Journal of Operations & Production Management*, vol. 38, no. 4, pp. 997-1019, 2018, doi:10.1108/IJOPM-11-2017-0717

[4] F. Chiocchio, S. Grenier, T. A. O'Neill, K. Savaria, J. D. Willms, "The Effects of Collaboration on Performance: A Multilevel Validation in Project Teams," *International Journal of Project Organisation and Management*, vol. 4, no. 1, pp. 1-37, 2012, doi:10.1504/IJPOM.2012.045362

[5] P. Northouse, "Transformational Leadership," in A. Bruckner, D. Simons, eds., *Leadership: Theory and Practice*, 4th ed., Sage, 2007, pp. 175-206.

https://doi.org/10.21872/2023IISE_2893

Proceedings of the IISE Annual Conference & Expo 2023
K. Babski-Reeves, B. Eksioglu, D. Hampton, eds.

Approximation of Optimal Control Surfaces for the Bass Model with Stochastic Dynamics

Abstract ID: 2893

Gabriel Nicolosi, Harold and Inge Marcus Department of Industrial and Manufacturing Engineering, Penn State University, University Park, PA and Christopher Griffin, Applied Research Lab at Penn State, University Park, PA

Abstract

The Bass diffusion equation is a well-known and established modeling approach for describing new product adoption in a competitive market. This model also describes diffusion phenomena in various contexts: infectious disease spread modeling and estimation, rumor spread on social networks, prediction of renewable energy technology markets, among others. Most of these models, however, consider a deterministic trajectory of the associated state variable (*e.g.*, market-share). In reality, the diffusion process is subject to noise, and a stochastic component must be added to the state dynamics. The stochastic Bass model has also been studied in many areas, such as energy markets and marketing. Exploring the stochastic version of the Bass diffusion model, we propose in this work an approximation of (stochastic) optimal control surfaces for a continuous-time problem arising from a 2×2 skew symmetric evolutionary game, providing the stochastic counter-part of the Fourier-based optimal control approximation already existent in the literature.

Keywords
Optimal control, Fourier approximation, evolutionary game, learning, stochastic Bass model.

1. Introduction and Relevant Literature

With its origins in the marketing literature, the Bass diffusion equation is a classical modeling approach initially developed to describe the underlying dynamics of new product adoption in a competitive market [1]. Its usefulness in modeling nonlinear phenomena is evident by its later application in diverse areas: infectious disease spread modeling and estimation [2], product sales forecasting [3], renewable energy technology diffusion [4], among others. The Bass model has also been used in the realm of non-deterministic problems: electricity markets [5] and sales forecasting [6] are just two examples. In the context of digital product maintenance, [7] derived a deterministic Bass-constrained optimal control problem (OCP) from a skew-symmetric game (a game with skew-symmetric payoff matrix), for which a solution was found by means of the classical Pontryagin's Maximum Principle (PMP), a classical result in the theory of optimal control that provides necessary conditions for control optimality. More recently, [8] proposed a backpropagation inspired Fourier-based approximation procedure for the same problem, approximating the solution found in [7] and generalizing it over a predefined range of system initial conditions. The work in [8] falls under the wide umbrella of global orthogonal collocation in numerical methods for OCPs, a well-established and active field of research. See [9] for an overview. Using the same approach as in [8], we construct approximating optimal control surfaces for the stochastic variation of the OCP arising from a 2×2 skew-symmetric evolutionary game, constrained by the Bass equation. In order to introduce stochasticity to the original problem, noise is added to the state dynamics in two different forms: additive noise to the fully observable state $x(t)$, and additive noise to the observer $y(t)$. This approach of adding noise is consistent with the literature of stochastic control and realistically models the underlying uncertainty involved in the diffusion process . Further details are discussed in Section 2, in which both deterministic and stochastic versions of the problem in question are introduced. Section 3 introduces the Fourier-based control parameterization used to generate the approximating optimal control and state surfaces. Some results from computational experiments are shown in Section 4. Finally, the general context of this work and future directions are briefly discussed in Section 5.

2. Preliminaries and Problem Construction

Consider the evolutionary dynamics of a population consisting of $n_1 + n_2 = n$ individuals, wherein these two groups represent non-consumer (n_1) and consumer (n_2) partitions of the total population. Defining $x_i = \frac{n_i}{n}$ as the percentages of individuals in each group, the system of differential equations describing the evolution of each group and the underlying market-share dynamics is

https://doi.org/10.21872/2023IISE_2893

$$\begin{aligned}\dot{x}_1 &= -\rho x_1 x_2 \\ \dot{x}_2 &= \rho x_2 x_1\end{aligned} \tag{1}$$

where ρ is the interaction payoff between both population partitions, giving rise to the skew-symmetric pay-off matrix $A = \left(\begin{smallmatrix} 0 & -\rho \\ \rho & 0 \end{smallmatrix}\right)$. We note that Equation (1) is an instance of the so-called replicator dynamics [10]. Since $x_1 + x_2 = 1$, Equation (1) is reduced to $\dot{x} = \rho x(1-x)$, which, in its turn, is a special case of the Bass diffusion equation from [1], *i.e.*:

$$\dot{x} = p(1-x) + q(1-x)x \tag{2}$$

for when the so-called coefficient of innovation $p = 0$. Now, consider an extraneous mechanism $u = u(t)$ that seeks to alter the dynamics shown in Equation (2), such as product maintenance effort aiming increased revenue stream through client base expansion. The Bass dynamics then becomes

$$\dot{x} = \pi(u)x(1-x), \tag{3}$$

where $\pi(u) = \beta u - \xi$ is a linear function of u, which is positive if and only if $\beta u > \xi$, *i.e.*, the intervention effect on the market-share dynamics is only positive against the next best alternative ξ. Following the example in [7], a company wishes to minimize a continuous performance index from the launching of a digital product at $t = 0$ up until the end of its life-cycle at $t = T$:

$$\min_u \int_0^T (Cu^2 - \alpha u x)\, dt, \tag{4}$$

where Cu^2 is the quadratic cost of maintenance effort and $\alpha u x$ is the linear revenue stream. If we require the solution of Equation (4) to respect Equation (2) departing from a given $x(0) = x_0$, we have a OCP with a deterministic Bass dynamics constraint.

2.1 The Deterministic Bass Equation Constrained OCP

The OCP of interest, with dynamics described by a deterministic Bass equation, is then formulated as:

$$\begin{aligned} &\min_u \int_0^T (Cu^2 - \alpha u x)\, dt \\ s.t. \quad &\dot{x} = x(1-x)(\beta u - \xi) \\ &x(0) = x_0 \\ &u(t) \geq 0. \end{aligned} \tag{5}$$

The solution to this OCP can be found analytically by applying the PMP like done in [7]. If instead an optimal control solution surface is desired for a range of initial conditions $x_0 \in X_0$, the procedure introduced by [8] can be applied, yielding approximations based on a two-dimensional Fourier series, dependent both on time t and initial condition x_0. The goal of this paper, however, is to provide an approximation in the taste of [8] to the stochastic variation of OCP (5), introduced next.

2.2 The Stochastic Bass Equation Constrained OCP

Building on OCP (5), we propose two stochastic model variations for the problem introduced above. The first model considers that the state dynamics is subject to additive white noise $\xi(t)$, giving a stochastic Bass model described by

$$\dot{x} = x(1-x)(\beta u - \eta) + \sigma\xi(t), \tag{6}$$

where $\sigma \in \mathbb{R}$. The second variation assumes deterministic dynamics as in [8], which in its turn cannot be observed without the presence of noise. Defining the observed state as $y(t)$, the noisy (observed) trajectory is formulated as

https://doi.org/10.21872/2023IISE_2893

$$y(t) = x(t) + \sigma\xi(t). \tag{7}$$

In what follows, we present the optimal control problem (OCP) formulation for both model dynamics variations presented in Equations (6) and (7). In light of the first stochastic model described above, the OCP under stochastic dynamics becomes

$$\begin{aligned} \min_u \quad & E\left[\int_0^T (Cu^2 - \alpha x u)\, dt\right] \\ s.t. \quad & \dot{x} = x(1-x)(\beta u - \eta) + \sigma\xi(t) \\ & x(0) = x_0 \\ & u \in L_2([0,T]), \end{aligned} \tag{8}$$

where the expectation operator E is needed in the objective functional in order to address the nondeterministic trajectory $x(t)$. The second OCP is defined in the same manner, however observing the trajectory, derived from Equation (6), through Equation (7). We also note that Equation (6) can be rewritten as

$$dx = x(1-x)(\beta u - \eta)\, dt + \sigma\, dW, \tag{9}$$

which is consistent with the notation in the literature of stochastic differential equations, and dW comes from the fact that the Brownian motion $W(t)$ has its time derivative defined as a white noise process, that is, $\dot{W}(t) = \xi(t) \sim \mathcal{N}(0,\sigma)$. Next, we introduced the control approximation $u \approx \hat{u}(t,x_0)$ and formulate the stochastic OCP, such as in (8), as a nonlinear programming (NLP) problem to be solved for the Fourier coefficients of the control approximation.

3. Optimal Control Surface Approximation

Our goal is to find a control approximation $\hat{u}(t,x_0)$ that solves the OCP (8) for any arbitrary initial condition in a predetermine set X_0. For this, we adapt the approximation scheme proposed in [8]. A two-variable cosine Fourier series is used to approximate the controllers $u(t,x_0)$:

$$u(t,x_0) \approx \hat{u}(t,x_0) = \sum_{m=0}^{M}\sum_{n=0}^{N} a_{mn} \cos\left(\frac{m\pi t}{T}\right) \cos\left(\frac{n\pi x_0}{\mathcal{X}_0}\right) \tag{10}$$

where a_{mn} are Fourier coefficients, with $M, N \in \mathbb{N}$, T is the same terminal time from (8) and $\mathcal{X}_0$ is the range within which the initials conditions vary in X_0. The objective functional in (8) is then transformed into a finite dimensional optimization problem, where the trajectory x satisfies its original dynamics constraint, given an initial condition, wherein u is approximated as in Equation (10).

3.1 Stochastic Dynamics

Substituting the approximation (10) into (6), we obtain

$$\dot{x} = x(1-x)(\beta\hat{u} - \eta) + \sigma\xi(t) \tag{11}$$

with $x(0) = x_0$, that is, the original dynamics constraint in OCP (8) wherein u is approximated as in Equation (10). For the stochastic constrained problem addressed in this work, the non-deterministic component of (11) must be taken into consideration. Therefore, for a given initial condition, if $\varphi(t)$ satisfies Equation (6) without the presence of noise $\xi(t)$, then $\widetilde{\varphi}(t) = \varphi(t) + \sigma W(t)$ satisfies its stochastic counterpart, where $\sigma W(t)$ is the Brownian motion affecting the state trajectory. The new objective functional becomes:

$$E\left[\int_0^T \left(Cu^2 - \alpha\left(\varphi + \sigma W(t)\right)u\right)\, dt\right]. \tag{12}$$

However, the new objective functional is equivalent to its deterministic counterpart. To show this, Equation (12) is rewritten as:

https://doi.org/10.21872/2023IISE_2893

$$\int_0^T (Cu^2 - \alpha\varphi u)\, dt - \alpha\sigma u\, E\left[\int_0^T W(t)\, dt\right] = \int_0^T (Cu^2 - \alpha\varphi u)\, dt, \tag{13}$$

since, on average, the definite integral of Brownian motion $W(t)$ is zero. Therefore, in this case, the objective functional remains unchanged from the deterministic one. Thus, the optimization procedure can be used exactly as in [8].

3.2 Stochastic State Observer

Now we address the case in which the state trajectory is observed as described by Equation (7). For this, we substitute the right-hand side of Equation (7) into our objective functional, obtaining:

$$E\left[\int_0^T (Cu^2 - \alpha(\varphi + \sigma\xi(t))u)\, dt\right] = \int_0^T (Cu^2 - \alpha\varphi u)\, dt - \alpha\sigma u\, E\left[\int_0^T \xi(t)\, dt\right] \tag{14}$$

By definition, $\dot{W}(t) = \xi(t)$, therefore

$$\int_0^T \xi(t)\, dt = \int_0^T \dot{W}(t)\, dt = W(T) - W(0) = W(T) \tag{15}$$

Thus, since $E[W(T)] = 0$, once again, the objective functional reduces to the deterministic case. We conclude that both in this case, and as in (3.1), the control approximation (10) is robust to the types of noise considered therein, providing the same approximation capabilities previously demonstrated in [8].

3.3 Nonlinear Programming Problem

Using the control approximation from Equation (10), we construct the following problem:

$$\begin{aligned} \min_u \quad & E\left[\int_0^T (Cu^2 - \alpha x u)\, dt\right] \\ s.t.\ & \dot{x} = x(1-x)(\beta\hat{u} - \eta) + \sigma\xi(t) \\ & x(0) = x_0 \\ & u(t, x_0) \approx \sum_{m=0}^{M}\sum_{n=0}^{N} a_{mn}\cos\left(\frac{m\pi t}{T}\right)\cos\left(\frac{n\pi x_0}{L}\right). \end{aligned} \tag{16}$$

We note that a similar formulation is valid for the case considered in Subsection (3.2). Like discussed in Subsection (3.1), if φ satisfies the flow determined by the deterministic component of the Bass dynamical constraint for a given initial condition x_0, then the OCP (16) becomes the following unconstrained NLP:

$$\min_{\mathbf{a}} \quad J(\mathbf{a};\, x_0) = \int_0^T \left(C\hat{u}(\mathbf{a})^2 - \alpha\varphi(\mathbf{a};\, x_0)\hat{u}(\mathbf{a})\right) dt. \tag{17}$$

Finally, the approximating surface of optimal controllers is found by solving

$$\min_{\mathbf{a}} \quad \sum_{x_0 \in X_0} J(\mathbf{a};\, x_0), \tag{18}$$

which can be done by a first order method like gradient descent, taking place in the space of Fourier coefficients **a**.

4. Computational Results

In this section we present computational results for the two cases considered in (3.1) and (3.2), obtained by solving the optimization problem in (18). For these experiments, we choose the following parameters values: $\alpha = 2, C = 1, \beta = \frac{1}{2}$, $\xi = \frac{1}{4}$, and $\sigma = 0.1$. Figure 1 shows, for both cases considered in (3.1) and (3.2), respectively, results for two choices of number of Fourier coefficients: $M = N = 3$ and $M = N = 5$. To illustrate the behavior of sample approximating optimal trajectories around the deterministic one, Figure 2 shows, for two fixed values of initial conditions, five samples of optimal state trajectory approximations obtained using the approximating optimal control.

https://doi.org/10.21872/2023IISE_2893

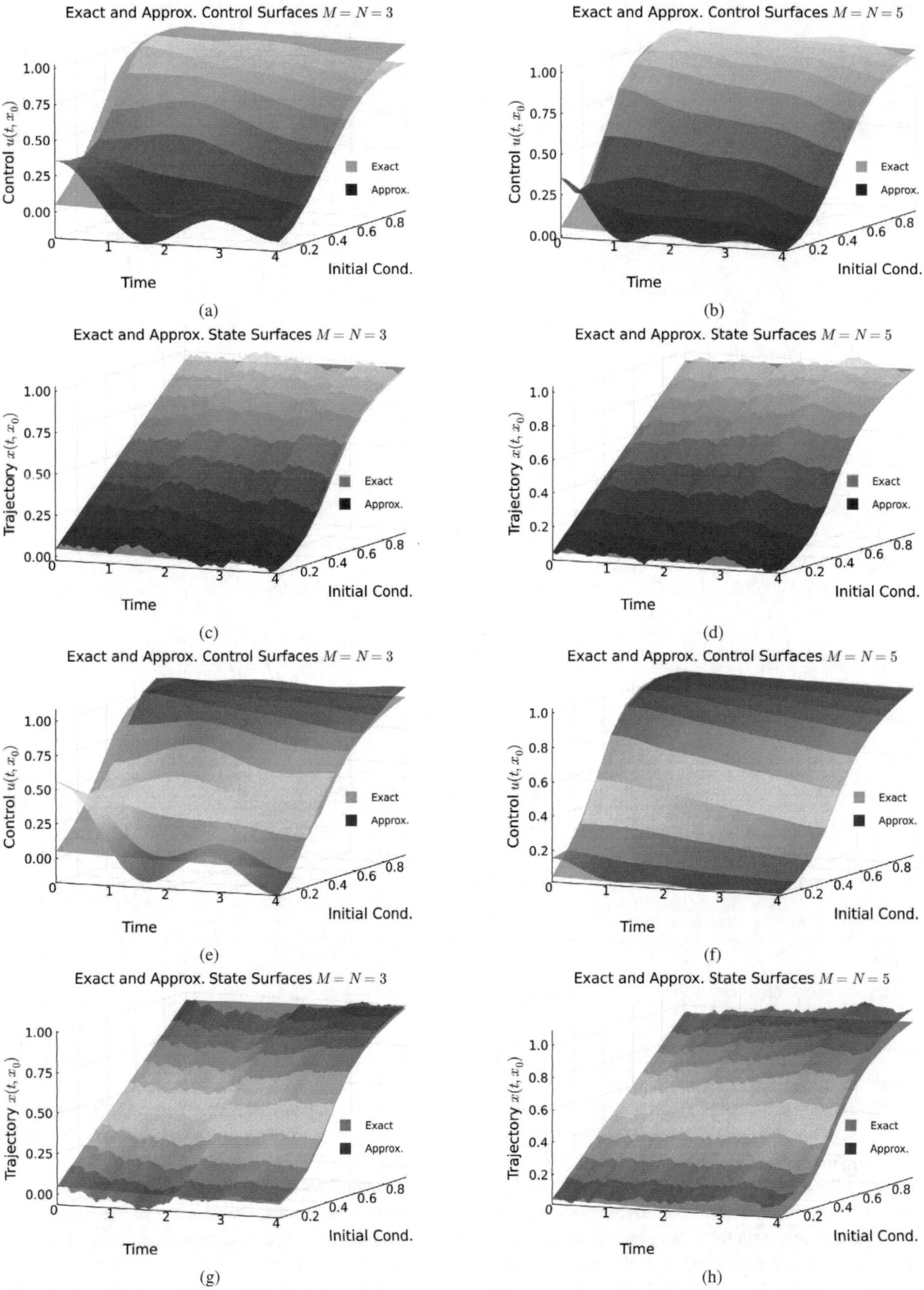

Figure 1: Optimal control surface approximations and (sample) optimal state trajectory surfaces for two choices of Fourier coefficients: $M = N = 3$ and $M = N = 5$ for the case of stochastic dynamics $((a) - (d))$ and noisy observer $((e) - (h))$ for $\sigma = 0.1$.

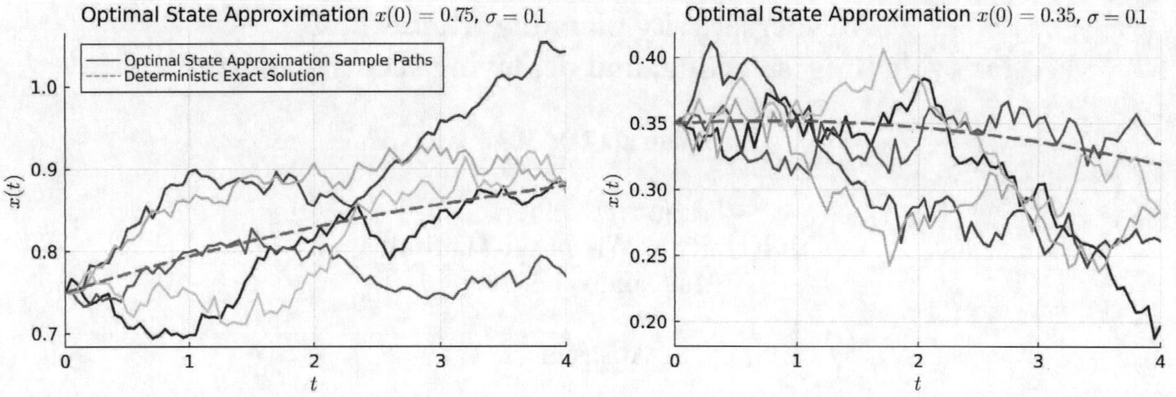

(a) (b)

Figure 2: Deterministic optimal state trajectory and five sample paths of optimal state approximations with fixed initial conditions $x(0) = 0.75$ (a) and $x(0) = 0.35$ (b), for $\sigma = 0.1$.

5. Conclusions and Future Directions

This work presented a Fourier-based approximation for a family of optimal controls arising from a stochastic variation of the OCP presented in [8]. Our analysis and computational results showed that the control approximation is robust to the two variations of noise introduced in the dynamics of the system's state variable and that it drives the state trajectory up or down depending on the initial state $x(0)$, which means that past a certain proportion threshold, the optimal control, instead of driving the market-share up, slowly brings it down toward an equilibrium, which is otherwise not feasible in the absence of a control input. Future work will challenge some assumptions of the current method, such as the separability of state and control in the dynamics, which guarantees one unique set of Fourier coefficients in the parameterization of both control and state. The interpretability of the underlying descent in the space of Fourier coefficients will also be investigated, aiming for the potential extraction of theoretical properties of the intrinsic OCP from the coefficients convergence to the approximated optimal solution.

References

[1] Frank M Bass. "A new product growth for model consumer durables". In: *Management science* 15.5 (1969), pp. 215–227.

[2] Enes Eryarsoy et al. "A novel diffusion-based model for estimating cases, and fatalities in epidemics: The case of COVID-19". In: *Journal of Business Research* 124 (2021), pp. 163–178.

[3] Zhi-Ping Fan, Yu-Jie Che, and Zhen-Yu Chen. "Product sales forecasting using online reviews and historical sales data: A method combining the Bass model and sentiment analysis". In: *Journal of business research* 74 (2017), pp. 90–100.

[4] Rajesh Kumar and Arun Agarwala. "Renewable energy technology diffusion model for techno-economics feasibility". In: *Renewable and Sustainable Energy Reviews* 54 (2016), pp. 1515–1524.

[5] CH Skiadas and AN Giovanis. "A stochastic Bass innovation diffusion model for studying the growth of electricity consumption in Greece". In: *Applied stochastic models and data analysis* 13.2 (1997), pp. 85–101.

[6] Johan Grasman and Marcel Kornelis. "Forecasting product sales with a stochastic Bass model". In: *Journal of Mathematics in Industry* 9 (2019), pp. 1–10.

[7] James Fan and Christopher Griffin. "Optimal digital product maintenance with a continuous revenue stream". In: *Operations Research Letters* 45.3 (2017), pp. 282–288.

[8] Gabriel Nicolosi, Terry Friesz, and Christopher Griffin. "Approximation of optimal control surfaces for 2×2 skew-symmetric evolutionary game dynamics". In: *Chaos, Solitons & Fractals* 163 (2022), p. 112535.

[9] John T Betts. *Practical Methods for Optimal Control Using Nonlinear Programming*. Vol. 36. SIAM, 2020.

[10] Peter Schuster and Karl Sigmund. "Replicator dynamics". In: *Journal of theoretical biology* 100.3 (1983), pp. 533–538.

https://doi.org/10.21872/2023IISE_3037

Proceedings of the IISE Annual Conference & Expo 2023
K. Babski-Reeves, B. Eksioglu, D. Hampton, eds.

A cybersecurity planning framework for evaluating, selecting, and deploying security controls

Abstract ID: 3037

Laura A. Albert
University of Wisconsin-Madison
Madison, WI 53706

Abstract

Organizations should proactively evaluate and address cybersecurity threats. However, there is not a clear set of guidelines and rules to help organizations make informed decisions surrounding cybersecurity planning. Additionally, cybersecurity practices must keep pace as new threats and technologies emerge. This research seeks to fill this gap by introducing a cybersecurity planning framework composed of three phases that consist of risk assessment, security control selection, and security control deployment. We discuss how risk analysis and integer programming can be used in this framework to help organizations make proactive planning decisions that balance multiple criteria such as risk, cost, and resource utilization.

Keywords
Cybersecurity Planning; Risk Analysis; Integer programming

1. Introduction
Protecting critical cyber-infrastructure requires a layered approach that involves threat identification, the proactive adoption of security controls and defenses, and better response and recovery strategies. Cybersecurity threats are adaptive and persistent, and therefore, defenses must constantly adapt and improve to keep pace with new risks. There is a growing need to protect cyber-physical systems using risk management techniques [1, 2]. However, many organizations find it challenging to keep up with cybersecurity best practices and new defenses given that they operate in resource-constrained environments. Government and industrial standards have been developed to inform planning efforts (e.g., [3 – 5]). These standards provide guidance into security planning, although they cannot be directly used for risk assessment. As a result, many organizations could benefit from the development of analytical tools to support risk-based cybersecurity planning.

In this paper, we introduce a cybersecurity planning framework to help organizations evaluate the security controls they have in place and prioritize additional security controls to implement and deploy. We take a high-level systems perspective to provide insight and guidance into the overall process. The framework we introduce supports risk-based decision-making using an organization's limited resources, and it supports proactive planning, not real-time detection, response, or recovery. The framework is composed of three phases that consist of risk assessment, security control selection, and deployment. In the first phase, we utilize a questionnaire, with questions developed in conjunction with experts in industry corresponding to security controls used for a specific industry. We develop a scoring mechanism based on risk analysis to evaluate an organization's questionnaire answers to inform risk planning decisions. Then, we outline how methods based on integer programming can identify a set of additional security controls to adopt, and we discuss how an organization can deploy the selected security controls. The questionnaire and the scores were developed with input from subject matter experts (SMEs) from industry who support the commercial real estate sector, and the approach can be the basis for other markets.

This paper is organized as follows. In Section 2, we introduce and overview the cybersecurity planning framework. Section 3 introduces the risk assessment questionnaire and its scoring mechanism. Section 4 investigates how to recommend security controls to manage risk, and Section 5 describes how to manage the phased roll-out of security controls. In Section 6, we offer concluding comments.

https://doi.org/10.21872/2023IISE_3037

2. Cybersecurity Planning Framework

In this section, we overview phases of a cybersecurity planning framework to help organizations assess risk and select security controls to protect against a variety of threats. Cybersecurity planning falls into the "Protect" function in NIST's Cybersecurity framework [2]. The functions—identify, protect, detect, respond, and recover—organize basic cybersecurity activities at their highest level. The "Protect" function captures the selection and deployment of cybersecurity safeguards to limit or contain the impact of a potential event, and it is composed of several phases. Our framework focuses on three phases related to planning: (1) risk assessment, (2) security control selection, and (3) deployment. Figure 1 provides a schematic of the three phases in the cybersecurity planning framework. Next, we briefly introduce these three phases.

Figure 1: Cybersecurity planning framework schematic

2.1. Risk assessment

There are several government and industrial cybersecurity standards that are used to guide planning decisions (e.g., [3 – 5]). Over time, many organizations have adopted these standards as mandatory controls. However, cybersecurity standards were not developed as risk management tools, and compliance with standards does not necessarily equate to risk-reduction. In addition, complying with all standards is resource-intensive for most organizations, which can deplete organizations' limited security resources [6]. There is a need to understand how various security controls can reduce risk so organizations can prioritize security controls in a sustainable way as well as balance risk reduction with other criteria, such as cost and resource utilization. Risk assessments to support cost-effective cybersecurity decision-making have met limited success due to the challenges in assessing risk [7, 8], accounting for human behavior [8], and deploying security controls in resource-constrained environments [6].

In recent years there has been increased interest in risk-based approaches to support cybersecurity planning [7, 9]. Evaluating risk involves identifying potential threats (vulnerabilities), estimating the conditional probability that each vulnerability is exploited, and identifying the expected consequences [1, 9]. Risk assessments rely on subject matter expert (SME) elicitation and subjective risk assessments. In industry, extensive questionnaires are often given to owners, vendors, integrators, and others who construct cyber defenses. The questionnaires are designed to proactively address cybersecurity issues by performing security risk assessments, although there are no industry-wide standard questionnaires that would allow for a more efficient use of SME time and consistent risk assessments.

2.2. Security control selection

There is a growing body of literature that studies how to mitigate cybersecurity risk through the strategic prioritization of security controls [9]. Enayaty-Ahanger et al. [10] provide a survey of optimization models for cyber-infrastructure security, including research that informs planning decisions. They note the importance of employing analytical tools to shed light on how to use limited resources to inform risk-based planning decisions in resource-constrained settings. A simple approach is to estimate the risk reduction associated with a security control and to prioritize security controls in rank order based on their cost-effectiveness [9]. More sophisticated approaches prioritize the selection of security controls by considering factors such as cost, overlapping capabilities, uncertainty in security control effectiveness, and the role of adaptive adversaries [11 – 13]. Attack modeling is important for cybersecurity protection models. Attack modeling often utilizes attack graphs to characterize potential attacks, where each path represents a series of exploits to achieve attack goals. Zheng et al. [11] introduce integer and stochastic programming models to select a portfolio of security controls subject to a security budget based on the linkages between the security controls and exploits in a set of attack graphs. This model was extended to consider two objectives, the first that maximizes the overall risk reduction and the second that maximizes the weighted number of attacks secured past a risk threshold [12]. Other research studies how to select a cost-effective set of security controls that maximally delay adversarial attacks that are attempted by multiple adversaries, given that there is uncertainty regarding how effective security controls may be [13].

2.3. Deployment

When new security controls are developed, they need to be implemented and operationalized in a timely manner by security personnel. However, there is a dearth of literature considering this issue despite its importance for achieving high levels of security. The importance of deployment is illustrated by the Equifax breach that was caused by the vulnerability Apache Struts CVE-2017-5638 in 2017 [14]. A patch for the vulnerability was released on March 7, 2017, but Equifax left the vulnerability unpatched until July 29, 2017 [14]. It later became evident that Equifax's IT systems had been breached and that the sensitive, personal data of 148 million of their customers had been exposed. The Equifax breach highlights the need for the phased roll-out of security controls, since merely having a patch to address a critical vulnerability is not adequate for protecting critical systems. To effectively manage risk, organizations need to deploy security controls, such as patches, and inform how security personnel incorporate these protection strategies into employee workflows. However, recent industry surveys indicate that deploying security controls is resource intensive [6], and resource limitations can delay or prevent security controls from being implemented. These issues have not yet been adequately addressed in the literature. Doing so would lift the assumptions made in the security control selection phase that security controls are either deployed or not deployed (a binary decision) and that these decisions are made in a single time period. Instead, deployment would inform the choice and implementation of security procedures, including how to deploy security controls. Optimal deployment, as opposed to ad hoc deployment, can match resources to risk, thereby offering the potential to reduce risk without increasing cost.

In the following three sections, we describe how analytical methods drawing upon risk analysis and integer programming can help guide the assessment, selection, and deployment of security controls.

3. Risk Assessment Questionnaire

We describe a risk assessment questionnaire scoring mechanism developed in conjunction with SMEs from industry. The questionnaire was developed to provide a consistent set of rules and guidelines for cybersecurity planning in the commercial real estate industry from a third party. This industry was used for proof-of-concept, with the idea that the questionnaire could be adapted to serve as the basis for a risk assessment for other markets. The questionnaire was motivated by the need for organizations to efficiently identify and manage cybersecurity risk. The vendors of products connected to building systems frequently fill out product questionnaires to assess security risk and to inform planning decisions. The questionnaires have similarities, but are phrased and scored differently, leading to inconsistent feedback. The questionnaires are also time-consuming and burden organizations. The overall goal of a standard questionnaire implemented by a third party is to improve cybersecurity standards in a way that helps organizations better manage risk by providing consistent and regular feedback.

In the questionnaire, security controls were considered across 11 categories of protection that are non-overlapping and complete. This allows us to assume independence across multiple categories of protection when assessing risk. The questionnaire consists of a total of 295 questions across the 11 categories, each of which corresponds to a single security control and its implementation. A total of 226 questions (76.6% of the total) require yes/no responses to simplify the implementation, with a "yes" response corresponding to a best practice, typically based on government or industrial standards [3 – 5]. Of the remaining questions, 57 questions (19.3% of the total) are multiple choice and 12 are free form (4.1% of the total). The multiple choice questions correspond to different types of implementation of a security control. Examples include how frequently passwords must be updated or how long video footage is saved.

Our main contribution is a scoring mechanism for the questionnaire. Performing a risk assessment is particularly challenging when implementing a questionnaire across multiple organizations. We could not directly estimate distributions for conditional probabilities and consequences, since those would be organization and setting specific. Instead, we adopted a simplified process that scores each question more generally based on typical implementations of security controls in the commercial real estate industry. We initially assumed each security control is independent to assess risk in an additive way. This assumption is lifted for security control selection in Section 4. Then, we elicited a weight for each question from three cybersecurity SMEs from industry. The weight for each question reflects the reduction in the expected consequences of a successful attack associated with each security control, where the weight approximately captures the reduction in risk (the change in the product of the conditional probability of success and expected consequences) for a typical organization. The three SMEs initially identified three tiers of weights: *preferred*, *urgent*, and *critical*. Further analysis indicated a need for an additional tier: *super-critical*. Weights were elicited for

https://doi.org/10.21872/2023IISE_3037

these four tiers, and these weights were refined after evaluating sample questionnaire answers. The final weights are 1 = *preferred*, 2 = *urgent*, and 5 = *critical*, and 10 = *super-critical*. Note that the weights and the assignment of weights to questions might differ for other markets.

SME elicitation was used to score the answer to each question. All yes-no questions (76.6% of all questions) were given a value of 0 for a no answer and 1 for a yes answer, and other questions (23.4% of all questions) were scored based on SME input to take on values between 0 and 1 based on the proportion of risk reduction that the particular security control implementation would achieve. We provide a summary of the questions that inform the risk assessment.

The questionnaire is scored by summing the points for the questions in each category, weighing the point total by the category weight, and then summing the weighted points in each category. Each category was initially assigned a category weight of 1 to quantify the impact of each question equally. Later, the SMEs recognized a need to re-weigh the categories to account for the different number of questions (security controls) in each category. To achieve this, three SMEs first estimated the total fraction of risk addressed by the security controls in aggregate in each category (these values added to 1.0). These values were rescaled based on the total number of points associated with the questions in each category to yield the (unscaled) category weights. The category weights were scaled such that the total number of points is between 0 and 100, with 100 being the ideal score.

Table 1 summarizes the questionnaire and its scores. It reports the scaled category weights, the distribution of the number of questions in each category across the four tiers, and the total weighted points in each category. The questionnaire yields scores in each category as well as an overall score to communicate risk to the organization that completes it. Additionally, alternative questionnaires could be developed for different stakeholders, such as owners, vendors, and integrators, to reflect their responsibility in managing risk.

Table 1: Questionnaire summary

Category	Category weight	Number of questions in each tier				Number of questions	Total weighted points
		Preferred	Urgent	Critical	Super Critical		
Company information	**0.40**	0	5	0	0	**5**	**4**
Physical Security	**0.06**	0	2	11	5	**18**	**6.5**
Network Topology	**0.08**	2	13	11	4	**30**	**10**
Authenticate & Admin	**0.05**	7	1	27	7	**42**	**11**
Remote Access, Mobility	**0.09**	8	12	2	4	**26**	**7.25**
System Endpoint	**0.11**	3	14	9	2	**28**	**10.5**
Data Center	**0.06**	2	6	22	3	**33**	**9**
Vulnerability Management	**0.10**	4	1	11	3	**19**	**9**
Data Storage	**0.07**	5	20	11	2	**38**	**8.5**
Policies	**0.16**	1	2	5	4	**12**	**11**
Bus Continuity	**0.05**	7	6	17	5	**35**	**7.75**
Training	**0.61**	9	0	0	0	**9**	**5.5**
Total						**295**	**100**

In the following two sections, we overview how industrial engineering methods based on integer programming could use the questionnaire results to select and deploy security controls.

4. Security Control Selection

The questionnaire score evaluates a portfolio of security controls. If this portfolio is inadequate for managing risk, a remediation plan can recommend security controls to adopt to reach an acceptable risk score. In this section, we outline how models from the literature can inform the selection of a set of security controls that is cost-effective and maximizes risk reduction. One simple way to create a remediation plan is to recommend the adoption of all security controls above a certain level (e.g., all *critical* and *super-critical* controls) at their highest levels. However, this

https://doi.org/10.21872/2023IISE_3037

approach is based on compliance and not risk analysis, and as a result, it may over-burden already resource-constrained organizations and may not lead to implementations that reduce actual risk to acceptable levels. To overcome these limitations, we recommend approaches based on integer programming and risk analysis.

Previous research on budgeted maximal multiple coverage models [11] introduces new integer and stochastic programming models to maximize overall risk reduction under uncertainty subject to a security budget. They account for dependencies between security controls by including multiple choice constraints and by modeling how security controls map to specific vulnerabilities in attack graphs. We propose adapting the modeling approach by [11] to use in conjunction with questionnaires. First, a mapping of questions from the questionnaire to security control standards can capture the dependencies between security controls, which in turn can be used as inputs for the integer programming model. The objective in [11] seeks to maximize overall risk reduction, not how to systematically reduce risk to an acceptable level for as many potential attacks as possible. This objective can be trivially adapted to identify a minimum cost set of security controls to reach a pre-specified overall questionnaire score (i.e., risk level).

Another approach is to prioritize security controls in such a way that an acceptable score is achieved in each questionnaire category. To do so, we can adapt the model the approach by Schmidt et al. [12] who introduce a bi-objective integer programming model that balances overall risk reduction (the same objective as in [11]) with a second objective that seeks to maximize the (weighted) number vulnerabilities whose risk levels are past an acceptable level as defined by a risk threshold. This second goal can be used to identify a cost-effective set of security controls to achieve an acceptable level in each category. For example, the analysis may recommend switching from a SMS-based multi-factor authentication (MFA) to MFA using an authentication app as one of the security controls.

5. Deployment

We recognize that selecting security controls is not sufficient for risk management—the security controls must be deployed. The problem of how to effectively implement a phased roll-out of security procedures remains an open problem in the literature. Rather than simply deciding which security controls to implement (a series of binary decisions), a phased roll-out of security controls dictates how limited resources are used to implement these controls over time. The resources can correspond to personnel time (with different types of personnel) and security budgets over deployment time horizons. For example, implementing MFA using an authentication app consists of selecting a vendor, checking compliance requirements, planning for a variety of access needs, educating users, distributing an app, implementing the new requirement, and responding to alerts over an extended time horizon. Without a deployment plan that takes limited resources into account, the proposed security controls may not be feasible with respect to an organization's resources, which in turn may prevent some security controls from being implemented. This could undermine planning efforts and lead to an organization accepting higher than anticipated levels of risk.

To support deployment decision-making, we propose a new approach that draws upon integer programming models in the scheduling literature to deploy security controls across multiple time periods. One approach is to model this problem as an extension of the Resource Constrained Project Scheduling Problem (RCPSP) [15]. Traditionally, the RCPSP seeks to schedule a set of *jobs* (security controls) with an objective that minimizes the makespan of the project. Each job requires resources of different types for deployment, and there may be precedence relationships between jobs. The RCPSP could use the output from the security control selection phase (see Section 4) as an input. The RCPSP is a well-studied problem in the literature with several variants. One such variant to consider is the multi-mode RCPSP (MRCPSP), where each job can be completed in one of several *modes*, which provide alternate resource/duration options for each job [15]. In the deployment domain, this may reflect different deployments of the same security control from different vendors or different implementations of the control that may reflect how often a task must be performed (e.g., every 30 days versus every 180 days). The output of this modeling approach could identify different security control implementation patterns and provide for the flexibility in the scheduling out of each control by dividing each control into a sequence of jobs to be completed.

6. Conclusions

Protecting cyber-physical systems against dynamic threats in resource-constrained systems requires analytical tools to support risk management and decision-making. This research recognizes that risk assessment ultimately requires a detailed plan that may require multiple phases. We take a systems approach and outline three such phases: risk assessment, security control selection, and deployment. A risk assessment can be used to identify areas where additional protection is required. Security control selection prioritizes solutions to help manage risk, and deployment helps the selected security controls be rolled out over time. When these phases are coordinated, they can be used to construct effective security defenses using limited security resources.

We recognize the opportunity to use industrial engineering methods drawing from risk analysis and integer programming to support decision making across all three phases in the cybersecurity planning framework. However, there are other possible approaches that could be utilized. Other aspects of cybersecurity—including detection, response, and recovery—could benefit from robust study using industrial engineering principles to give decision-makers a suite of tools to use for proactive planning.

Acknowledgements

This work was funded by the National Science Foundation Award 2000986. The views and conclusions contained in this document are those of the author and should not be interpreted as necessarily representing the official policies, either expressed or implied, of the National Science Foundation. The author would like to gratefully acknowledge David Bunzel, Min Kyriannis, and two other anonymous industry experts for their input and feedback.

References

[1] The White House, "Executive Order on Securing the Information and Communications Technology and Services Supply," Office of the Press Secretary, Washington, D.C., 2019.

[2] National Institute of Standards and Technology (NIST). "Framework for improving critical infrastructure cybersecurity," version 1.1. Technical report, NIST, Washington, D.C., 2018.

[3] National Institute of Standards and Technology (NIST). "Security and privacy controls for federal information systems and organizations," NIST Special Publication 800-53, Revision 4, Washington, D.C., 2013.

[4] R. Ross, V. Pillitteri, K. Dempsey, M. Riddle, G, Guissanie. "Protecting Controlled Unclassified Information in Nonfederal Systems and Organizations," National Institute of Standards and Technology Report SP 800-171 Rev. 2, Washington, D.C., 2020.

[5] The Center for Internet Security. "The CIS Critical Security Controls for Effective Cyber Defense," Version 6.0, East Greenbush, NY, USA, October 15, 2015.

[6] R. Stevens, J. Dykstra, W. Knox Everette, J. Chapman, G. Bladow, A. Farmer, K. Halliday, and M.L. Mazurek. "Compliance Cautions: Investigating Security Issues Associated with US Digital-Security Standards," in *Proceedings of the Network and Distributed Systems Security (NDSS) Symposium*, San Diego, CA, USA, February 23-26, 2020, doi: 10.14722/ndss.2020.24003.

[7] G. D. Wyss, J. F. Clem, J. L. Darby, K. Dunphy-Guzman, J. P. Hinton and K. W. Mitchiner, "Risk-based cost-benefit analysis for security assessment problems," in *Proceedings of the 44th Annual 2010 IEEE International Carnahan Conference on Security Technology*, San Jose, CA, USA, October 5-8, 2010, pp. 286-295, doi: 10.1109/CCST.2010.5678687.

[8] N.M. Scala, A.C., Reilly, P.L. Goethals, M. Cukier. "Risk and the five hard problems of cybersecurity," *Risk Analysis*, vol. 39, no. 10, pp. 2119-2126, 2019, doi: 10.1111/risa.13309.

[9] D.W. Hubbard, and R. Seiersen. *How to measure anything in cybersecurity risk*. Hoboke, NJ, USA: John Wiley & Sons, 2016.

[10] F. Enayaty-Ahangar, L.A. Albert, E. DuBois, "A survey of optimization models and methods for cyberinfrastructure security." *IISE Transactions*, vol. 53, no. 2, pp. 182-198, 2020, doi: 10.1080/24725854.2020.1781306.

[11] K. Zheng, L.A. Albert, J.R. Luedtke, E. Towle, "A budgeted maximum multiple coverage model for cybersecurity planning and management." *IISE Transactions*, vol. *51, no.* 12, pp. 1303-1317, 2019, doi: 10.1080/24725854.2019.1584832.

[12] A. Schmidt, L.A. Albert, K. Zheng. "Risk management for cyber-infrastructure protection: A bi-objective integer programming approach," *Reliability Engineering & System Safety*, vol. 205, 2021, no. 1, pp. 107093, 205, doi: 10.1016/j.ress.2020.107093.

[13] K. Zheng, L.A. Albert. "Interdiction models for delaying adversarial attacks against critical information technology infrastructure." *Naval Research Logistics,* vol. 66, no. 5, pp. 411-429, 2019, doi: 10.1002/nav.21859.

[14] Electionic Privacy Information Center (EPIC), "Equifax data breach," website, 2022. [Online]. Available: https://archive.epic.org/privacy/data-breach/equifax/ [Accessed December 15, 2022].

[15] Kolisch, Rainer, Arno Sprecher, and Andreas Drexl. "Characterization and generation of a general class of resource-constrained project scheduling problems," *Management science*, vol. 41, no. 10, pp. 1693-1703, 1995, doi: 10.1287/mnsc.41.10.1693.

https://doi.org/10.21872/2023IISE_3104

Proceedings of the IISE Annual Conference & Expo 2023
K. Babski-Reeves, B. Eksioglu, D. Hampton, eds.

Experiential Learning in Engineering Education: A Case Study in Probability and Statistics Class Project at the University of Texas at Arlington

Abstract ID: 3104

Yiran Yang[1], Jing Zhao[2], Martin Wallace[3], Ryan Paul Hulla[4], Mohammad Ahnaf Shahriar[1]
[1] Department of Industrial, Manufacturing, and Systems Engineering, University of Texas at Arlington, Arlington, TX 76019
[2] Department of Industrial Engineering, Penn State Behrend, Erie, PA 16563
[3] University of Texas at Arlington Libraries, Arlington, TX 76019
[4] Center for Research on Teaching and Learning Excellence, University of Texas at Arlington, Arlington, TX 76019

Abstract

Probability and statistics are the fundamental knowledge that is required by many college majors such as industrial engineering, data science, computer engineering, and computer science, not only in the University of Texas at Arlington (UTA) but also in most 4-year colleges. Classes that cover probability and statistics are offered at both undergraduate and graduate levels. To enhance the students' active learning in probability and statistics classes, especially their understanding of discrete and continuous probability distributions as well as hypothesis testing, a case study is conducted at UTA where an experiential course project is designed and implemented, as a part of the UTA Libraries' Experiential Learning Faculty Facilitators Program. Specifically, two experiential data collection processes are designed: (1) the collection of Nylon 12 particle dimensions (e.g., diameter, circularity, etc.) using a microscope, and (2) the collection of time intervals between students arriving at the UTA Central Library. The impact of this experiential project on students' learning is evaluated through a series of predesigned survey questions. The main hypothesis investigated in this research is that students' self-perceptions of their competencies (in analyzing data, selecting appropriate software, exploring data, assembling team, collaborating in teams, managing knowledge, and applying knowledge in a new situation) will increase by participating in the designed experiential project.

Keywords
Engineering education; Experiential learning; Experiential project; Probability and statistics

1. Introduction
Kolb's Experiential Learning Theory [1] defines experiential learning as "the process whereby knowledge is created through the transformation of experience. Knowledge results from the combination of grasping and transforming experience." Research studies have shown that different experiential learning methods could be helpful in STEM education in many aspects such as increasing students' problem-solving ability [2] and enhancing the participation of underrepresented undergraduates [3]. Over the years, experiential learning techniques have been adopted in industry for workplace training [4]–[6] as well as in classrooms for education in disciplines such as marketing [7], sociology [8], international business [9], business communication [10], accounting [11], mechanical engineering [12], mechanics [13], mechatronics [14], and civil engineering and construction management [15]. In the literature, previous studies have been performed to investigate how to design and improve experiential learning in STEM education, e.g., Oklahoma State University in the course of fire protection and safety engineering technology [16] and Swiss Federal Institute of Technology Lausanne in the design, manufacture, sell, and maintenance of airplane diesel engines [17].
University of Texas at Arlington (UTA), as a Carnegie R-1 University and the 4th institution in the state of Texas to achieve designation as a Texas Tier One University, is dedicated to partnering with faculty and the community to integrate experiential learning in the curricula, focusing on curricular design that utilizes the UTA Libraries' spaces, creative technologies, and resources with an emphasis on hands-on learning and reflection. This paper represents a collaborative effort by UTA faculty, staff, students, and collaborators in the design and development of experiential course projects in probability and statistics classes. This research study was performed in a graduate-level class for

https://doi.org/10.21872/2023IISE_3104

students with a data science major, but the project design can be extended to other undergraduate and graduate level probability and statistics classes for all engineering majors.

2. Problem Description

In the probability and statistics class, the class project usually has two parts: (1) data collection and descriptive analysis, and (2) hypothesis testing. The main objectives of the project are to help students acquire skills in (1) applying learned concepts in class to solving numerical problems, and (2) collecting data and performing descriptive and inferential statistical analyses. Specifically, students are asked to collect two types of data samples, i.e., Data Set I which contains at least 100 observations for a continuous random variable from a population that is suspected to be normally distributed, and Data Set II which contains the time intervals between consecutive occurrences of an event for more than 100 occurrences. Students have the freedom of choosing topics of their interest, and they often collect data from online open sources. Previous interviews with students have identified a few different reasons for the predominant online data collection: higher efficiency, convenience, availability to large data sets, and health concerns due to COVID-19. Meanwhile, the class instructor has observed that sometimes, students get overwhelmed by the excessive amount of data and start to choose data based on availability instead of data characteristics and/or research interest. Therefore, new experiential course projects are designed in the research study. Specifically, 70 students enrolled in the class are put into 5-person 14 groups, where 7 groups will collect Data Set I experientially from an engineering research lab and collect Data Set II online, and the other 7 groups will collect Data Set I online and collect Data Set II experientially at UTA Central Library. In experiential Data Set I, students are asked to use a microscope to measure the particle dimensions (including length, width, circularity, etc.) of Nylon 12 power (randomly selected) used in selective laser sintering 3D printer; In experiential Data Set II, students are asked to record the time interval between students entering the library and going to different directions. Students are asked to finish the survey questions upon the completion of their projects at the end of the semester.

3. Methodology

For this research study, a reflective survey was designed and validated by researchers at the University of Texas at Arlington, developed specifically to measure competencies acquired through experiential learning [18], [19]. Seven categories of questions are designed to evaluate different aspects of students' performance, shown in Table 1. Responses that the participants need to choose from include: 1. Does not describe me; 2. Describes me slightly well; 3. Describes me moderately well; 4. Describes me very well; and 5. Describes me extremely well.

Table 1. A summary of survey questions

Category	Question
A. Analyze data	A.1 How well does this statement describe you? I can analyze data.
	A.2 I can analyze data using probability tools such as descriptive statistical analysis and goodness of fit analysis.
	A. 3 I can identify questions that can be answered or problems that can be solved by analyzing data.
	A.4 I can question assumptions in hypothesis tests drawn from data analysis.
	A.5 I can make conclusions in hypothesis tests based on data analysis.
	A.6 When working on future projects, what process will you use to analyze data? What criteria will you use to evaluate the data? (Optional)
B. Explore data	B.1 How well does this statement describe you? I can explore an idea, question, and/or problem and potential solutions using data.
	B.2 Please explain your response to the above question. How did this course affect your answer?
	B.3 I can collect input from stakeholders and peers.
	B.4 I can research existing relevant data and ideas.
	B.5 Imagine that you need to determine potential ways of solving a work-related problem. How would you approach brainstorming and researching different ideas and evaluating the potential solutions? What role would your coworkers and/or supervisors play in determining the best path forward? (Optional)
C. Appropriaten ess of	C.1 How well does this statement describe you? I can assess the availability and appropriateness of statistical analysis software, tools, and methods to solve a problem.
	C.2 Please explain your response to the above question. How did this course affect your answer?

Software, Tools, and Methods	*C.3 I can research various statistical analysis software, tools, and methods to determine limitations and suitability for a specific application.*
	C.4 I can choose the most appropriate statistical analysis software, tools, and methods (physical, digital, and rhetorical) for the job.
	C.5 I can acquire the necessary statistical analysis software, tools, and methods.
	C.6 I can investigate alternate statistical analysis software, tools, and methods when a desired one is not available.
	C.7 I can fabricate necessary tools, reimagine material choices, develop alternate workflows, and/or revise project scope when alternative tools or materials are not available.
	C.8 Imagine you were recently hired and your new boss asks you to create a prototype for a new project. How will you determine the most appropriate tools to use? (Optional)
D. Assemble Effective Teams	D.1 How well does this statement describe you? I can assemble effective teams.
	D.2 Please explain your response to the above question. How did this course affect your answer?
	D.3 I can recognize opportunities to collaborate with others who provide diverse experiences and perspectives.
	D.4 I can gauge the costs & benefits of "Doing-it-Yourself" (DIY) or "Doing-it-Together" (DIT).
	D.5 I can recruit team members with diverse skills appropriate for specific project requirements.
	D.6 I can join a team where my skills are sought and valued.
	D.7 I can solicit advice, knowledge, and specific skills from experts.
	D.8 Imagine that you need to assemble or join a highly effective team in a job-related scenario. What criteria will you use to select team members, or to identify a team where your skills will be valued? (Optional)
E. Collaborate Effectively	E.1 How well does this statement describe you? I can collaborate effectively with team members and stakeholders. E.2 Please explain your response to the above question. How did this course affect your answer?
	E.3 I can listen to others.
	E. 4 I can learn from and with others.
	E.5 I can communicate respectfully and clearly with team members and stakeholders.
	E.6 I can follow through on team commitments and responsibilities.
	E.7 I can practice accountability both personally and with team members.
	E.8 I can appraise contributions to the success of the team.
	E.9 Imagine that you are preparing for a job-related team project. What collaboration techniques would you recommend to your team? What are some ineffective collaboration practices that you would avoid? (Optional)
F. Knowledge Management	F.1 How well does this statement describe you? I can employ effective knowledge management practices.
	F.2 Please explain your response to the above question. How did this course affect your answer?
	F.3 I can restate technical and statistical jargon for the layperson.
	F.4 I can document steps clearly with sufficient detail for others to follow and replicate.
	F.5 I can use version control to manage project outputs and documentation.
	F.6 I can preserve project outputs and documentation for long-term access.
	F.7 Imagine that you have to finish writing a technical manual about how to use a 3D printer at your company. Another employee began writing the manual a year ago, but has retired. You have access to the 3D printer. Describe some of the steps you might take to accomplish this task.
G. Apply Knowledge Gained into Other Disciplines	G.1 How well does this statement describe you? I can apply knowledge gained into other disciplines, workforce, and community situations.
	G.2 Please explain your response to the above question. How did this course affect your answer?
	G.3 I can teach skills and share insights with other makers.
	G.4 I can recognize and cultivate transferrable skills.
	G.5 I can transfer knowledge, skills, and methods of inquiry across disciplines and activities.
	G.6 I can familiarize myself with the skillsets of others.
	G.7 I can connect those seeking to learn something with those who have relevant experience.
	G.8 Imagine you are working on a group project with team members from a variety of

https://doi.org/10.21872/2023IISE_3104

	disciplines. The project is too labor intensive to simply have everyone work on the aspects that most align with their existing skills. How might you cultivate a group skillshare that would honor everyone's potential contributions? (Optional)

(Note: Italic questions are reflection questions.)

4. Results

A total of 63 participants answered the questions listed in Table 1. The survey results are analyzed as shown in Table 2. Specifically, hypothesis tests are performed to check the mean and variance of the responses collected for reflection questions (i.e., reflect back at the beginning of the semester vs. at the end of the semester) using paired t-test for mean difference and the Bonett's method for standard deviation difference.

Table 2. Results of paired t-test and variance test

c	Q.	**Estimation for Mean Different**			**Paired T-Test**		**Test of Two Variances**	
		Mean	**StDEV**	**95% CI for Mean Difference**	**T-Value**	**P-Value**	**95% CI for SD Ratio**	**P-Value**
A	A.2	-1.540	1.175	(-1.836, -1.244)	-10.40	<0.001	(1.084, 1.930)	0.010
	A.3	-1.460	1.119	(-1.742, -1.178)	-10.36	<0.001	(1.126, 1.955)	0.004
	A.4	-1.603	1.009	(-1.857, -1.349)	-12.62	<0.001	(1.222, 2.227)	<0.001
	A.5	-1.651	1.034	(-1.911, -1.390)	-12.67	<0.001	(1.187, 2.084)	0.001
B	B.3	-1.492	0.982	(-1.739, -1.245)	-12.06	<0.001	(1.063, 1.971)	0.019
	B.4	-1.524	1.120	(-1.806, -1.242)	-10.80	<0.001	(0.935, 1.841)	*0.132*
C	C.3	-1.429	1.120	(-1.774, -1.210)	-10.58	<0.001	(0.909, 1.640)	*0.222*
	C.4	-1.540	1.148	(-1.829, -1.251)	-10.65	<0.001	(0.893, 1.513)	*0.289*
	C.5	-1.444	1.104	(-1.722, -1.166)	-10.39	<0.001	(0.896, 1.558)	*0.260*
	C.6	-1.444	1.104	(-1.722, -1.166)	-10.39	<0.001	(0.769, 1.474)	*0.740*
	C.7	-1.460	1.133	(-1.746, -1.175)	-10.23	<0.001	(0.824, 1.431)	*0.568*
D	D.3	-1.175	1.476	(-1.546, -0.803)	-6.32	<0.001	(0.814, 1.625)	*0.629*
	D.4	-1.190	1.330	(-1.525, -0.855)	-7.10	<0.001	(0.683, 1.320)	*0.545*
	D.5	-0.968	1.545	(-1.357, -0.579)	-4.98	<0.001	(0.719, 1.247)	*0.543*
	D.6	-1.016	1.621	(-1.424, -0.608)	-4.97	<0.001	(0.745, 1.362)	*0.800*
	D.7	-0.968	1.425	(-1.327, -0.609)	-5.39	<0.001	(0.670, 1.291)	*0.474*
E	E.3	-0.952	1.099	(-1.229, -0.676)	-6.88	<0.001	(1.589, 2.519)	<0.001
	E.4	-0.841	1.066	(-1.110, -0.573)	-6.27	<0.001	(1.430, 2.453)	<0.001
	E.5	-0.778	1.039	(-1.039, -0.516)	-5.94	<0.001	(1.342, 2.710)	<0.001
	E.6	-0.841	1.081	(-1.113, -0.569)	-6.18	<0.001	(1.329, 2.760)	<0.001
	E.7	-0.810	1.014	(-1.065, -0.554)	-6.34	<0.001	(1.132, 2.292)	0.005
	E.8	-0.730	0.987	(-0.979, -0.482)	-5.87	<0.001	(1.438, 2606)	<0.001
F	F.3	-1.159	1.081	(-1.431, -0.887)	-8.51	<0.001	(0.880, 1.555)	*0.332*
	F.4	-1.127	1.024	(-1.385, -0.869)	-8.74	<0.001	(1.308, 2.010)	0.024
	F.5	-1.063	1.061	(-1.331, -0.796)	-7.96	<0.001	(0.919, 1.652)	*0.192*
	F.6	-1.190	1.189	(-1.490, -0.891)	-7.94	<0.001	(1.237, 2.202)	<0.001
G	G.3	-1.095	1.132	(-1.380, -0.810)	-7.68	<0.001	(1.266, 1.897)	<0.001
	G.4	-1.095	1.073	(-1.366, -0.825)	-8.10	<0.001	(1.133, 2.066)	0.003
	G.5	-1.063	1.045	(-1.327, -0.800)	-8.08	<0.001	(1.244, 2.088)	<0.001
	G.6	-1.000	0.984	(-1.248, -0.752)	-8.07	<0.001	(1.212, 2.081)	0.001
	G.7	-1.079	1.021	(-1.336, -0.822)	-8.39	<0.001	(1.166, 2.178)	0.002

(*Note*: Italic numbers are the P-values that are larger than 0.05.)

As shown in Table 2, P values from the paired t-test are all statistically significant ($p < 0.001$), supporting the rejection of the null hypothesis that the responses collected before and after the project have an equal mean. In other words, a statistically significant mean difference was observed in students' self-perception in terms of reflection back to the beginning of the semester compared to the end of the semester. Following a similar analysis, statistically significant variance difference was observed for most questions in categories A, E, F, and G. For example, the variance increase in E suggests that although on average, students are more confident in collaborating with others effectively, their responses have great variations. The average mean increases among different competencies are plotted in Figure 1 (a) and the average scores among different competencies are plotted in Figure 1 (b). The significant variance changes for certain sub-competencies are plotted in Figure 2.

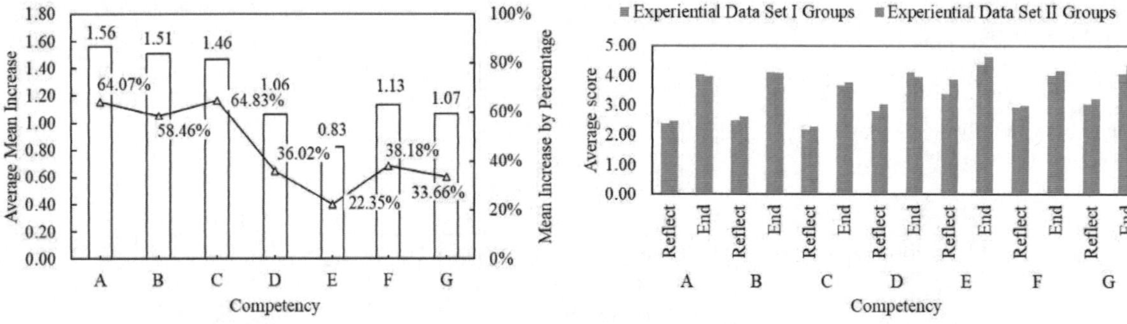

(a) (b)

Figure 1. (a) Average mean increases, and (b) average scores among competencies

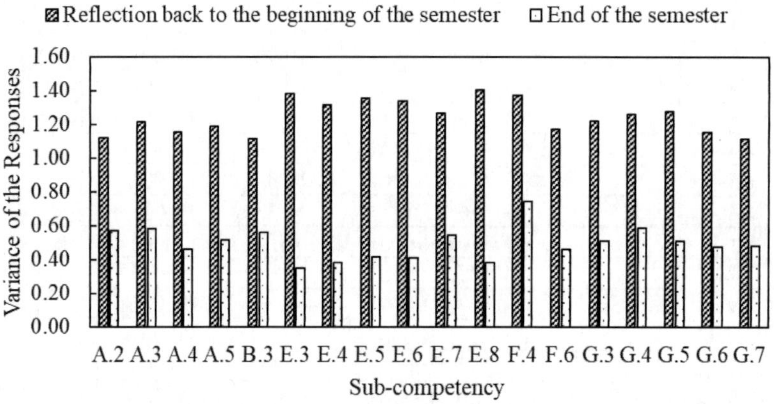

Figure 2. The significant variance changes among sub-competencies

5. Conclusions

This research study was conducted to investigate the effectiveness of the designed experiential projects in enhancing students' self-perceptions of their skill competencies, for a probability and statistics class at UTA. Survey results show great promise that an experiential learning course can increase students' abilities to analyze and explore data, manage knowledge, and apply knowledge in a new situation. Results indicated students had greater variance on their opinions in self-improvement of selecting appropriate software and assembling teams, which suggests a potential direction of improvement in future classes. The experiential projects will be shared with colleagues/collaborators to be used in a broader community and improved when more feedback is received. To extend this work, a comparison between students' self-perceptions of their competencies and their group peers' perceptions of their competencies will be conducted to identify any potential correlations; a pre/post-test of theoretical-practical knowledge in probability and statistics will also be conducted. In addition, other forms of experiential projects have already been designed and ready to be implemented in UTA's probability and statistics classes, such as data collection in an virtual reality environment.

Acknowledgements

The authors (Dr. Yang and Mr. Wallace) would like to acknowledge the support from University of Texas at Arlington (UTA) Libraries' Experiential Learning Faculty Fellowship (2022-2023). We would also like to acknowledge the work of our colleague Morgan Chivers. Mr. Chivers, UTA Libraries FabLab Librarian, led the design and validation of the survey tool used in this research study.

References

[1] R. R. Sims, "Kolb's Experiential Learning Theory: A Framework for Assessing Person-Job Interaction," *Acad. Manag. Rev.*, vol. 8, no. 3, pp. 501–508, 1983.

[2] I. F. Lestari, "Experiential learning using STEM approach in improving students' problem solving ability," in *Journal of Physics: Conference Series*, 2021, p. 012005.

[3] M. R. Goldberg, R. A. Cooper, D. Ding, and A. Koontz, "Using Experiential Learning to Inspire, Educate, and Empower Underrepresented Undergraduates in STEM," in *2011 ASEE Annual Conference & Exposition*, 2011, pp. 22.1615.1-22.1615.18.

[4] L. H. Lewis and C. J. Williams, "Experiential learning: Past and present," *New Dir. Adult Contin. Educ.*, vol. 1994, no. 62, pp. 5–16, 1994, doi: 10.1002/ace.36719946203.

[5] D. A. Nembhard and M. V. Uzumeri, "Experiential learning and forgetting for manual and cognitive tasks," *Int. J. Ind. Ergon.*, vol. 25, no. 4, pp. 315–326, 2000, doi: 10.1016/S0169-8141(99)00021-9.

[6] T. J. Krupnik, C. Shennan, W. H. Settle, M. Demont, A. B. Ndiaye, and J. Rodenburg, "Improving irrigated rice production in the Senegal River Valley through experiential learning and innovation," *Agric. Syst.*, vol. 109, pp. 101–112, 2012, doi: https://doi.org/10.1016/j.agsy.2012.01.008.

[7] D. D. Gremler, K. D. Hoffman, S. M. Keaveney, and L. K. Wright, "Experiential Learning Exercises in Services Marketing Courses," *J. Mark. Educ.*, vol. 22, no. 1, pp. 35–44, Apr. 2000, doi: 10.1177/0273475300221005.

[8] M. C. Wright, "Getting More out of Less: The Benefits of Short-Term Experiential Learning in Undergraduate Sociology Courses," *Teach. Sociol.*, vol. 28, no. 2, pp. 116–126, Feb. 2000, doi: 10.2307/1319259.

[9] P. Paul and K. Mukhopadhyay, "Experiential Learning in International Business Education," *J. Teach. Int. Bus.*, vol. 16, no. 2, pp. 7–25, Apr. 2005, doi: 10.1300/J066v16n02_02.

[10] P. M. Saunders, "Experiential learning, Cases, and Simulations in Business Communication," *Bus. Commun. Q.*, vol. 60, no. 1, pp. 97–114, Mar. 1997, doi: 10.1177/108056999706000108.

[11] L. Gittings, R. Taplin, and R. Kerr, "Experiential learning activities in university accounting education: A systematic literature review," *J. Account. Educ.*, vol. 52, p. 100680, 2020, doi: https://doi.org/10.1016/j.jaccedu.2020.100680.

[12] S. F. Pamungkas, I. Widiastuti, and Suharno, "Kolb's experiential learning for vocational education in mechanical engineering: A review," *AIP Conf. Proc.*, vol. 2114, no. 1, p. 30023, Jun. 2019, doi: 10.1063/1.5112427.

[13] S. M. Holzer and R. H. Andruet, "Experiential Learning in Mechanics with Multimedia," *Int. J. Eng. Educ.*, vol. 16, no. 5, pp. 372–384, 2000.

[14] M. K. Habib, F. Nagata, and K. Watanabe, "Mechatronics: Experiential learning and the stimulation of thinking skills," *Educ. Sci.*, vol. 11, no. 2, pp. 1–22, 2021, doi: 10.3390/educsci11020046.

[15] Z. Jingxiao, X. Haiyan, S. Klaus, X. Bo, L. Hui, and S. Martin, "Integrated Experiential Learning–Based Framework to Facilitate Project Planning in Civil Engineering and Construction Management Courses," *J. Prof. Issues Eng. Educ. Pract.*, vol. 145, no. 4, p. 5019005, Oct. 2019, doi: 10.1061/(ASCE)EI.1943-5541.0000421.

[16] R. J. Agnew, "EXPERIENTIAL LEARNING IN HIGHER EDUCATION FOR THE 21ST CENTURY VIRGINIA R . CHARTER , Oklahoma State University TIMOTHY A . WILSON , Oklahoma State University," vol. 14, no. 11.

[17] G. Regev, D. C. Gause, and A. Wegmann, "Experiential learning approach for requirements engineering education," *Requir. Eng.*, vol. 14, no. 4, pp. 269–287, 2009, doi: 10.1007/s00766-009-0084-x.

[18] M. Wallace, R. Hulla, M. Chivers, and A. Alexander, "Comparing Teamwork & Collaboration Competencies between a Technology in Art Education course and an Engineering Project Management Course," 2020.

[19] M. Chivers, M. Wallace, R. Hulla, A. Alexander, and J. Cantu, "Comparing Teamwork & Collaboration Competencies between a Technology in Art Education course and an Engineering Project Management course," 2023, [Online]. Available: In Preparation

https://doi.org/10.21872/2023IISE_3119

Proceedings of the IISE Annual Conference & Expo 2023
K. Babski-Reeves, B. Eksioglu, D. Hampton, eds.

The Edge-based Contiguous p-median Problem: Formulations and Path-based Insights

Zeyad Kassem and Adolfo R. Escobedo
School of Computing and Augmented Intelligence
Arizona State University, Tempe, Arizona, 85281

Abstract

This paper introduces the edge-based contiguous p-median problem to partition a road network into a given number of compact and contiguous territories. The problem incorporates actual path-based node-to-edge distances, and is thereby especially suited to last mile logistics applications. Two binary programming models are introduced to solve the problem, and they differ in how they enforce contiguity. The first utilizes cut set-based contiguity constraints, which are exponential in number. Hence, this model is paired with an iterative branch and bound algorithm with a cut generation scheme (B&B&Cut) that generates only a small number of these constraints. The second model utilizes shortest-path contiguity (SPC) constraints, which are polynomial in number. In addition, we demonstrate that the SPC constraints represent supervalid inequalities of the edge-based p-median problem, meaning that they cut off integer feasible but non-optimal solutions. The two solution approaches associated with the introduced models are tested on road networks with up to 1735 nodes and 2226 edges; the shortest path-based formulation attains speed ups in computational time up to almost 4x relative to the computational time of solving the first optimization model via B&B&Cut.

Keywords
Edge-based p-median with contiguity, shortest-path contiguity, logistics, combinatorial optimization, supervalid inequalities

1. Introduction
Over the last few years, there has been a global acceleration in the trend to transition to an e-commerce business model. This has led to increased interest in last mile freight logistics, which is the final leg of a delivery service in a consumer supply chain network where the consignment is sent directly to the consumer or to a certain collection point from a transportation hub [3]. Last mile logistics is the least efficient part of the supply chain, with costs contributing up to 28% of the total transportation cost. In developed economies, last mile logistics can contribute up to 30% of vehicle miles of travel and up to 50% of greenhouse gas (GHG) emissions [3].

Different studies have shown that traffic density and closeness of the warehouses to the service area are among the most dominant factors for changing the operational and environmental costs [16]. Traffic density can be leveraged to reduce the unit cost by increasing the consolidation of shipments to customers among the different routes and road segments [2]. Increasing the consolidation of shipments can lead to the utilization of larger and more cost-effective vehicles, which can help distribute the fixed costs among more units and also lower the variable costs [2]. Discrete location-allocation models could enforce these considerations by determining the optimal location of warehouses and customer allocations to these warehouses, while also satisfying additional practical criteria such as contiguity from the warehouses to each of the customers they serve.

Location-allocation models have been studied extensively and are prevalent in both the public and private sectors [8]. In general, these models aim to simultaneously locate a set of facilities and/or services (e.g., schools, hospitals, and warehouses) and to allocate customers or communities to the chosen locations with the aim of optimizing one or several decision criteria [13]. Two main types of discrete location-allocation models have been considered in the literature: p-median models and p-center models. p-median models locate p facilities from among a discrete set of locations, and their goal is to minimize the sum of the weighted distances from each facility to its assigned nodes. p-center problems locate p facilities, where the goal is to minimize the maximum distance from each facility to its assigned center. p-center models are typically utilized in the public sector, and their applications include locating public facilities, schools, and emergency systems such as fire stations and ambulance depots [12]. On the other hand,

https://doi.org/10.21872/2023IISE_3119

p-median models are said to be better suited for private companies [11]. Most of the literature on p-median problems concentrates on node-based partitioning models, meaning that the customers and facilities are situated on the nodes of the underlying territory graph. However, the edge-based variant of this problem, which can better reflect the locations of customers (i.e., streets) in e-commerce applications, has received little attention. Models most closely related to the featured edge-based problem are edge-based districting models [7, 9]. Although they share certain aspects in common, the latter enforce additional criteria such as work balance—that is, each facility must serve approximately the same number of customers—which may conflict with the desire to achieve compact service areas. On the other hand, contiguity constraints used in districting models, which to our knowledge have not been considered for edge-based p-median models, could help enhance the service quality of p-median problem solutions.

This paper introduces a new variant of the p-median problem, namely the edge-based contiguous p-median problem (ECpM). Models introduced to solve these problems are suited for the subsequent planning of efficient road-based routing for e-commerce companies. Specifically, this paper introduces a binary programming formulation that partitions a road network into a fixed number of compact and contiguous territories by locating centers and allocating roads to these centers, with the objective of minimizing the total distance traveled within each territory. To achieve contiguity, we propose two binary programming models, which differ on how they enforce contiguity. The first requires an exponential number of cut set-based constraints, and the second entails a polynomial number of shortest path-based constraints. The former is solved via B&B&Cut, and the latter with off-the-shelf methods.

The remainder of the paper is structured as follows. Section 2 provides background and a brief overview of related works. Section 3 presents two formulations for the problem, and presents additional theoretical insights. Section 4 describes computational studies. Finally, Section 5 concludes the paper.

2. Background and Related Work

Certain criteria are needed to design territories that lead to efficient last mile logistics services including contiguity and compactness [15]. A territory is considered to be *contiguous* if travel between any two points within a territory can happen without leaving the territory. A territory is considered to be *compact* if its shape is nearly circular or approximately square-shaped, non-distorted, without holes, and with smooth boundaries [4]. In general, compact territories lead to shorter travel distances within each territory.

3. The Edge-based Contiguous p-Median (ECpM) Problem

This section defines the ECpM problem formally, and it presents a binary programming model with cut set-based contiguity constraints. The latter constraints are exponential in number, and their full specification requires prohibitive computational resources. Therefore, this model is solved with a B&B&Cut strategy that typically generates only a small subset of these constraints, as needed by each specific instance. We then propose an alternative approach that utilizes shortest-path contiguity constraints, which are polynomial in number. In addition, we demonstrate that the SPC constraints represent supervalid inequalities of the edge-based p-median problem, meaning that they cut off integer feasible but non-optimal solutions.

3.1 Problem Definition

The aim of the ECpM problem is to partition a road network into a specified number of compact and contiguous territories, onto which some type of last mile logistics service is to be presumably carried out along the roads of each territory. We proposed two binary programming models, which differ in how they enforce contiguity. The objective function aims to minimize the dispersion value, which is the sum of node-to-edge distances from the center node of each service area to all of its allocated edges.

The underlying road network is modeled as an undirected connected planar graph $G = (V, E)$. The dispersion value between node $i \in V$ and edge $(j,k) \in E$, $d_{i,(j,k)}$, is defined as follows.

Definition 3.1. Let $d_{i,(j,k)}$ be equal to the sum of positive edge lengths $l_{(m,n)}$ that constitute the shortest path from node i to node j or k (depending on which is the tail node in the shortest path). The distance $d_{i,(j,k)}$ is calculated as

$$d_{i,(j,k)} = \min\left\{ \sum_{(m,n)\in SP_{i,j}} l_{(m,n)}, \sum_{(m,n)\in SP_{i,k}} l_{(m,n)} \right\}, \tag{1}$$

where $SP_{i,j}$ is the shortest path from node i to node j. Note that the shortest path between every pair of nodes in V can be efficiently computed using Dijkstra's algorithm [1]. The above definition has been previously utilized in the districting context in the works of García-Ayala et al. [7] and Kassem and Escobedo [9].

Definition 3.2. Let $SP_{i,(j,k)}$ be the edges $(l,m) \in E$ that make up the shortest path from node i to edge (j,k), which is either $SP_{i,j}$ or $SP_{i,k}$ (depending on which is the tail node in the shortest path).

https://doi.org/10.21872/2023IISE_3119

3.2 Cut Set-based Formulation for ECpM

This subsection presents the first binary programming formulation for ECpM. The model aims to partition E into a specified number (p) compact and contiguous districts. An exponential number of cut-set based contiguity constraints are included to prevent any disconnections between the edges located within the same territory. These constraints were first formulated by García-Ayala et al. [7], who also explained that it is possible to solve their model by carrying only a small number of contiguity constraints, specifically via an iterative branch and bound algorithm with a cut generation scheme (B&B&Cut). The below formulation is solved with an analogous approach.

Sets and Indices:

$i,j,k \in V$	Set of nodes (corresponding to street crossings or dead ends of the underlying road network).
$(j,k) \in E$	Set of edges (corresponding to a road or street of the underlying road network).
$S \subset E$	Subset of edges.
$V(S) \subset V$	Set of nodes that are incident to any edge in S.
$\sigma(S)$	Cut set of S (i.e., $\sigma(S) := \{(i,j) \in E \mid i \in V(S), j \in V \setminus V(S)\}$).

Parameters:

$d_{i,(j,k)}$	Shortest-path distance from node i to edge (j,k) (see Definition 3.1).
p	Number of districts.

Binary Variables:

$$w_i = \begin{cases} 1 & \text{if node } i \text{ is a center node,} \\ 0 & \text{otherwise.} \end{cases}$$

$$x_{i,(j,k)} = \begin{cases} 1 & \text{if edge } (j,k) \text{ is assigned to center node } i, \\ 0 & \text{otherwise.} \end{cases}$$

$$\min \sum_{i \in V} \sum_{(j,k) \in E} d_{i,(j,k)} x_{i,(j,k)} \tag{2}$$

subject to

$$\sum_{i \in V} x_{i,(j,k)} = 1 \qquad \forall (j,k) \in E \tag{3}$$

$$\sum_{i \in V} w_i = p \tag{4}$$

$$x_{i,(j,k)} \leq w_i \qquad \forall i \in V, \forall (j,k) \in E \tag{5}$$

$$\sum_{s \in \sigma(S)} x_{is} - \sum_{s \in S} x_{is} \geq x_{i,(j,k)} - |S| \qquad \forall i \in V, \forall (j,k) \in E, S \subset E \setminus \sigma(\{(j,k)\}) \tag{6}$$

$$x_{i,(j,k)} \in \mathbb{B} \qquad \forall i \in V, \forall (j,k) \in E \tag{7}$$

$$w_i \in \mathbb{B} \qquad \forall i \in V \tag{8}$$

Objective function (2) maximizes compactness by minimizing the dispersion value, which is defined as the sum of node-to-edge distances from each center node i to its assigned edges. Constraint (3) ensures that each edge $(j,k) \in E$ is allocated to exactly one center node. Constraint (4) ensures that exactly p center nodes must be selected. Constraint (5) allows edge (j,k) to be allocated to node i ($x_{i,(j,k)} = 1$) only if i is a center node ($w_i = 1$). Constraint (6) ensures that any edge (j,k) allocated to a center node i should be adjacent to other edges allocated to the same center node. Let $S \subset E \setminus \sigma(\{(j,k)\})$ be any subset of edges not adjacent to edge (j,k) and allocated to center node i. If edge (j,k) is not assigned to center node i ($x_{i,(j,k)} = 0$), the constraint is always satisfied. Moreover, if there is one or more edges in set S that are not assigned to center node i, then the value of second term is strictly less than $|S|$ and the constraint is always satisfied. Constraint (6) is activated only when all edges in S are assigned to center node i and edge (j,k) is assigned to center node i ($x_{i,(j,k)} = 1$). In this case, the value of the first term must be greater than or equal to 1. This means that at least one edge in the cut set of set S must be assigned to center node i. Applying these constraints for all possible subsets S ensures that all edges within a service area are connected.

https://doi.org/10.21872/2023IISE_3119

3.3 Compact Formulation of ECpM with Shortest-path Contiguity Constraints

There have been several attempts to model shortest-path contiguity in spatial optimization problems. Zoltners et al. [17] introduce the concept of hierarchical adjacency trees for the sales territory alignment problem with predetermined centers. For each predetermined center node, a set of shortest paths is determined from the center to all other nodes with predecessor-successor relationships of the nodes along each shortest path recorded. Contiguity is enforced by ensuring that no node is allocated to a center node unless at least one of its immediate preceding nodes along the shortest path is also allocated to the same center node [17]. Cova et al. [5] derive the shortest path contiguity-n constraints for single-region site search problems. These constraints allow each spatial unit to be reached from the root along only one of its n shortest paths; to that end, the first, the second, and up until nth shortest paths are computed from each spatial unit to the root, where n is reasonably small. Mehrotra et al. [14] developed and implemented a special version of shortest path contiguity-n constraints to a political districting problem, where only the first shortest path is considered. More recently, Farughi et al. [6] introduced a distinct set of shortest-path contiguity constraints for designing health care districts. The constraints ensure that if nodes j and k are allocated to center node i, then all nodes along the $SP_{j,k}$ must also be assigned to center node i [6]. A common feature of all these shortest-path constraints is that they are defined exclusively for node-based territorial planning models, except for [5] where the spatial unit is a raster, which is nevertheless equally representable as a node.

To the best of our knowledge, there are no shortest-path contiguity constraints defined for edge-based location-allocation models. This work seeks to fill this research gap—throughout the paper, we refer to the shortest-path contiguity constraints developed herein as SPC, for convenience. The main intuition behind the SPC constraints is that, for any connected planar graph with more than one edge, if edge (j,k) is assigned to center node i, then all edges along the shortest path $SP_{i,(j,k)}$ must also be assigned to i. If implemented on all edges, this condition guarantees contiguity within each district, as it forces all the edges along $SP_{i,(j,k)}$ to be assigned to center node i. That is, there must be a path between any two edges assigned to the same center node.

The next theorem demonstrates that the SPC constraints represent *supervalid inequalities* of the edge-based p-median problem, abbreviated as EpM (i.e., ECpM without contiguity). That is, adding the SPC constraints may remove feasible but non-optimal solutions to the EpM problem.

Theorem 3.1. Let $G=(V,E)$ be an undirected connected planar graph with $|E|\geq 2$ and assume that the dispersion value is defined as the sum of node-to-edge distances $d_{i,(j,k)}$, determined according to Definition 3.1. If edge (j,k) is assigned to center node i ($x_{i,(j,k)}=1$), then there is an optimal solution to EpM where all edges $(l,m)\in SP_{i,(j,k)}$ are assigned to i.

Proof. Assume, by contradiction, that (j,k) is assigned to i in an optimal solution to ECpM, but $(l,m)\in SP_{i,(j,k)}$ is not allocated to node i. Without the loss of generality, let edge $(l,m)\in SP_{i,(j,k)}$ be allocated to center node i', and let node l be the tail node on the $SP_{i,(l,m)}$ and $SP_{i',(l,m)}$.The following two cases are considered:
Case 1: $d_{i',(l,m)}<d_{i,(l,m)}$. We have that

$$d_{i,(j,k)} = d_{i,(l,m)} + d_{l,(j,k)} \tag{9a}$$

$$> d_{i',(l,m)} + d_{l,(j,k)}. \tag{9b}$$

Equation (9a) corresponds to the distance from node i to edge (j,k), whereas (9b) corresponds to the distance from i' to (j,k). Since $d_{i',(l,m)}<d_{i,(l,m)}$ holds, then $d_{i,(j,k)}>d_{i',(j,k)}$. This contradicts our assumption that center node i is the optimal center for edge (j,k).
Case 2: $d_{i',(l,m)}=d_{i,(l,m)}$. In this case, we obtain the equation

$$d_{i,(j,k)} = d_{i,(l,m)} + d_{l,(j,k)} = d_{i',(l,m)} + d_{l,(j,k)} = d_{i',(j,k)}. \tag{10a}$$

Equation (10a) corresponds to the distance from node i to edge (j,k), which is equal to the distance from i' to (j,k), since $d_{i',(l,m)}=d_{i,(l,m)}$. This implies that there is an alternative optimal solution where (l,m) is allocated to node i without changing the dispersion value. □

Since $SP_{i,(j,k)}$ is a contiguous path, allocating the edges along the $SP_{i,(j,k)}$ to center node i if edge (j,k) is allocated to i guarantees contiguity within each district. This means that we can add SPC constraints to ECpM instead of cut set-based contiguity constraints (6) to enforce contiguity within each district. It is possible to implement the above concept in multiple logically equivalent ways. Next, we specify three different options.

$$\text{(SPC-1)} \qquad |SP_{i,(j,k)}|x_{i,(j,k)} \leq \sum_{(l,m)\in SP_{i,(j,k)}} x_{i,(l,m)} \;\; \forall i \in V, \forall (j,k) \in E \qquad (6')$$

In this case, if $x_{i,(j,k)} = 1$, then constraint (6′) enforces $\sum_{(l,m)\in SP_{i,(j,k)}} x_{i,(l,m)} \geq |SP_{i,(j,k)}|$, where $|SP_{i,(j,k)}|$ denotes the number of edges on the shortest path from node i to edge (j,k). If $x_{i,(j,k)} = 0$, the left hand side of constraint (6′) is equal to zero and the constraint becomes redundant.

$$\text{(SPC-2)} \qquad x_{i,(j,k)} \leq x_{i,(l,m)} \;\; \forall i \in V, \forall (j,k) \in E, \forall (l,m) \in SP_{i,(j,k)} \qquad (6'')$$

Here, if $x_{i,(j,k)} = 1$, then constraint (6″) enforces $x_{i,(l,m)} = 1 \; \forall (l,m) \in SP_{i,(j,k)}$. If $x_{i,(j,k)} = 0$, then the left hand side of constraint (6″) is equal to zero and the constraint becomes redundant.

$$\text{(SPC-3)} \qquad x_{i,(j,k)} \leq x_{i,(l,m)} \;\; \forall i \in V, \forall (j,k) \in E, (l,m) \in SP_{i,(j,k)}, (l,m) \in \sigma(\{(j,k)\}) \qquad (6''')$$

In this case, if $x_{i,(j,k)} = 1$, constraint (6‴) enforces $x_{i,(l,m)} = 1$, where (l,m) is the preceding edge on the shortest path from edge (j,k) to center node i. If $x_{i,(j,k)} = 0$, the left hand side of constraint (6‴) is equal to zero and the constraint becomes redundant.

Adding SPC-1 and SPC-3 requires adding $O(|V||E|)$ constraints, while adding SPC-2 requires adding $O(|V|^2|E|)$ constraints. In our computational tests, we incorporate SPC-3 constraints into our compact ECpM model, which requires adding a relatively lower number of constraints compared to SPC-2 constraints and has a tighter linear relaxation compared to SPC-1 constraints.

4. Results

The optimization models were coded in Python 3.7 and solved using CPLEX 20.10. The computational tests were performed on a high performance computing node equipped with 126 GB of memory and an Intel Xeon E5 v4 Processor running at 2.40 GHz. The nine problem instances are derived from three real-life road networks located in Denmark [10]. The following improvement ratio in computational time is calculated as

$$\text{Improvement Ratio} = \frac{\text{Time to solve cut set-based EC}p\text{M}}{\text{Time to solve compact EC}p\text{M with SPC-3}}. \qquad (11)$$

Table 1: Computational Results (in seconds) for Solving Cut Set-based ECpM via B&B&Cut and Compact ECpM with SPC Constraints

Instance	$\|V\|$	$\|E\|$	p	Cut Set B&B&Cut	SPC-3 Formulation	Improvement Ratio
1	1094	1447	2	887	1,396	0.64
2	1094	1447	50	481	521	0.92
3	1094	1447	100	815	481	1.70
4	1564	1805	2	5,758	3,791	1.52
5	1564	1805	50	6,124	1,821	3.36
6	1564	1805	100	6,432	1,393	4.62
7	1735	2226	2	8,905	6,638	1.34
8	1735	2226	50	9,695	3,135	3.09
9	1735	2226	100	10,463	3,128	3.34

Table 1 shows the instance number, number of nodes ($|V|$), number of edges ($|E|$), number of districts (p), time to solve cut set-based ECpM via an iterative branch and bound algorithm with a cut generation scheme (Cut Set B&B&Cut), time to solve compact ECpM with SPC-3 constraints (SPC-3 Formulation), and the improvement ratio. Based on the results in Table 1, solving SPC-3 Formulation outperformed Cut Set B&B&Cut in seven out of the nine instances, even though the latter adds only a small number of contiguity constraints. Cut Set B&B&C performs better than SPC-3 Formulation on the smallest instances (with $|V| = 1094$ and $|E| = 1447$). For the relatively small road network associated with these instances, the larger model size caused by the addition of SPC-3 constraints had a bigger effect than the resulting reduction in the feasibility space. The SPC-3 constraints had a bigger impact on the larger instances. In fact, SPC-3 Formulation solved all larger instances faster than EpM, that is, ECpM without contiguity constraints. The latter fact supports the usefulness of the SPC constraints as supervalid inequalities for the EpM, and it represents a new effective technique for solving edge-based p-median models.

https://doi.org/10.21872/2023IISE_3119

5. Conclusion

This paper introduces and derives a binary programming formulation for the edge-based contiguous p-median (ECpM) optimization problem. In addition, it derives new shortest-path contiguity constraints that can scale the solution of large-scale instances of this problem—with speed ups of up to 4x for the instances tested in this work. In the future, we will perform more comprehensive computational tests and explore additional network insights to further reduce the computational effort of ECpM problems.

References

[1] Ravindra K Ahuja, Thomas L Magnanti, and James B Orlin. *Network Flows*. Prentice Hall, New Jersey, 1988.

[2] Sibel A Alumur, James F Campbell, Ivan Contreras, Bahar Y Kara, Vladimir Marianov, and Morton E O'Kelly. Perspectives on modeling hub location problems. *European Journal of Operational Research*, 291(1):1–17, 2021.

[3] Techane Bosona. Urban freight last mile logistics—challenges and opportunities to improve sustainability: a literature review. *Sustainability*, 12(21):8769, 2020.

[4] Alexander Butsch, Jörg Kalcsics, and Gilbert Laporte. Districting for arc routing. *INFORMS Journal on Computing*, **26**(4):809–824, 2014.

[5] Thomas J Cova and Richard L Church. Contiguity constraints for single-region site search problems. *Geographical Analysis*, 32(4):306–329, 2000.

[6] Hiwa Farughi, Madjid Tavana, Sobhan Mostafayi, and Francisco J Santos Arteaga. A novel optimization model for designing compact, balanced, and contiguous healthcare districts. *Journal of the Operational Research Society*, 71(11):1740–1759, 2020.

[7] Gabriela García-Ayala, José Luis González-Velarde, Roger Z Ríos-Mercado, and Elena Fernández. A novel model for arc territory design: Promoting eulerian districts. *International Transactions in Operational Research*, **23**(3):433–458, 2016.

[8] Jaroslav Janáček and Marek Kvet. Sequential approximate approach to the p-median problem. *Computers & Industrial Engineering*, 94:83–92, 2016.

[9] Zeyad Kassem and Adolfo R Escobedo. Models and network insights for edge-based districting with simultaneous location-allocation decisions. *IISE Transactions*, pages 1–13, 2022.

[10] Lone Kiilerich and Sanne Wøhlk. New large-scale data instances for carp and new variations of carp. *INFOR: Information Systems and Operational Research*, 56(1):1–32, 2018.

[11] Andreas Klose and Andreas Drexl. Facility location models for distribution system design. *European journal of operational research*, 162(1):4–29, 2005.

[12] Vladimir Marianov and Charles ReVelle. Siting emergency services. *Facility Location: a survey of applications and methods*, 1:199–223, 1995.

[13] Vladimir Marianov, Daniel Serra, and Z Drezner. Location problems in the public sector. *Facility location: Applications and theory*, 1:119–150, 2002.

[14] Anuj Mehrotra, Ellis L Johnson, and George L Nemhauser. An optimization based heuristic for political districting. *Management Science*, 44(8):1100–1114, 1998.

[15] M Cândida Mourão and Leonor S Pinto. An updated annotated bibliography on arc routing problems. *Networks*, **70**(3):144–194, 2017.

[16] Erica Wygonik and Anne V Goodchild. Urban form and last-mile goods movement: Factors affecting vehicle miles travelled and emissions. *Transportation Research Part D: Transport and Environment*, 61:217–229, 2018.

[17] Andris A Zoltners and Prabhakant Sinha. Sales territory alignment: A review and model. *Management Science*, 29(11):1237–1256, 1983.

https://doi.org/10.21872/2023IISE_3137

Proceedings of the IISE Annual Conference & Expo 2023
K. Babski-Reeves, B. Eksioglu, D. Hampton, eds.

Impact of Replanning on A*-HOP Performance to Solve Stochastic CTP

Neel Chanchad
University of Arkansas
Fayetteville, Arkansas

Ashlea Milburn
University of Arkansas
Fayetteville, Arkansas

Abstract

After a disaster, uncertainty regarding the status of the possibly disrupted road network makes the delivery of emergency supplies such as food and water to critical locations challenging. This situation can be modeled as a variant of the Canadian Traveler Problem (CTP). In the CTP, the goal is to find the shortest path from a source to a destination along a network comprised of edges that may or may not be available. The status of an edge as available or not can only be ascertained by visiting a node incident to it. In the stochastic variant of CTP, a disruption probability p_e for each edge e is assumed to be available. The literature contains only a few algorithm implementations and computational studies for this NP-hard problem, and these algorithms differ according to how often re-planning is carried out. For example, an agent can re-plan at every node to choose the next node to visit, or an agent can plan a sequence of nodes to visit and only re-plan when the discovery of disrupted edges renders it necessary. In this paper we analyze the relationship between these two re-planning strategies and solution quality via a computational study comprised of CTP test instances from the literature. Initial results indicate that re-planning more frequently can be helpful when the proportion of disrupted edges is higher than expected but may lead to lower quality solutions when the proportion of disrupted edges is lower than expected.

Keywords
Stochastic, A*, Canadian Traveler Problem, Path-planning

1. Introduction

Disasters such as hurricanes, floods, and earthquakes can cause severe damage to infrastructure leading to blocked or destroyed routes, landslides, and washouts [1, 2]. This can impede transportation infrastructure such as roads, airports and ports, making it challenging to transport relief aid and personnel to affected areas. Furthermore, power outages and communication failures can impede coordination efforts and make it difficult to disseminate information about the status of roads and transportation systems. In the wake of a disaster, the road network can be severely compromised, creating challenges for organizations such as the Federal Emergency Management Agency (FEMA) to deliver essential goods and supplies to affected communities [1]. The inability to navigate blocked or damaged roads can lead to delays and increased travel time due to detours, which can have dire consequences for those who are in need of basic necessities such as food, water, and shelter. To mitigate these challenges, it is important to rely on algorithms that can efficiently plan routes based on probabilistic information about road blockages as exact status is unknown during the immediate disaster response phase. These algorithms should also possess the capability to replan routes in real-time, as first responders and other agents encounter blockages and acquire more knowledge about the road network. A sophisticated approach is required to ensure that goods and supplies can reach those in need as quickly and efficiently as possible. This problem of determining the most efficient path between a source and a destination is referred to as the Canadian Traveler Problem (CTP) in the literature [3]. The objective of the agent in single source single destination CTP is to minimize the travel cost required to reach the destination from the source

https://doi.org/10.21872/2023IISE_3137

node. In the CTP, the agent is equipped with information about the graph structure $G = (N, E)$, where N represents the set of nodes, including a source node v_s and destination node v_d, and E represents the set of edges. Additionally, the agent has access to information regarding the edge cost (c_e) and probability of edge blockage (p_e) for each edge $e \in E$. The true state of the network is referred to as the weather w, where $w \subseteq E$ is the set of edges that have not failed. However, the exact status of an edge e as blocked or available (i.e., whether $e \in w$ or $e \in E \setminus w$) remains unknown until the agent reaches one of its endpoints.

A study by Alseth [4] proposed an A* based algorithm known as A*-HOP to address the CTP for both single and multiple agents. The A*-HOP algorithm employs an initial path-planning strategy towards the destination while the agent is stationed at the source node. Then, the agent's traversal of the network adheres to this planned path until a blocked edge on the planned path is encountered, at which point re-planning is initiated. Note that in this case, re-planning is essential, as the agent's next move along the planned path is infeasible because the edge is blocked. This study investigates whether re-planning more frequently than only when necessary to avoid a blocked edge can improve agent paths. We observe that new information is obtained by the agent every time the agent arrives to a previously unvisited node, whether a blocked edge is discovered or not. That is, the uncertainty regarding the statuses of all edges incident to the node as either blocked or available is resolved upon arrival to the node. Thus, we explore whether re-planning at every new node, when some uncertainty inherent in the problem is resolved, leads to more informed and better decisions. This paper proposes a modified version of the A*-HOP algorithm with more frequent replanning (R), referred to as the A*-HOP-R algorithm, and limits the scope of the investigation to single agent CTP scenarios.

The subsequent portion of this manuscript is organized as follows: In section 2, we delve into the existing literature pertaining to the CTP. This is followed by a description of our solution methodology in section 3. The experimental details and corresponding results are presented in section 4 and 5 respectively. Finally, we conclude the paper with section 6, which encompasses our conclusions and potential avenues for future research.

2. Literature

The term CTP was first defined by Papadimitriou and Yannakakis [3] and it was demonstrated that the problem is PSPACE-complete. Since its introduction, many variants of the CTP have been explored. In this paper, we review only those papers focused on providing computational results, versus theoretical results, for single and multi-agent stochastic CTP variants.

Eyerich et al. [5] introduced the first computational results for stochastic single agent CTP. The author introduced four distinct policies, namely Optimistic (OMT), Hindsight Optimization (HOP), Optimistic Rollout (ORO), and Upper Confidence Bounds applied to Trees (UCT). The OMT policy is based on the free space assumption, meaning that the agent assumes every edge in the network to be available with probability of 1 until the agent observes the edge to be blocked, after which the probability is updated to 0. The remaining HOP, ORO, and UCT policies rely on rollouts, also referred to in the literature as scenario-based planning.

In HOP, rollouts are used to calculate the expected cost of reaching the destination from neighboring nodes. For each rollout, the agent assumes perfect information and calculates the shortest path cost to reach the destination using any shortest path algorithm. The expected cost is calculated as the average cost across all rollouts for neighboring nodes, and the agent uses this expected cost to select the next node to visit. In contrast, the ORO policy uses the OMT policy for each rollout, applying the free space assumption rather than assuming perfect information. Both in HOP and ORO policies, rollouts are independent, meaning that the decisions made by the agent during each rollout are not influenced by previous rollouts. The UCT policy, however, seeks to balance exploration and exploitation. Unlike the other policies, the rollouts in UCT are not independent. The results demonstrated that the UCT policy outperforms the other policies.

In a multi-agent setting, Shiri and Salman [6] employed the OMT policy and utilized Dijkstra's algorithm to plan paths for agents starting from the same source and heading towards the same destination. To diversify the paths of the agents, an iterative penalty method was implemented, which entailed penalizing the edge costs along the planned path of each agent. Subsequently, the penalized edge costs were used to generate paths for the following agents. Alseth [4] aimed to enhance the results obtained by Shiri and Salman [6] and employed a rollout-based HOP policy with A* algorithm to generate paths for agents. The expected cost was calculated using the same method as described in Eyerich et al. [5]. The expected cost served as a heuristic value for the A* algorithm. The iterative node penalty method was used to

https://doi.org/10.21872/2023IISE_3137

diversify the paths, which involved penalizing the expected cost of the nodes involved in the planned path of each agent instead of penalizing the edge costs as done in Shiri and Salman [6]. Both algorithms plan the paths for the agents at the source and continue to be followed until a blocked edge is discovered in any agent's path, after which re-planning is performed. The A*-HOP approach was found to be more effective than the OMT-based algorithm proposed by Shiri and Salman [6], leading us to choose A*-HOP for further analysis to examine the impact of more frequent re-planning.

3. Methods

In A*-HOP proposed by Alseth [4], the agent uses the A* algorithm [7] to plan an entire path from the agent's current node to the destination, and the agent follows this planned path until a blocked edge in the planned path is discovered. The objective of A* algorithm is to find the shortest path between a source and a destination. The A* algorithm selects the path that minimizes the function mentioned in equation 1:

$$f(n) = g(n) + h(n). \tag{1}$$

The term $g(n)$ refers to the path cost to reach intermediate node n from source node v_s and $h(n)$ refers to the heuristic function that provides an expected cost to reach the destination v_d from n. To obtain the expected cost $h(n)$, the HOP policy mentioned in Eyerich et al. [5] is used. In HOP, sample weathers called rollout weathers are generated. Each rollout weather $r \in \{1,2,3,...,R\}$ is assumed to be known with perfect information. Dijkstra's algorithm is used to determine the cost of reaching the destination from every node $C_r(n)$, using only those edges available in r. The average of these costs across all rollout weathers for each node is calculated using equation 2 and is supplied as the expected cost to the A* algorithm, thereby enabling the creation of a path:

$$h(n) = \frac{\sum_{r\in\{1,2,3,...,R\}} C_r(n)}{R}. \tag{2}$$

In this paper, we modify the algorithm in Alseth [4] as follows: instead of re-planning only when an agent encounters a blocked edge in their planned path, we implement re-planning every time an agent arrives to a previously unvisited node and gains new information regarding the status of edges incident to it. This approach aims to provide more frequent opportunities for decision-making with the expectation that more information will lead to more informed decisions. Figure 1 shows the difference in the two approach and phases an agent undergoes whenever the agent reaches a new node. The first phase is sensing, in which the agent evaluates and clarifies the status of the adjacent edges. Subsequently, the agent engages in the re-planning phase utilizing an A* algorithm to devise a path from the current node to the destination. For A*-HOP, the re-planning phase is only executed if the planned path contains one or more blocked edges. For A*-HOP-R, the re-planning phase is performed at every new node irrespective of whether agent's planned path contains blocked edge or not. Finally, the agent proceeds to the moving stage, in which it commences travel towards the next node as outlined in the planned path generated by the A* algorithm.

4. Experimental Details

Our experimentation is conducted on Delaunay graphs [5], analyzing graphs of sizes 20, 50, and 100 nodes. To ensure robustness and generalizability of our findings, we employed 10 different graphs for each size and for each of these graphs, we used 1000 distinct weathers (note that weather is synonymous with the term test instance). All graphs used were taken from [5], where the cost c_e and blockage probability p_e of each edge was generated using an uniform [1,50] and [0,1) distribution respectively. For each test instance, 3000 rollout weathers are utilized. The graphs, weathers, and rollout weathers used in our study are identical to those employed in Alseth [4], to ensure a fair and unbiased analysis. There are total of 30,000 test instances in this computational study.

5. Results

Table 1 provides a comparison of the average cost and performance between A*-HOP and A*-HOP-R over 1000 instances for each of the 30 graphs. The first column lists the name of the graphs, while the second and third columns indicate a 95% confidence interval for average path cost to destination across 1000 instances obtained using A*-HOP and A*-HOP-R respectively. For example, for graph 20-1, the 95% confidence interval on average path cost across

https://doi.org/10.21872/2023IISE_3137

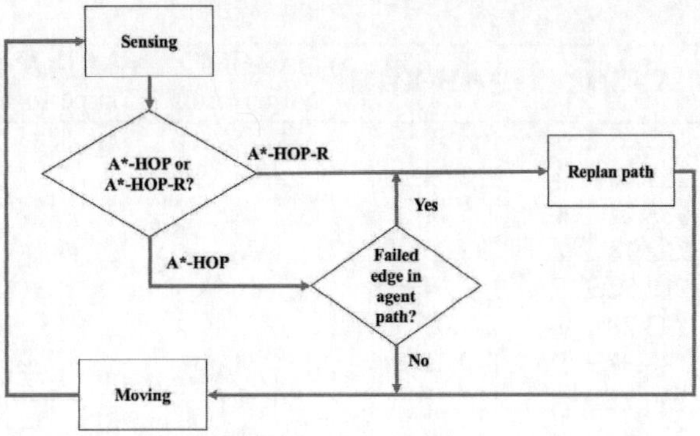

Figure 1: Agent Flow Chart

1000 instances is 163.67 ± 5 for A*-HOP and 163.59 ± 5 for A*-HOP-R. The fourth and fifth columns of Table 1 provide the numbers of weathers in which each algorithm outperforms (i.e., obtains a path of lower cost than) the other. Finally, the last column provides the number of weathers in which both algorithms produced a path of equal cost. For example, for graph 20-1, the two algorithms produce a path of equal cost for 605 out of 1000 test instances and across the remaining 395 test instances, A*-HOP produces the better path in 207 instances and A*-HOP-R produces the better path in 188 instances. The solutions in these 188 instances represent new best-known solutions in the literature.

To enable summary comparisons across all graphs of the same size, averages are provided for columns two and three and totals are provided for columns four through six in Table 1. For example, in the 11th row, across all graphs of size 20, the average cost of paths produced by A*-HOP and A*-HOP-R are 156.45 units and 156.94 units respectively. Out of 10,000 test instances on graphs of size 20, the two algorithms produce a path of equal cost for 8908 test instances and across the remaining 1092 test instances, A*-HOP produces the better path in 656 instances and A*-HOP-R produces the better path in 436 instances. The solutions in these 436 instances represent new best-known solutions in the literature.

To determine whether differences in A*-HOP and A*-HOP-R solutions are statistically significant at the 0.05 level, we first define the difference d_i as the path cost produced by A*-HOP-R for instance i minus the path cost produced by A*-HOP for the same instance. Then we conduct two paired t-tests on the mean difference, μ_d, where the hypothesized value of μ_d is zero. In the first paired t-test, the alternative hypothesis is $H_1^1 : \mu_d > 0$, indicating significantly lower path costs produced by A*-HOP. In the second paired t-test, the alternative hypothesis is $H_1^2 : \mu_d < 0$, indicating significantly lower path costs produced by A*-HOP-R. Bold values in the second and third columns of Table 1 identify those graphs where the null hypothesis is rejected in favor of H_1^1 and H_1^2, respectively. From the table, we can observe that A*-HOP produces paths with significantly lower costs in 10 out of 30 graphs (e.g., 20-2, 20-3, etc.). A*-HOP-R only produces paths with significantly lower costs in 1 out of 30 graphs, graph 50-7. For the remaining graphs, there is no statistically significant performance difference between the two algorithms.

Using A*-HOP-R to re-plan more frequently did enable us to find new best-known solutions for 2142 out of 30,000 test instances (436 for graphs of size 20, 900 for graphs of size 50, and 806 for graphs of size 100). For those new best-known solutions, the average percent improvements in path cost for graph sizes 20, 50, and 100 are 16.2%, 16.8%, and 9.1% respectively. It is also noteworthy to examine the maximum improvement observed in these graphs. The maximum improvements identified in new best-known solutions for graph sizes 20, 50, and 100 are 69.0% (368 units of distance), 78.8% (600 units of distance), and 62.7% (781 units of distance), respectively. The remaining best-known solutions for these test instances are produced by A*-HOP in [4]. It is important to note the only solutions for this set of test instances available in the literature are those in [4]. While the graph structures of the test instances were first introduced in [5], the exact weathers from [5] are not available for comparison.

It can be observed that as the size of the graph increases from 20 to 50, the percentage of instances in which A*-HOP-R yields a new best-known solution rises from 4.4% to 9.0%. However, when the graph size increases from 50 to 100,

https://doi.org/10.21872/2023IISE_3137

Table 1: Average cost for 20, 50 and 100 node graph for different policies with 95% confidence interval

Graph	A*-HOP	A*-HOP-R	A*-HOP outperforms	A*-HOP-R outperforms	Tie
20-1	163.67 ± 5	163.59 ± 5	207	188	605
20-2	**151.26** ± **4**	152.53 ± 4	82	38	880
20-3	**135.24** ± **6**	135.71 ± 6	62	4	934
20-4	232.78 ± 7	233.09 ± 7	78	71	851
20-5	112.92 ± 6	113.04 ± 6	16	10	974
20-6	133.28 ± 4	133.03 ± 4	34	18	948
20-7	**148.87** ± **5**	149.81 ± 5	31	7	962
20-8	136.72 ± 5	136.87 ± 5	5	2	993
20-9	174.87 ± 4	175.43 ± 4	40	18	942
20-10	**174.88** ± **6**	176.30 ± 6	101	80	819
Avg\Tot	**156.45** ± **2**	156.94 ± 2	656	436	8908
50-1	**215.11** ± **8**	226.54 ± 9	328	295	377
50-2	378.49 ± 8	379.37 ± 8	165	146	689
50-3	267.46 ± 8	268.13 ± 8	84	55	861
50-4	239.26 ± 7	239.32 ± 7	66	35	899
50-5	230.01 ± 7	230.96 ± 7	112	32	856
50-6	**239.95** ± **9**	244.35 ± 9	162	75	763
50-7	211.54 ± 7	**210.44** ± **7**	44	40	916
50-8	296.68 ± 9	297.62 ± 9	409	111	480
50-9	**240.99** ± **10**	243.15 ± 10	144	65	791
50-10	259.89 ± 7	260.42 ± 7	71	46	883
Avg\Tot	**257.94** ± **3**	260.03 ± 3	1585	900	7515
100-1	**284.14** ± **7**	284.70 ± 7	37	9	954
100-2	156.01 ± 7	156.07 ± 7	13	17	970
100-3	**423.77** ± **14**	432.12 ± 14	284	155	561
100-4	322.87 ± 8	322.40 ± 8	32	47	921
100-5	**366.53** ± **14**	370.97 ± 14	261	196	543
100-6	413.72 ± 10	413.72 ± 11	236	108	656
100-7	394.95 ± 14	395.30 ± 14	72	57	871
100-8	291.16 ± 10	291.79 ± 10	177	76	747
100-9	272.69 ± 11	272.19 ± 11	78	56	866
100-10	350.10 ± 9	351.19 ± 9	146	85	769
Avg\Tot	**327.59** ± **4**	329.04 ± 4	1336	806	7858

this percentage decreases to 8.1%. The influence of graph size on the proportion of instances in which travel cost is reduced for both algorithms remains ambiguous.

While there are some test instances for which A*-HOP-R yielded improvements over solutions produced by A*-HOP, overall these results suggest that the strategy of re-planning more frequently, each time new information is gained, is not as effective in reducing average path cost as we had hypothesized. In an attempt to provide some guidance regarding in what situations to recommend one algorithm over another, we compared the set of test instances in which A*-HOP performs better to the set of instances in which A*-HOP-R performs better, excluding those test instances in which both algorithms perform equally well from our analysis. Our initial findings suggest that A*-HOP-R tends to perform well for more severely-disrupted instances. That is, when the proportion of failed edges in the real weather exceeds the expected number of failed edges according the edge failure probability distributions, replanning more frequently (via A*-HOP-R) tends to be beneficial. On the other hand in less severely disrupted instances, where

https://doi.org/10.21872/2023IISE_3137

relatively fewer edge failures are discovered compared with what is expected, replanning only when necessary (via A*-HOP) is preferred. However, we temper these observations with the caveat that a thorough analysis would required to fully substantiate these initial findings.

6. Conclusion and Future Work

In this study, the proposed algorithm A*-HOP-R was a modification of A*-HOP, in which re-planning was implemented at every node where new information was obtained by the agent. The purpose of this modification was to investigate whether more frequent re-planning would lead to better-informed decisions and ultimately result in a lower travel cost. The computational experiment was conducted on Delaunay graphs and revealed that the original algorithm, A*-HOP produced significantly better results than A*-HOP-R for 10 out of 30 graphs, and A*-HOP-R produced significantly better results than A*-HOP for 1 out of 30 graphs. In addition, A*-HOP-R was able to identify new best-known solutions for 2142 out of 30,000 test instances. Although A*-HOP-R may not appear to be promising on average, the fact that 7.1% of the instances showed improvement suggests that the algorithm is promising in certain cases. Further analysis is required to determine the characteristics of these instances and understand which algorithm is more appropriate under different weathers. The results section 5 provides some direction towards achieving this understanding.

In order to determine a clear conclusion, it is necessary to formulate a hypothesis and undergo thorough testing. This represents a substantial area for future investigation. Since the algorithm was tested exclusively on Delaunay graphs, further examination and comparison of both algorithms across various graph types can offer greater insights into their respective performances. Examining the performance of these algorithms in stochastic k-CTP variations, where the number of failed edges is limited, may also prove intriguing. Furthermore, the effects of more frequent re-planning could be evaluated in multi-agent scenarios. Lastly, the impact of re-planning more frequently can be tested when multiple agents are available for traversing the graph to reveal edge status.

Acknowledgements

This material is based upon work supported by the National Science Foundation under Grant 1554412.

References

[1] FEMA. FEMA Preliminary Damage Assessment Guide. Technical report, FEMA, May 2020. URL https://www.fema.gov/sites/default/files/2020-07/fema_preliminary-disaster-assessment_guide.pdf.

[2] UNU-EHS. Infrastructure-Damage. URL https://interconnectedrisks.org/impacts/infrastructure-damage.

[3] Christos H Papadimitriou and Mihalis Yannakakis. Shortest paths without a map. *Theoretical Computer Science*, 84(1):127–150, 1991.

[4] Andrew Alseth. *l-CTP: Utilizing Multiple Agents to Find Efficient Routes in Disrupted Networks.* University of Arkansas, 2020.

[5] Patrick Eyerich, Thomas Keller, and Malte Helmert. High-quality policies for the canadian traveler's problem. In *Twenty-Fourth AAAI Conference on Artificial Intelligence*, 2010.

[6] Davood Shiri and F Sibel Salman. Online optimization of first-responder routes in disaster response logistics. *IBM Journal of Research and Development*, 64(1/2):13–1, 2019.

[7] Peter E Hart, Nils J Nilsson, and Bertram Raphael. A formal basis for the heuristic determination of minimum cost paths. *IEEE transactions on Systems Science and Cybernetics*, 4(2):100–107, 1968.

https://doi.org/10.21872/2023IISE_3156

Proceedings of the IISE Annual Conference & Expo 2023
K. Babski-Reeves, B. Eksioglu, D. Hampton, eds.

Uncertainty Quantification for Healthcare Data

Abstract ID: 3156

Hala Ghali, Sang Won Yoon, Daehan Won
Systems Science and Industrial Engineering Department
Binghamton University, Binghamton, NY 13902

Abstract

Technology has been of great assistance in the medical field; however, the adoption of machine learning and artificial intelligence is still experimental. The main issue that is halting this leap in medical technology is the trustworthiness of the product of artificial intelligence-powered machines. This research aims to create a framework that reviews and quantifies the uncertainty of machine learning models. For this purpose, healthcare data will be utilized to provide a machine learning and optimization-based solution for a medical-based issue. There exist multiple branches in healthcare that can highly benefit from the integration of machine learning-based automation and decision support, in this research the focus will be organ donation and transplant outcomes. The organ donation data used will be extracted synthetically and will be used to carry a regression task. The estimation of uncertainty will be calculated using a novel elaborate algorithm that includes Bayesian Neural Networks, Monte Carlo Dropout, and L2 regularization . As part of the methodology, the prediction of uncertainty will be executed on three parts, including the prediction of epistemic uncertainty, aleatoric uncertainty, and combined uncertainties. This study is developed to help understand the type of uncertainties that healthcare data and models presents, and subsequently how to predict them efficiently in order to increase confidence in machine learning outcomes in healthcare.

Keywords
Uncertainty Quantification, Healthcare Data, Machine Learning, Regression Analysis.

1. Introduction

Given the complex problems that healthcare presents, great hope, and emphasis are tied to the new technologies to help solve them. The main requirements for ML or any AI tools or applications is big data, because the larger the data, the more possibilities to find patterns and valuable findings. There is no shortage in terms of data in healthcare compared to many other sectors, however, that comes with its challenges. Therefore, a need for preprocessing techniques has arisen to deal with such issues. Uncertainty quantification constitute a solution to deal with accuracy of methods and results which improve healthcare related outcomes. The uncertainty can have many sources which can be grouped into two major categories, aleatoric and epistemic uncertainty [1]. Aleatoric uncertainty is referred to the inherent uncertainty in a given system and thus is harder to overcome by more data collection or experimentation. It is a stochastic type of variation, that can only be represented through a probability distribution. Epistemic uncertainty is solely emanating from the actual model and can be traced to a lack of knowledge about this latter. This specific uncertainty can be manageable by gaining more information about the system as a whole [2]. In this paper, a description of the problem will be presented followed by an in-depth literature review of methodologies used to address uncertainty quantification in healthcare. Also, a clear description of the methodologies and data profiles will be discussed in the third section. The next part will include analysis of the results and will showcase the performance of the methods applied. Finally, a conclusion will include a summary of the main findings and provide insights into future work.

2. Problem Description

The quality of any machine learning model or the basis of any AI application is highly reliant on the quality of data. Since it is hard to control the quality of the data that constitutes the base of the model, more techniques have been put into place as part of the preprocessing to address those problems. Unfortunately, the uncertainty component that comes from the data itself, its variability, and problems are not addressed specifically but as part of the accuracy of the model itself. There exists no significant emphasis on the study of the uncertainty that the data and model bring to the results

https://doi.org/10.21872/2023IISE_3156

and how it can be first quantified and understood to be late on mitigated. The machine learning known procedure trusts the applied algorithm to deal with uncertainty in terms of a black box, which is later on quantified using performance metrics based on training and testing of the data as final reporting measures.
Given the importance and the volume of the effect of these results and findings, it is of utmost importance that the results be reported accurately, and the performance of the model be studied thoroughly through all its steps. The margin of error in healthcare and medicine does not leave a gap for unjustified uncertainty, therefore the quantification of uncertainty should be integrated into the machine learning process from beginning to end.

3. Related Research

Unlike other sectors, dealing with uncertainty in healthcare comes with a specific set of challenges. Multiple techniques, both recent and spanning the last 2 decades have been useful in quantifying uncertainty and its role in machine learning applications in healthcare. Researchers have recently explored the role of uncertainty in a marking event in the healthcare sector specifically and humanity in general and that is COVID-19. Uncertainty quantification through Bayesian inference was used to verify facts about the pandemic [3], and also to estimate the uncertainty in deep learning models that are used to help with the diagnosis [4]. Similar research in immunology but related to HIV has also used Bayesian inference to analyze specific patterns in data [5], and the use of hierarchical methods for uncertainty estimation in dynamic differential equations [5]. For coronary heart disease, a model was devised to assess patient risk [6] where both linear discriminant analysis and adaptive neuro-fuzzy systems were optimized to control uncertainty in a Korean dataset which resulted in a high predicting rate of coronary disease.

Rule induction techniques have been used in various research in the late 90s and 2000s [7], [8], [9] . For instance, it was utilized in research through an induction rule that allowed the algorithm to extract all necessary patterns needed for medical diagnosis. The prediction model was a success and allowed the exploration of three datasets and also provided statistical measures as part of the output [10]. In efforts to locate the noise in breast cancer imaging, Monte Carlo simulation was employed [11], [12]. In chemotherapy, monte Carlo simulation was also used in uncertainty quantification for research that predicts the dose of radiotherapy needed for specific treatment [13] The same monte Carlo simulation methodology was promoted by [11] where a big database encompassing all organs and tissues of both genders up to 15 years old was generated to facilitate the dose used in scanner procedures.
Other research has explicitly leveraged the power of the Dempster-Shafer theory (DST) as a way to eliminate as much uncertainty as possible when dealing with medical data, especially what relates to subjectivity [14]. Analogous research has been conducted to mitigate the uncertainty in medical models and data through imprecise probability methods that oversee both types of uncertainty; aleatoric and epistemic [14].

4. Methodology

In addition to the description of the data characteristics used in this analysis. This section will also include a detailed summary of the methods. Figure 1 is a high-level depiction of the methodology that will be discussed.

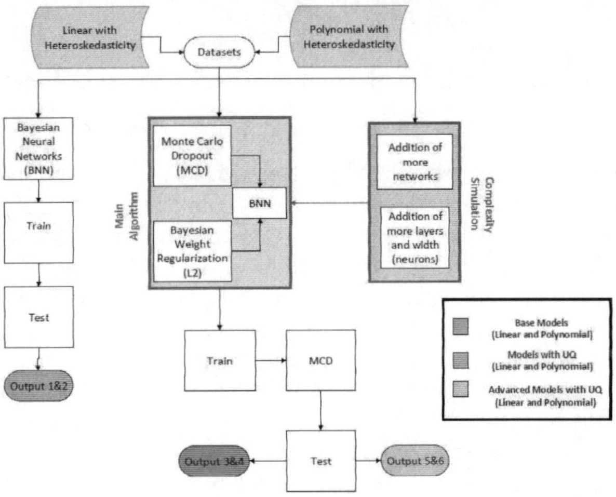

Figure 1: A high-level description of the initial performed analysis

https://doi.org/10.21872/2023IISE_3156
Proceedings of the IISE Annual Conference & Expo 2023
K. Babski-Reeves, B. Eksioglu, D. Hampton, eds.

4.1. Data Description

In order to experiment with the concepts that are addressed in this section, a series of organ donation and transplant synthetic data is developed to test the methodologies. To better explain these concepts a heteroskedastic model is established for the regression task. Heteroskedasticity is present when the data violates the homoscedasticity assumption [15]. In other words, this data will lack homogeneity in the variance and the latter will vary greatly among the random variables [15]. Heteroskedasticity refers to the phenomenon where the variance of a variable's error term is not constant across all values of the predictor variable(s). In other words, the spread of the residuals (the difference between the observed and predicted values) is not the same for all values of the predictor variable(s). This can lead to biased parameter estimates and incorrect inferences if not considered [16].

Uncertainty quantification and prediction are affected by heteroskedasticity in a number of ways. For example, when making predictions, heteroskedasticity can lead to a larger spread of predicted values at certain points in the predictor space, resulting in increased uncertainty. This can also lead to overconfidence in predictions made in regions where the error variance is lower. In terms of uncertainty quantification, heteroskedasticity can lead to underestimation of the true uncertainty of the model's parameters. This is because traditional methods for assessing uncertainty, such as calculating the standard error of the estimates, assume that the error variance is constant. When this assumption is violated, the standard errors will be too small, leading to overly narrow confidence intervals [16].

4.2. Monte Carlo Dropout

The Monte Carlo Dropout (MCD) method is a technique for approximating Bayesian inference in neural networks. It is based on the idea of adding dropout regularization to a neural network during both training and inference[17]. Dropout is a regularization technique that randomly drops out (i.e., sets to zero) a certain proportion of neurons during training. This helps to prevent overfitting by encouraging the network to learn multiple independent representations of the data. The Monte Carlo part of the method comes from the idea of sampling from the posterior distribution of the weights, given the data [17]. In the MCD method, dropout is applied to the network during inference by randomly dropping out neurons with a certain probability defined by the user at each forward pass. This results in a different network architecture at each forward pass, effectively averaging over different architectures. The final output is then obtained by averaging the output of the network over multiple forward passes with dropout. Mathematically, the MC Dropout method can be formulated as follows in Equation (1) [17]:

Let x be the input data, and y be the corresponding output. Let $p(x, y)$ be the posterior distribution of the weights of the network given the data. We can approximate this distribution using a point estimate of the weights, $\widehat{w}$, obtained by maximizing the likelihood of the data. To obtain a more robust estimate of the output, we can perform multiple forward passes of the network with dropout, each with a different dropout mask.
The final output is obtained by averaging the output of the network over multiple forward passes with dropout:

$$y_{pred} = \frac{1}{T} \sum_{i=1}^{T} y_i \tag{1}$$

Where T is the number of samples and y_i is the output of the network with the ith dropout mask.

4.3. Bayesian Neural Networks

Bayesian Neural Networks (BNNs) are a type of neural network that incorporates Bayesian principles, specifically Bayesian inference, into the network architecture. They are used to model the uncertainty in the weights of the neural network, which is important for making predictions in real-world applications where data is often noisy or uncertain [18]. The mathematical formulation of a BNN involves defining a prior probability distribution over the weights of the network, and then using Bayesian inference to compute the posterior distribution given the data. The prior distribution represents our belief about the weights before we see the data, and the posterior distribution represents our updated belief after we have seen the data [18]. One common approach to defining the prior distribution is to use a Gaussian distribution with a mean of 0 and a fixed standard deviation. This is known as a "weight prior" and it can be written as shown in Equation (2) [18]:

$$p(w) = N(w; 0, \sigma^2) \tag{2}$$

Where w is the weight vector, $N(w; 0, \sigma^2)$ is the Gaussian distribution, and σ^2 is the variance of the distribution.

https://doi.org/10.21872/2023IISE_3156

5. Results

In this section a thorough description of the experimental results will be presented. The results of the synthetic data extracted that was described above are used to map a simple regression task. MCD, BNNs and L2 regularization is used to predict the different types of uncertainty.

5.1. Epistemic Uncertainty Prediction

In order to predict epistemic uncertainty, this model is implemented through neural networks and different number of nodes were tested and 6 layers were considered. The main methodology that will be used to estimate uncertainty in this section is Monte Calo Drop out (MCD) is a highly effective tool to approximate variational inference. This model is primarily used for deep learning to help with computer vision [19]. Monte Carlo Dropout is a method for estimating uncertainty in deep learning models. It involves using dropout, a regularization technique, during the forward pass of a neural network at test time. During training, dropout is applied to the input of certain layers by randomly setting a proportion of the input units to zero. This helps to prevent overfitting by forcing the network to learn multiple independent representations of the input data. This also makes the output stochastic. At test time, dropout is not applied, and the network makes a single prediction for each input. The Monte Carlo Dropout method involves using dropout during the forward pass of the network at test time, but instead of making a single prediction, the network makes multiple predictions with different dropout masks. These predictions are then averaged to produce a final prediction, and the variance of the predictions can be used to estimate the uncertainty of the model [19].
This developed model has a specific architecture that takes a number of parameters. It takes a number of hidden units in each layer of the model. For this specific instance, the model will have three hidden layers with 10, 20 and 30 units respectively. The number of epochs is set at 500 with a batch size of 10 and experiments with 5 different number of neurons ranging from 5 to 20. The probability of dropping out a unit during training is between 0 and 1. Dropout is a regularization technique that helps prevent overfitting by randomly setting a fraction of input units to zero during training. L2 regularization coefficient is also specified. L2 regularization adds a penalty to the model's loss function to discourage large weights, which can help prevent overfitting. The dropout probability that has been used for training and testing this study is of (p=10%). The L2 regularization is defined as follows in Equation (3):

$$L(\theta, p) = \frac{(y-\hat{y})^T(y-\hat{y})}{2N} + \frac{1-p}{2N}\ \|\theta\|_2^2 \quad (3)$$

The activation function for this neural network is hyperbolic tangents, and the dropout layers are inserted between the hidden layers to apply dropout regularization. The MCD output used to sample epistemic uncertainty for a number of times denoted with T and the expected output is an average of the output of all models during that specific run. Equations (4) and (5) denotes this process.

$$E(y) = \frac{\sum_{t=1}^{T}\hat{y}}{T} \quad (4)$$

$$Var(y) = \frac{\sum_{t=1}^{T}\hat{y}(x)^T\,\hat{y}(x)}{T} - E(y)^T\,E(y) \quad (5)$$

Figure 2 presents a comparison between the original model and the developed model with MCD for epistemic uncertainty quantification. It is clear that the original model does not provide any feedback towards uncertainty or its quantity across different parts of the data. However, in the second graph, it is clear that the epistemic uncertainty is

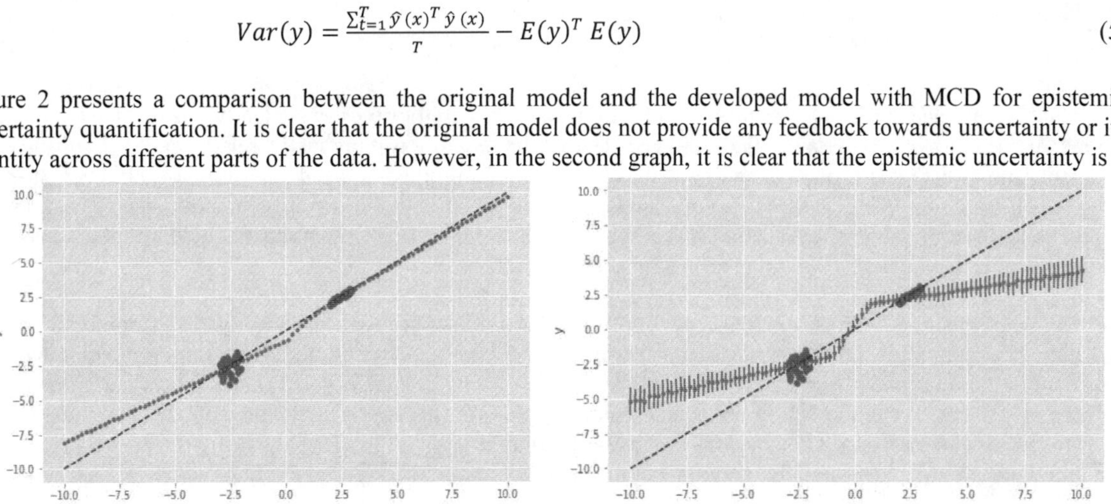

Figure 2: The predicted epistemic uncertainty (right) vs. the original model (left)

https://doi.org/10.21872/2023IISE_3156

significant in the endings because of data missing. The portion between the two scatter points is the same across the two graphs because the uncertainty is not related to epistemic uncertainty.

5.2. Aleatoric Uncertainty Prediction

In order to estimate aleatoric uncertainty, we can use a direct prediction of heteroskedastic aleatoric uncertainty. We encourage the model to display aleatoric uncertainty through including aleatoric variance $\sigma(x_i)^2$ in the Equation (6) and this prediction is regularize using the log of the variance represented as $log\ \sigma(x_i)^2$.

$$L(\theta, p) = \frac{\sum_{i=1}^{N}(y_i - \hat{y}_i)^2}{2N\sigma(x_i)^2} + \frac{1}{2} log\ \sigma(x_i)^2 \tag{6}$$

while the loss function that is implemented is demonstrated in Equation (7).

$$L(\theta, p) = \frac{\sum_{i=1}^{N} exp\ (-s_i)(y_i - \hat{y}_i)^2}{2N} + \frac{1}{2} s_i \tag{7}$$

Keeping the previous methodology for the neural network and the MCD the results are shown in the graph in Figure (3). It shows how the aleatoric uncertainty is higher between the two data points clusters to the contrary of epistemic uncertainty.

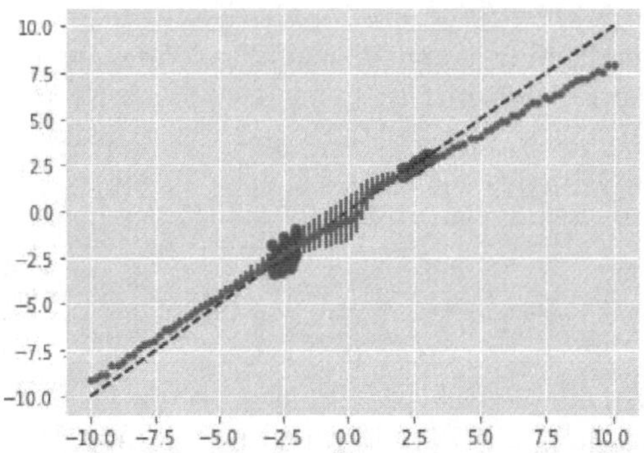

Figure 3: The predicted aleatoric uncertainty in the linear regression data

5.3. Combined Uncertainty Prediction

The MCD method allows for estimating both uncertainties in one model through a slight change in its mathematical equation represented in Equation (13).

$$L(\theta, p) = \frac{\sum_{i=1}^{N}(y_i - \hat{y}_i)^2}{2N\sigma(x_i)^2} + \frac{1}{2} lo\ g\ \sigma(x_i)^2 + \frac{1-p}{2N} \parallel \theta \parallel_2^2 \tag{8}$$

Given that the absence of additional data creates more epistemic uncertainty we can evidently, see that the areas where that is the case epistemic uncertainty flares up. For aleatoric uncertainty is more model related and it is indicative of the quality of the data that is at hand and how does it epitomize the model at hand, and more data collection is not going to help with improving this specific uncertainty. The combined aleatoric and epistemic can also be useful if the purpose is to quantify the general uncertainty, but it is more useful to examine them separately to be able to make the right decision about the model. This representation is shown in Figure 4.

The running of the model comes with the expense of time that the user needs to be aware of. The formulation of this algorithm has been done in a way to minimize run time. However, for the algorithm that is combining both uncertainties, the run time is higher similarly to aleatoric uncertainty quantification Table 1 shows the difference of run time in seconds. It is important to note that the dataset used for this experimentation is fairly small, so the user needs to take this into account when dealing with a higher dataset.

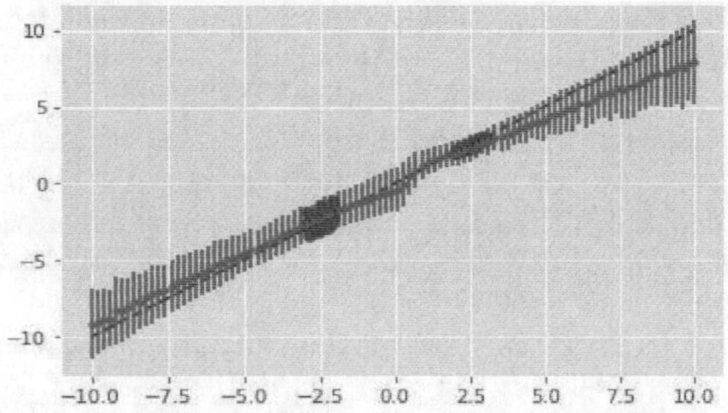

Figure 4: The predicted combined uncertainty standard deviation

Table 1: Comparison of run times of all the models

Model	Training time (s)
Original	33.605
Epistemic	37.952
Aleatoric	94.679
Combined uncertainties	95.758

6. Conclusions

In this paper, a comprehensive review of uncertainty quantification in healthcare data was carried out. A literature review to map the main methodologies and research gap was explored which provided motivation for the analysis described in later sections. An analysis of organ donation and transplant synthetic data was utilized in a simple regression task to discover uncertainty quantification of the data. Both types of uncertainties were predicted using a combination of MCD and BNN with L2 regularization methods. Combined aleatoric and epistemic uncertainties were predicted as well, and the performance of algorithm was compared for all three predictions. Predicting combined uncertainties required more run time and memory power than running epistemic or the original model. These results highly affect the trustworthiness of machine learning models and is highly informative for ways to improve data used for a better outcome. For future work, adapting this algorithm for a classification task can be highly useful in this context.

References

[1] A. der Kiureghian and O. Ditlevsen, "Aleatory or epistemic? Does it matter?," *Structural Safety*, vol. 31, no. 2, pp. 105–112, Mar. 2009, doi: 10.1016/j.strusafe.2008.06.020.

[2] V. Volodina and P. Challenor, "The importance of uncertainty quantification in model reproducibility," *Philosophical Transactions of the Royal Society A: Mathematical, Physical and Engineering Sciences*, vol. 379, no. 2197, p. rsta.2020.0071, May 2021, doi: 10.1098/rsta.2020.0071.

[3] H.-C. Lin *et al.*, "Bayesian inference of nonylphenol exposure for assessing human dietary risk," *Science of The Total Environment*, vol. 713, p. 136710, Apr. 2020, doi: 10.1016/j.scitotenv.2020.136710.

[4] B. Ghoshal and A. Tucker, "Estimating Uncertainty and Interpretability in Deep Learning for Coronavirus (COVID-19) Detection," Mar. 2020, [Online]. Available: http://arxiv.org/abs/2003.10769

[5] Y. Huang, H. Wu, and E. P. Acosta, "Hierarchical Bayesian inference for HIV dynamic differential equation models incorporating multiple treatment factors," *Biometrical Journal*, vol. 52, no. 4, pp. 470–486, Aug. 2010, doi: 10.1002/bimj.200900173.

[6] J.-G. Yang, J.-K. Kim, U.-G. Kang, and Y.-H. Lee, "coronary heart disease optimization system on adaptive-network-based fuzzy inference system and linear discriminant analysis (ANFIS–LDA)," *Pers Ubiquitous Comput*, vol. 18, no. 6, pp. 1351–1362, Aug. 2014, doi: 10.1007/s00779-013-0737-0.

[7] H.-C. Chou, C.-H. Cheng, and J.-R. Chang, "Extracting drug utilization knowledge using self-organizing map and rough set theory," *Expert Syst Appl*, vol. 33, no. 2, pp. 499–508, Aug. 2007, doi: 10.1016/j.eswa.2006.05.020.

[8] M. Ningler, G. Stockmanns, G. Schneider, H.-D. Kochs, and E. Kochs, "Adapted variable precision rough set approach for EEG analysis," *Artif Intell Med*, vol. 47, no. 3, pp. 239–261, Nov. 2009, doi: 10.1016/j.artmed.2009.07.004.

https://doi.org/10.21872/2023IISE_3156

[9] Erekat, A., Madathil, S. C., & Khasawneh, M. (2018). Interventional Radiology Procedure Scheduling Using Self-Adaptive Simulated Annealing Optimization Algorithm. In IIE Annual Conference. Proceedings (pp. 1102-1107). Institute of Industrial and Systems Engineers (IISE).

[10] S. Tsumoto, "Extraction of Experts' Decision Rules from Clinical Databases Using Rough Set Model," *Intelligent Data Analysis*, vol. 2, no. 3, pp. 215–227, Jul. 1998, doi: 10.3233/IDA-1998-2305.

[11] C. Lee, K. P. Kim, D. J. Long, and W. E. Bolch, "Organ doses for reference pediatric and adolescent patients undergoing computed tomography estimated by Monte Carlo simulation," *Med Phys*, vol. 39, no. 4, pp. 2129–2146, Mar. 2012, doi: 10.1118/1.3693052.

[12] K. Bush, R. Townson, and S. Zavgorodni, "Monte Carlo simulation of RapidArc radiotherapy delivery," *Phys Med Biol*, vol. 53, no. 19, pp. N359–N370, Oct. 2008, doi: 10.1088/0031-9155/53/19/N01.

[13] X.-L. Sun, H. Wang, X.-K. Li, G.-H. Cao, Y. Kuang, and X.-C. Zhang, "Monte Carlo computer simulation of a camera system for proton beam range verification in cancer treatment," *Future Generation Computer Systems*, vol. 102, pp. 978–991, Jan. 2020, doi: 10.1016/j.future.2019.09.011.

[14] R. Alizadehsani *et al.*, "Handling of uncertainty in medical data using machine learning and probability theory techniques: a review of 30 years (1991–2020)," *Ann Oper Res*, 2021, doi: 10.1007/s10479-021-04006-2.

[15] P. J. Rosopa, M. M. Schaffer, and A. N. Schroeder, "Managing heteroscedasticity in general linear models.," *Psychol Methods*, vol. 18, no. 3, pp. 335–351, 2013, doi: 10.1037/a0032553.

[16] H. White, "A Heteroskedasticity-Consistent Covariance Matrix Estimator and a Direct Test for Heteroskedasticity," Econometrica, vol. 48, no. 4, p. 817, May 1980, doi: 10.2307/1912934.

[17] C. Blundell, J. Cornebise, K. Kavukcuoglu, and D. Wierstra, "Weight Uncertainty in Neural Networks," May 2015.

[18] D. P. Kingma and M. Welling, "Auto-Encoding Variational Bayes," Dec. 2013.

[19] S. Gruber, P. D. Ullrich, and K., "UNIVERSITY HEIDELBERG SEMINAR REPORT EXPLAINABLE MACHINE LEARNING What Uncertainties Do We Need in Bayesian Deep Learning for Computer Vision?," 2018.

https://doi.org/10.21872/2023IISE_3187

Proceedings of the IISE Annual Conference & Expo 2023
K. Babski-Reeves, B. Eksioglu, D. Hampton, eds.

Stakeholder analysis using cognitive mapping concept in construction projects

Abstract ID: 3187

Ying Thaviphoke, Ph.D.
Miguel Garcia
College of Arts and Sciences, School of Engineering
Texas A&M International University, Laredo, TX 78041

Abstract

Construction projects are complex systems. They involve complex activities, including multiple stakeholders. Hence, there is a need for a well-thought-out stakeholder analysis to ensure the success of a project. It is because each stakeholder has different needs toward the project. Internally, construction projects require teams of specialists to develop ideal solutions. However, external stakeholders can make an impact on the project as well. A stakeholder analysis provides the importance of the needs of different stakeholders and how each stakeholder should be treated throughout a project life cycle. Getting a decent outcome from a stakeholder analysis can also mitigate the risk of failure – not meeting budget, time, and scope expectations. This paper proposes the cognitive mapping concept as a tool for stakeholder analysis in construction projects. A cognitive map is comprised of nodes and arrows. Nodes represent statements or needs, while arrows provide directions and the magnitude of the relationships among nodes. Together with stakeholder analyses (e.g., stakeholder engagement priority, stakeholder typology, and classification), the cognitive mapping concept adds value to a stakeholder analysis in a construction project. This paper provides the application of the cognitive mapping concept for a better stakeholder analysis effort.

Keywords

Stakeholder analysis, Cognitive mapping, Construction project

1. Introduction

Construction projects involve various stakeholders who individually express their needs, values, and expectations about the project. These values or expectations are sometimes in conflict with each other. The conflicting values and expectations may lead to failures if they are inappropriately managed. It is crucial to find a form of consensus among stakeholders to ensure the project's success. The individual view is explained in the personal construct theory [1]. The theory argues that people try to "make sense" of the environment to "manage and control" those surroundings. Stakeholders in construction projects are no different. One of the challenges that the project manager might face is evaluating stakeholder needs and expectations with respect to the main objectives of the project. The determination of priority needs should be managed properly. When there is a lack of understanding of the needs of stakeholders, the project might not be regarded as successful (i.e., deliver the desired outcome). Stakeholder analysis is important in problem-solving efforts [2].

Stakeholder analysis attempts "to develop a strategic view of the human and institutional landscape, the relationships between different stakeholders and the issues they care about most" [3, p. 526]. Construction projects always involve multiple stakeholders [4]–[6]). Therefore, to ensure the success of a project, stakeholder analysis can be considered one of the most important processes in construction projects. It is important because the outcome of the process would allow "project teams to understand the stakeholder environment and develop appropriate engagement strategies" [7, p. 294]. Various stakeholder analysis techniques have been mentioned in the literature [2], [3], and [8]. This paper proposes the cognitive mapping concept as a tool for stakeholder analysis in construction projects. The cognitive mapping concept has been applied in a group decision-making process [9]. Cognitive maps represent multiple concepts and their relationships and interactions respectively.

This paper is organized as followings. It first explores potential problems in stakeholder analysis. Then, the introduction of the cognitive mapping concept as a tool in stakeholder analysis. Next, an illustrative example of the application of the cognitive mapping concept is discussed to demonstrate the proposed approach. The results are analyzed based on a hypothetical scenario (a real estate project). Finally, conclusions, limitations, and future research ideas are presented.

https://doi.org/10.21872/2023IISE_3187

2. Problem Description

A stakeholder refers to any person that affects or can be affected by the system of interest. In other words, stakeholders are "any group or individual who can affect or is affected by the achievements of the organization's objective" [10]. As a concept, in construction projects, stakeholders can range from a construction worker to the CEO of the construction company to the clients.

A poor stakeholder analysis could lead to the project's failure. Mok, Shen, and Yang [6] mentioned some major pitfalls of poor stakeholder analysis in the construction of a Chinese opera building. The major causes for such pitfalls are unmet client requirements, failure to balance between scope and project resources, and lack of leadership team makeup, vision, charisma, and learning stance. Unmet client requirements are caused by the lack of communication and understanding between contractors and clients which reflects the final construction project lacking the requirements or requests of the client. Lack of leadership team makeup, vision, charisma, and learning is caused by not picking the right people to work on the project, not having a clear idea of the project's scope, and lacking the opportunity for contractors to study and learn about the scope given before starting the project.

Jergeas et al. [11] reported four issues found with poor stakeholder analysis in a project. First, stakeholders are not satisfied with the outcome of the project. Second, there are disruptions that affect both budgets and the schedule of the project. Failure to balance between scope and project budgets is caused by several reasons such as not completing the project within the budget, not getting approvals or permits for construction, and lack of esthetic or functionality [6]. Third, the project does not receive enough endorsement at the early stage. Finally, a poor stakeholder analysis might affect the potential future collaborations with the main stakeholders.

As mentioned in this section, stakeholder analysis is crucial to understand the stakeholder environment and developing appropriate engagement strategies. Therefore, the use of cognitive mapping will be helpful to illustrate and show the interaction between stakeholders in an organized and easy way to understand. In other words, there is a need to establish a consensus among stakeholders. Appropriate strategies can be developed after the consensus is reached.

3. Related Research

Stakeholder analysis methods categorize and analyze stakeholders with respect to their individual needs and expectations. Also, the impact that each stakeholder could make on a project [13]. One of the objectives of stakeholder analysis is to understand the interests and stances of various stakeholders [13]. It is fair to say that the outcome of a stakeholder analysis affects the entire operations of any project. Therefore, it is important to critically understand all stakeholders' needs [14]. The focus should be on those who have the potential to make an impact on the outcome of the project.

Mok and Shen [7] argue that stakeholders in construction projects can be analyzed from three aspects: Stakeholders' values and their interdependencies, relationships and interactions among stakeholders, and dynamics of stakeholders and issues. The authors apply input-output analysis and network theory as a tool in stakeholder analysis efforts. The method uses mathematical equations which take into account different factors such as perturbation, interdependency, and ineffectiveness of the different stakeholders and result in the stakeholder influence in the system.

Thaviphoke [9] investigates the outcome of the application of the cognitive mapping concept in a group decision-making process. Multiple stakeholders are asked to make a decision in a complex environment. It was found that the stakeholders who make decisions using cognitive maps show a higher level of problem-solving confidence than the stakeholders who did not use the cognitive mapping concept. The results of this experimental research imply that the cognitive mapping concept can be useful in a multiple stakeholders environment.

The cognitive mapping concept has been applied as a tool in stakeholder analysis [3] and [8]. The cognitive mapping concept is applied to capture the subjective thoughts of an individual. Villafranca and Loureiro [14, p. 1071] argue that "the cognitive map helps to structure complex situations with subjective information, and assumption". A cognitive map is a visual representation of an individual or a group's mental model. In other words, it represents a perception of an individual or a group toward a system of interest [16].

https://doi.org/10.21872/2023IISE_3187

4. Methodology

A cognitive map represents the mental model of an individual (i.e., how an individual perceives reality). Leonhardt Kjærgaard and Blegind Jensen [15, p. 1110] explain that "cognitive maps provide the possibility of representing users' thoughts, ideas, and critical comments about a technology and other related aspects." It should be noted that if one cognitive map represents an individual's view, a collection of cognitive maps (i.e., causal map) would represent a group's view [16]. As mentioned earlier, one of the objectives of a stakeholder analysis is to understand multiple stakeholders and generate a consensus. It is crucial to understand the way they perceive the objectives or the values of the project. According to the personal construct theory by George Kelly, stakeholders could view values differently from each other even if they are in the same project. Villafranca and Loureiro [14] provide a six-step iterative process to create a cognitive map from stakeholders. The authors indicate four advantages of using the cognitive mapping concept in stakeholder analysis: knowledge generation, leveled knowledge, exhaustiveness, and rationale capture. It is fair to say that these four aspects bring about consensus among stakeholders.

Although there are multiple stakeholders in construction projects, they can be categorized into three key stakeholders: Clients, consultants, and contractors [4]. These key stakeholders can be considered internal as opposed to external stakeholders (e.g., local community, government, NGO). Karlsen [5] identifies important stakeholders in a project (construction included). These stakeholders are, for example, clients, contractors, end users, consultants, suppliers, public authorities, etc. [6] also argue that clients, contractors, cost consultants, designers, end users, government agencies, and subcontractors are the main stakeholders in construction projects. Thaviphoke and Hester [8] applied the cognitive mapping concept to investigate a problem in a construction project. They found that the concept helped increase the level of problem understanding from stakeholders' different points of view. In a real estate scenario, it is fair to say that the end users are homebuyers.

Thaviphoke [9] argues that the cognitive mapping concept provides a bigger picture of the situation for stakeholders. In his experiment, following the Strategic Options Development Analysis I approach discussed in Pidd [17], individual cognitive maps were combined and turned into a causal map. Multiple stakeholders in a group then discussed and negotiated to try to establish a consensus using the causal map. This paper follows this approach. The next section presents the illustrative results of the application of the cognitive mapping concept as a tool in stakeholder management in construction projects.

5. Results

This paper presents an illustrative result of the cognitive mapping concept in stakeholder analysis. Regardless of the size, it is fair to say that most construction projects will have clients, contractors, consultants, end users, and public authorities as the main stakeholders [4], [6], and [5]. Therefore, in this paper, the results of the analysis are based on these main stakeholders. As mentioned earlier, homebuyers could be considered end users. Also, a real estate development company could be considered a client in a construction project. The illustrative results start with individuals' perceptions of the project followed by the merged map. All the maps are constructed using the freeware Mental Modeler software package (mentalmodeler.org), developed by a consortium of US universities and industry partners.

This hypothetical scenario is a real estate project in a residential area. The product of this project could range from a condominium, house, gated community, and so on. This paper focuses on the domain understanding stage of stakeholder analysis as discussed in a requirements management spiral process [18]. Figure 1 illustrates stakeholders' perceptions toward the outcome of the project.

https://doi.org/10.21872/2023IISE_3187

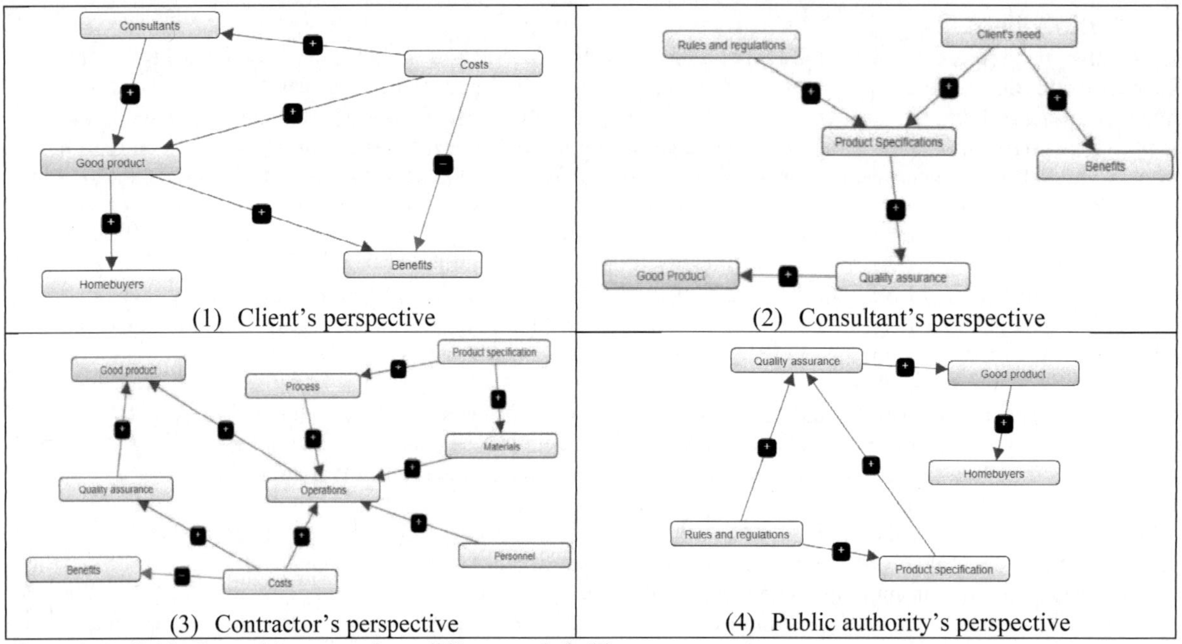

Figure 1: Individual cognitive maps from different stakeholders

A real estate development company is a client in this scenario. From the client's perspective (1), the ultimate goal of this stakeholder is to maximize its benefit. This stakeholder needs to reach out to another stakeholder, the consultant, to ensure that the cost is minimized, and the good product is delivered. The good product will then satisfy the end users which are homebuyers in this case. Satisfied homebuyers will then maximize the benefits. On the other hand, increasing costs can negatively impact the benefit of the project. The negative sign on the red arrow identifies the direct negative impact between concepts.

From the consultant's perspective (2), the goal of this stakeholder is also to maximize its benefit. However, this is not the same benefit shown in (1). A consultant firm can maximize its benefit through its service to the client. The main responsibility of this stakeholder is to ensure that the project will produce a good product. Product specification can be derived from the client's needs along with rules & regulations. This stakeholder also controls the quality of the product through quality assurance processes. It should be noted that "rules and regulations," "product specification," and "Quality assurance" are not present from the client's perspective (1). These three concepts are also present in the public authority's perspective (4) though. This is because the goal of public authority is to ensure the safety of citizens. As it shows no benefits from this stakeholder's perspective.

The busiest map in this scenario is the contractor's perspective (3). It is predictable because this is where the actual construction occurs. A contractor wants to maximize its benefits by having good operations to ensure a good product. Operations can be comprised of process, material, and personnel. While the personnel can be hired internally through a human resource department, the process and material need to follow the product specification. The main responsibility of this stakeholder is to manage the resources, time, and scope of the project. The more money that the contractor spends on the operation, the less benefit it will generate. However, the contractor cannot solely control the cost because it is governed by the product specifications.

These different perceptions are derived from each stakeholder. It can be noticed that all stakeholders see the "good product" as part of the project (green concept). However, they perceive it in different ways. These assumptions are based on the personal construct theory by Kelly and the experience of the author in construction projects. As mentioned earlier, if one cognitive map represents an individual's view, a collection of cognitive maps (i.e., a causal map) would represent a group's view. Figure 2 provides a merged map from four cognitive maps shown in Figure 1.

https://doi.org/10.21872/2023IISE_3187
Proceedings of the IISE Annual Conference & Expo 2023
K. Babski-Reeves, B. Eksioglu, D. Hampton, eds.

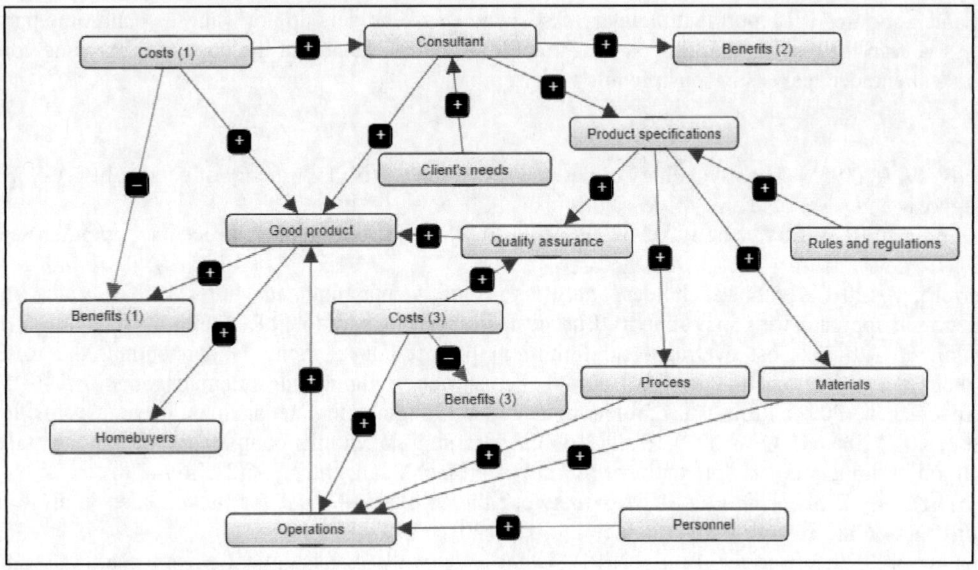

Figure 2: The stakeholders' merged map

Firstly, one of the most noticeable aspects that stands out in this merged map is the repetitive concepts – costs and benefits. As mentioned earlier, each stakeholder has different perceptions of the problem in front of them. Having a merged map will help clarify this ambiguity. As shown in Figure 2, it is now clear that "Cost 1" is different than "Cost 3," and "Benefit 1," "Benefit (2)" and "Benefit (3) are three different notions. This finding should act as a reference point for discussions or negotiations. For example, suppose that a project manager has to make a tough decision that could impact the benefit. Having a merged map will show which exact benefit would be affected by this tough decision.

Secondly, there is only one "good product" concept in the merged map which could be considered the ultimate goal of the project. As it is shown in Figure 1, a "good product" could be reached from different means because of different stakeholders' perceptions. This notion could help solve some of the problems mentioned earlier in this paper. Last but not least, the merged map (casual map) provides the big picture of a project. In other words, it creates a consensus among stakeholders. Regardless of the stakeholder role, one should have a general idea of what the project is about (the goal) and what the important factors are to achieve the goal. For example, without a causal map, a client might not know that there is a quality assurance component in the project at all. This notion might create some tensions between a client and a contractor if there are reports on increasing costs of quality assurance from the latter stakeholder.

Thirdly, the high-traffic concepts can be identified. High-traffic concepts are the ones that have a high number of interactions [16]. These concepts are usually critical in the problem of interest. From Figure 2, there are four high-traffic concepts. These are "Good product," "Operations," "Product Specifications," and "Consultant." These concepts have the potential to play vital roles in this project. All stakeholders should pay close attention to them to ensure the success of the project.

This study poses some limitations. The results are illustrative applications of the cognitive mapping concept during the stakeholder analysis process. The generalization of the results is limited to the subjective characteristics of a project. For example, the number of stakeholders, different specifications, different rules and regulations, or different project types. However, the results should apply to general construction projects.

6. Conclusions

This paper presents an illustrative application of the cognitive mapping concept in a stakeholder analysis in construction projects. The presented approach provides some insights to create a consensus among stakeholders. Also, the final product of the approach indicates some potential key concepts that could enhance the success rate of a project. Future research should address the limitations mentioned in the results section. This includes applying the cognitive mapping concept in a project with a different number of stakeholders, or different project types (e.g., transportation, other infrastructure). Also, stakeholders must be willing to participate in the cognitive map generation process.

https://doi.org/10.21872/2023IISE_3187

Villafranca and Loureiro [12] report that there are cases where stakeholders do not want to spend time participating in the process. Given the results from this paper, the authors are confident that the cognitive mapping concept is a useful tool in stakeholder analysis in construction projects.

References

[1] Kelly, G. A. (1955). The psychology of personal constructs. Vol. 1. A theory of personality. Vol. 2. Clinical diagnosis and psychotherapy. W. W. Norton.
[2] Hester, P. (2015). Analyzing stakeholders using fuzzy cognitive mapping. Procedia Computer Science, 61, 92-97.
[3] Ferretti, V. (2016). From stakeholders analysis to cognitive mapping and Multi-Attribute Value Theory: An integrated approach for policy support. European Journal of Operational Research, 253(2), 524-541.
[4] Doloi, H. (2013). Cost overruns and failure in project management: Understanding the roles of key stakeholders in construction projects. Journal of construction engineering and management, 139(3), 267-279.
[5] Karlsen, J. T. (2002). Project stakeholder management. Engineering management journal, 14(4), 19-24.
[6] Mok, K. Y., Shen, G. Q., & Yang, R. J. (2017). Addressing stakeholder complexity and major pitfalls in large cultural building projects. International Journal of Project Management, 35(3), 463-478.
[7] Mok, K. Y., & Shen, G. Q. (2016). A network-theory Based Model for Stakeholder Analysis in Major Construction Projects. Procedia Engineering, 164 (2016) 292 – 298.
[8] Thaviphoke, Y., & Hester, P. T. (2018). Chinese drywall: A Systemic Decision-making Approach. In Applications of Management Science (Vol. 19, pp. 119-153). Emerald Publishing Limited.
[9] Thaviphoke, Y. (2020). An investigation on the effectiveness of a problem structuring method in a group decision-making process (Doctor of Philosophy (PhD)). Old Dominion University, Norfolk, VA. Retrieved from https://digitalcommons.odu.edu/emse_etds/182
[10] Freeman, R. E. (1984). Strategic management: a stakeholder approach. Massachusetts: Pitman.
[11] Jergeas, G. F., Williamson, E., Skulmoski, G. J., & Thomas, J. L. (2000). Stakeholder management on construction projects. AACE international transactions, P12A.
[12] Mok, K. Y., Shen, G. Q., & Yang, J. (2015). Stakeholder management studies in mega construction projects: A review and future directions. International journal of project management, 33(2), 446-457.
[13] Schiller, C., Winters, M., Hanson, H. M., & Ashe, M. C. (2013). A framework for stakeholder identification in concept mapping and health research: a novel process and its application to older adult mobility and the built environment. BMC public health, 13(1), 1-9.
[14] Villafranca, B. C. L., & Loureiro, G. (2013). Stakeholder Analysis Process Using Cognitive Mapping. In Concurrent Engineering Approaches for Sustainable Product Development in a Multi-Disciplinary Environment: Proceedings of the 19th ISPE International Conference on Concurrent Engineering (pp. 1069-1080). Springer London.
[15] Kjærgaard, L., & Jensen, B. (2014). Using Cognitive Mapping to Represent and Share Users' Interpretations of Technology. Communications of the Association for Information Systems (Vol. 34) , Article 51, 1097-1114.
[16] Collins, A. J., Thaviphoke, Y., & Tako, A. A. (2022). Using Strategic Options Development and Analysis (SODA) to understand the simulation accessibility problem. Journal of the Operational Research Society, 1-22.
[17] Pidd, M. (2009). Tools for thinking: Modeling in management science. West Sussex, UK: John Wiley & Sons.
[18] Van Lamsweerde, A. (2009). Requirements engineering: From system goals to UML models to software (Vol. 10, p. 34). Chichester, UK: John Wiley & Sons.
[19] Stefan Olander (2007) Stakeholder impact analysis in construction project management, Construction Management and Economics, 25:3, 277-287, DOI: 10.1080/01446190600879125

https://doi.org/10.21872/2023IISE_3407

Proceedings of the IISE Annual Conference & Expo 2023
K. Babski-Reeves, B. Eksioglu, D. Hampton, eds.

The Effect of Different Optimization Strategies to Physics-Constrained Deep Learning for Soil Moisture Estimation

Jianxin Xie, The University of Tennessee Knoxville, Dr. Zheyu Jiang, Oklahoma State University, and Dr. Bing Yao, University of Tennessee Knoxville

Abstract ID: 3407

Abstract

Soil moisture is a key hydrological parameter that has significant importance to the human society and environment. Accurate modeling and monitoring of soil moisture in crop fields, especially in the root zone (top 100 cm of soil), is essential for improving agricultural production and crop yield with the help of precision irrigation and farming tools. Realizing the full sensor data potential depends greatly on advanced analytical and predictive domain-aware models. In this work, we propose a physics-constrained deep learning (P-DL) framework to integrate physics-based principles on water transport and water sensing signals for effective reconstruction of the soil moisture dynamics. We adopt three different optimizers, namely Adam, RMSprop, and GD, to minimize the loss function of P-DL during the training process. In the illustrative case study, we demonstrate the empirical convergence of Adam optimizers outperforms the other optimization methods in both mini-batch and full-batch training.

Keywords

Richards equation, soil moisture, physics-informed neural network, optimization

1. Introduction

Soil moisture is a vital hydrological state variable that has a significant impact on the global environment and human society. Detailed monitoring and modeling of soil moisture spatiotemporal dynamics are of crucial importance to numerous applications including freshwater allocation, weather forecasting, and natural disaster predictions (e.g., floods, landslides, and droughts).

The soil moisture dynamic is governed by a hydrological model in the form of a partial differential equation (PDE), called the Richardson-Richards equation (RRE) [1]. RRE encompasses Buckingham-Darcy law [2] that describes both saturated and unsaturated water flow in soil and continuity equation that helps to describe the water flux. RRE captures the nonlinear function relationship between three important soil moisture variables, i.e., the soil volumetric water content θ, the pressure head ψ , and hydrologic water conductivity K. The characteristic relations between ψ with θ and ψ with K can be characterized by the water retention curve and hydraulic conductivity function, respectively, to delineate the characteristics of water and solute movement in soils. Many parametric models that combine both careful experimental work and perceptive theoretical insights have been developed to describe these two soil hydraulic relations (also known as constitutive relationships).

In order to quantitatively simulate and visualize the soil moisture dynamics, scientists have developed mesh-based algorithms such as the finite element method, to numerically solve the Richardson-Richards equation. However, mesh-based methods involve the discretization of both the spatial and temporal domain of the soil moisture evolvement process. The computation complexity to solve the RRE at each time step is proportional to the number of discretized nodes within the targeted soil field, resulting in expensive computational costs for detailed modeling. More importantly, real-world sensor measurements cannot be readily assimilated into the FEM numerical procedure, leading to inferior applicability of mesh-based simulation in real-world practice.

It is worth noting that the movement of water and their interrelationships in the soil has already been well summarized in the RRE and various parametric constitutive models. To take full advantage of both physics knowledge and empirical sensor observation, Raissi et al. [3] built a physics-informed neural network (PINN) framework that integrates the well-established physics laws with deep learning to suppress the model dependence on training data. The efficacy of the PINN has already been verified in numerous physical systems, such as liquid flow simulation, elastodynamic problems, non-linear structural systems, cardiovascular flow modeling, and cardiac electrodynamics simulation [4].

https://doi.org/10.21872/2023IISE_3407

Previously, researchers have investigated the application of PINNs to model the soil moisture dynamics by engaging RRE. Notably, Banbai et al. [5] embedded RRE into PINN to inversely learn the soil moisture dynamics only from volumetric water content observations without engaging any pre-assumptions on soil hydraulic functions and realize a free-form representation of constitutive relationships.

Here, we propose a physics-constrained deep learning (P-DL) based on the PINN framework that accommodates the sensor measurements of the directly observed independent variable, i.e., the water pressure head ψ, in a three-dimensinal (3D) soil geometry. The van Genuchten model is engaged to provide the explicit nonlinear relations between the pressure head and the other variable in interest, which can also further facilitate the physics embodiment as the model constraint in the PINN. Moreover, the performance of the predictive modeling depends to a great extent on the proper selection of the optimization technique. In this study, we further investigate the effect of the three most commonly-used optimizers, i.e., Adaptive Moment Estimation (Adam), Root Mean Square Propagation (RMSprop), and Gradient Descent (GD) for both mini-batch and full-batch training in minimizing the loss function of P-DL that is designed to not only satisfactorily match the pressure head sensor measurements but also respect the RRE with the constitutive relations.

2. Research Methodology

2.1 Richardson-Richards Equation

In this paper, we engage non-linear RRE to describe the flow of water in 3D unsaturated homogeneous rigid soil and ignore the sink term [1]: $\frac{\partial\theta(\psi)}{\partial t} = -\nabla \cdot q$, where θ is the soil volumetric water content, ψ is pressure head, t denotes time, q represents the soil water flux density. This relation is also known as the continuity equation with respect to the mass balance of the soil water. The Buckingham-Darcy law [2] defines the relationship between q and ψ, which describes both saturated and unsaturated water flow in the soil: $q = -K(\psi)\cdot\nabla(\psi+z)$, where K is the hydraulic conductivity. The dynamics of soil water can be summarized in RRE using preceding relations:

$$\frac{\partial\theta(\psi)}{\partial t} = \nabla\cdot(K(\psi)\nabla(\psi+z)) \tag{1}$$

It is worth noting that the pressure head ψ is a primal variable that is dependent on t and spatial instances $s = [x,y,z]$. Thus, the left-hand side of Eq. (1) can be reformulated as $\frac{\partial\theta}{\partial\psi}\frac{\partial\psi}{\partial t}$ based on the chain rule. The function relationship of $\theta(\psi)$ and $K(\psi)$ are usually referred as water retention curves (WRCs) and hydraulic conductivity functions (HCFs), respectively, which are commonly specified by parametric models. Here, without loss of generality, we choose the commonly-used van Genuchten model [6] to characterize WRCs and HCFs:

$$\begin{aligned}\theta(\psi) &= \frac{\theta_s-\theta_r}{\left[1+(\alpha|\psi|)^n\right]^m}+\theta_r\\ K(\psi) &= K_s\frac{\left\{1-(\alpha|\psi|)^{n-1}[1+(\alpha|\psi|)^n]^{-m}\right\}^2}{[1+(\alpha|\psi|)^n]^{m/2}}\end{aligned} \tag{2}$$

where K_s, θ_s, θ_r are the saturated hydraulic conductivity, saturated volumetric moisture content, and residual moisture content, respectively. Parameters n, m, and α stand for curve-fitting soil hydraulic properties, where $m = 1-1/n$. These soil hydraulic parameters determine the soil properties of the field, the values of which are taken from [7].

2.2 Physics-constrained deep learning (P-DL)

The soil moisture dynamics that are characterized by the WRC and HCF majorly depend on the accurate modeling of $\psi(s,t)$. We engage a feedforward fully-connected DNN that is trained not only to satisfy the sensor measurement of pressure head ψ_m, but also to respect the underlying physics principles (RRE) to approximate the nonlinear relationships between the input spatiotemporal instances (s,t) and the decision variable, i.e., pressure head ψ. We model the spatiotemporal soil water pressure head distribution as: $[s,t] \xrightarrow{\mathcal{N}(s,t;\Theta_{NN})} \hat{\psi}(s,t)$, where $\mathcal{N}(s,t;\Theta_{NN})$ denotes the DNN model, Θ_{NN} stands for the neural network parameters. The DNN is constructed with an input layer composed of spatiotemporal coordinates $[s,t]$, multiple hidden layers to approximate nonlinear functional relationships, and one output layer for the prediction of $\hat{\psi}(s,t,\Theta_{NN})$. The RRE-based physics principles are further embedded into the DNN along with the sensor data constraint through a unique loss function defined as:

$$\mathcal{L}(\Theta_{NN}) = \mathcal{L}_D + \mathcal{L}_{RRE} \tag{3}$$

The total loss $\mathcal{L}(\Theta_{NN})$ consists of the following two key components:

1) **Data-driven loss** $\mathcal{L}_D$: The pressure head signals can be recorded at multiple locations on the horizontal plane (xy-plane) at different fixed depths using the soil moisture sensors. In this work, the pressure head is sampled for 30

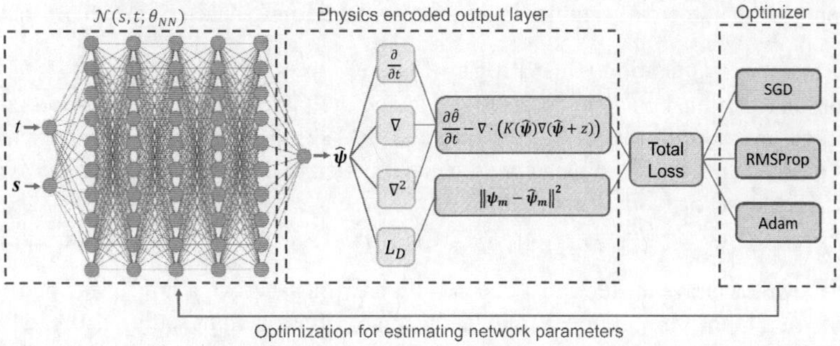

Figure 1: Illustration of the proposed P-DL framework for soil moisture prediction.

discrete time instances at 5 equally spaced depths in the vertical direction. As such, each sensor fetches a time series signal of pressure heads at a specific location in the soil geometry, denoted as $\boldsymbol{\psi}_m(s,t)$. The DNN needs to be trained to generate $\hat{\psi}$ that match the sensor measurements closely. Hence, the data-driven loss $\mathcal{L}_D$, enforcing agreement between the measured and predicted pressure head signals, is defined as:

$$\mathcal{L}_D = \frac{1}{N_m}\sum_{i=1}^{N_m}(\psi_m(s_i,t_i) - \hat{\psi}_m(s_i,t_i))^2 \tag{4}$$

where N_m denotes the number of spatiotemporal measurements.

(2) **RRE-based loss** $\mathcal{L}_{RRE}$: To further enhance the estimation accuracy and the model robustness, physics-based regularization is imposed over the spatiotemporal collocation points, $[s_i,t_i]$'s, which are randomly chosen from the spatiotemporal domain of the hydraulic process in the targe soil field. The physics constraint enforces the hydraulic model presented in RRE (see Eq. (1)) by encouraging the physics-based residuals to be close to zero. Specifically, RRE-based residuals are defined as:

$$r_\psi(s,t,\Theta_{NN}) := \frac{\partial\theta}{\partial t} - \nabla\cdot(K(\hat{\psi})\nabla(\hat{\psi}+z)) \tag{5}$$

It is worth noting that the gradient and partial derivative in Eq.(5) can be reformulated in terms of ψ and its derivatives:

$$r_\psi(s,t,\Theta_{NN}) := \frac{\partial\theta}{\partial\psi}\frac{\partial\psi}{\partial t} - \left(\frac{\partial K}{\partial\psi}\frac{\partial\psi}{\partial x}\frac{\partial\psi}{\partial x} + K\frac{\partial^2\psi}{\partial x^2} + \frac{\partial K}{\partial\psi}\frac{\partial\psi}{\partial y}\frac{\partial\psi}{\partial y} + K\frac{\partial^2\psi}{\partial y^2} + \frac{\partial K}{\partial\psi}\frac{\partial\psi}{\partial z}\left(\frac{\partial\psi}{\partial z}+1\right) + K\frac{\partial^2\psi}{\partial z^2}\right) \tag{6}$$

The gradient $\frac{\partial\theta}{\partial\psi}$ and $\frac{\partial K}{\partial\psi}$ can be directly computed according to the explicit function relationships in Eq. 2. The first and second derivatives of ψ can be easily computed using automatic differentiation, which is a widely used technique in deep learning and is more efficient, accurate, and reliable than numerical differentiation used in FEM-based modeling [8]. The physics-based constraint will then be realized by encouraging the values of $r_\psi(s,t;\Theta_{NN})$ to be close to zero. Thus, the RRE-based loss is defined as:

$$\mathcal{L}_{RRE} = \frac{1}{N_f}\sum_{i=1}^{N_f}(r_\psi(s_i,t_i;\Theta_{NN}))^2 \tag{7}$$

where N_f represents the total number of selected spatiotemporal collocation points in the hydrological process to encode the RRE into the DNN. The physics-based loss is further embedded into the overall loss function in Eq. (3) to respect both the data-driven loss and the underlying soil moisture system physics for the reliable modeling of the spatiotemporal soil moisture dynamics.

2.3 Optimization Techniques

2.3.1 Mini-batch/batch Gradient Descent (GD)

Mini-batch gradient descent can be viewed as a variation of the gradient descent algorithm that involves stochasticity. Batch GD uses the entire training dataset to update the neural network parameters at each iteration. In contrast, mini-batch GD reduces the loss by replacing the actual gradient that is calculated from the whole data set with an estimated counterpart computed from randomly selected subsets of the data, which allows faster updates for one iteration. Mini-batch GD can help to escape from the saddle point in virtue of the introduced stochasticity. However, due to the inherent variance, the path taken by the algorithm to reach the minimum is usually noisier than the typical gradient descent and with a slow convergence asymptotically. The updating rule is given as follow:

$$\Theta_t = \Theta_{t-1} - \eta\frac{1}{B}\sum_{i=1}^{B}\nabla_\Theta\mathcal{L}(s_i,t_i,\psi_i;\Theta) \tag{8}$$

where η is the learning rate, $\{s_i,t_i;\psi_i\}$ is one spatiotemporal coordinate and pressure head sensor measurement pair. For classic batch GD, B stands for the total number of the training datasets, i.e., $B = |\psi_m|$; whereas for mini-batch stochastic GD, B denotes the number of samples in a subset of the training datasets, i.e., $B > 1$ but $B < |\psi_m|$.

2.3.2 Root Mean Square Propagation (RMSProp)

One of the limitations of traditional gradient descent is engaging the same step size for each neural network parameter throughout the training process. To overcome this shortage, RMSProp is proposed as an extension to the gradient descent that allows the step size in each dimension to be automatically adapted based on a decaying moving average of partial gradients. The neural network is updated by:

$$E[g^2]_t = \beta E[g^2]_{t-1} + (1-\beta)\left(\delta\mathcal{L}/\delta\Theta\right)^2, \quad \Theta_t = \Theta_{t-1} - (\eta/\sqrt{E\left[g^2\right]_t})\delta\mathcal{L}/\delta\Theta \tag{9}$$

where $E[g^2]_t$ is the moving average at step t that depends on previous average $E[g^2]_{t-1}$ and current gradient $\frac{\delta\mathcal{L}}{\delta\Theta}$. β is a weight of past time step to the current update moving average, whose default setting is 0.9. Then the adaptive step size is calculated by dividing the step size η with an exponentially decaying average of squared gradients. The use of a decaying moving average allows the algorithm to discard early gradients and focus on the most recently observed partial gradients information during the progress of the search.

2.3.3 Adaptive Moment Estimation (Adam)

Adam is another method that computes adaptive learning rates for each parameter in the neural network [9]. In addition to the automatic learning rate adaption for each input variable by storing an exponentially decaying average of past squared gradients like RMSProp, Adam further smooths the search process by introducing an exponentially decaying moving average of past gradients m_t, which shares the similar idea as momentum that can help accelerate mini-batch GD in the relevant direction and dampens oscillations by also counting a fraction of the past time step gradients to the current update. The decaying average of the gradient history m_t and its squared version v_t at step $t$are computed as:

$$m_t = \beta_1 m_{t-1} + (1-\beta_1)g_t, \quad v_t = \beta_2 v_{t-1} + (1-\beta_2)g_t^2 \tag{10}$$

where β_1 and β_2 are the hyperparameters whose value is usually chosen as 0.9 for β_1, 0.999 for β_2. [9] stated that m_t and v_t are biased towards zero during the first few steps, especially when the decay rates β_1 and β_2 are small. They figured to cancel out these biases by computing bias-corrected versions of m_t and v_t:

$$\hat{m}_t = \frac{m_t}{1-\beta_1^t}, \quad \hat{v}_t = \frac{v_t}{1-\beta_2^t} \tag{11}$$

Then the neural network updating rule is formulated as:

$$\Theta_{t+1} = \Theta_t - \frac{\eta}{\sqrt{\hat{v}_t}+\varepsilon}\hat{m}_t \tag{12}$$

3. Experimental Results

We evaluate the performance of different gradient descent methods, i.e., mini-batch/batch GD, RMSProp, and Adam, to the proposed P-DL framework in a 3D soil moisture system to estimate the pressure head from sparse sensor measurement. The neural network is carried on TensorFlow-GPU with Python application programming interface (API). Note that the CPU used for the computation is Intel(R) Xeon(R) W-2265 CPU @ 3.50GHz. The GPU is NVIDIA RTX A4500. The soil geometry is a cubic shape formed by 20 nodes in x and y direction, and 10 nodes in z direction. The geometry information and the benchmark system dynamics are obtained from [7]. Note that the sensor measurement noise is inevitable in real-world practice. Thus, in this work, we add a noise level $\sigma_\varepsilon = 0.005$ to the simulation data. In other words, sensor observation is denoted as $\psi_m(s,t) = \psi(s,t) + \varepsilon(s,t)$, where the noise follows a Gaussian distribution as $\varepsilon(s,t) \sim \mathcal{N}(0, \sigma_\varepsilon^2)$. The soil moisture pressure head is measured at 15 randomly selected locations on horizontal xy-plane, and for each location, sensors are evenly distributed at 5 different depths. Thus, 75 time series data of pressure heads are collected and served as sensor observation, i.e., ψ_m, in the loss function $\mathcal{L}$ to train the P-DL. The temporal domain for 0.9 hours is uniformly discretized in 30 temporal instances. The modeling result will be quantified on the whole spatiotemporal domain of the soil using Relative Error (re), which is defined as:

$$re_\psi = \frac{\|\hat{\psi}(s,t) - \psi(s,t)\|_2}{\|\psi(s,t)\|_2} \qquad re_\theta = \frac{\|\hat{\theta}(s,t) - \theta(s,t)\|_2}{\|\theta(s,t)\|_2} \tag{13}$$

where $\hat{\psi}, \hat{\theta}$ and ψ, θ denote the predicted and true pressure head dynamics and water content, respectively. In the present investigation, a feedforward fully-connected DNN is constructed by incorporating the physics law in the P-DL approach to model the soil moisture dynamics. The DNN consists of five hidden layers with ten neurons in each layer. We randomly choose collocation points $N_f = 10{,}000$ among 120000 spatiotemporal instances from the soil moisture dynamic domain to encode RRE into the DNN. The neural network is carried on TensorFlow-GPU with Python application programming interface (API), and we found that the computation speed is around 6.5 times faster than running on CPU only. Note that the CPU used for the computation is Intel(R) Xeon(R) W-2223CPU @ 3.6GHz. The GPU is NVIDIA Quadro P2200

https://doi.org/10.21872/2023IISE_3407

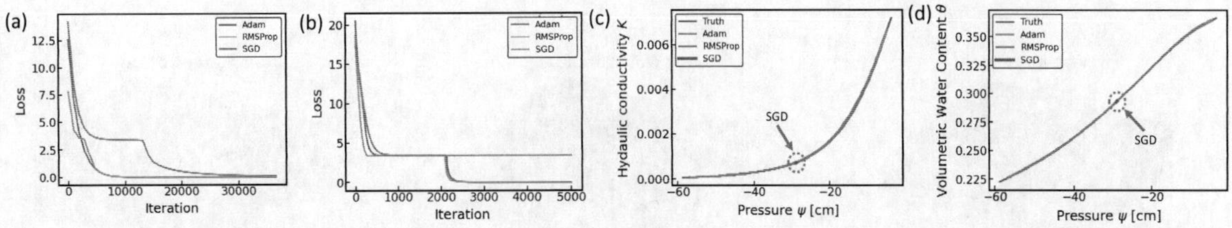

Figure 2: The comparison of training loss $\mathcal{L}$ of P-DL model using mini-batch (a) and full batch (b) with different optimization methods. (c) Hydraulic conductivity function (HCF), (d) Water retention curves (WRC) generated from the P-DL model and van Genuchten model.

Table 1: Comparison of *re* of ψ and θ using different optimization algorithms with mini-batch and full batch.

		GD	RMSProp	Adam
ψ	Mini-batch	0.0649	0.0106	0.0079
	Full batch	0.4251	0.0324	0.0049
θ	Mini-batch	0.0140	0.0029	0.0022
	Full batch	0.1319	0.0090	0.0009

3.1 Convergence and performance observations

In this subsection, we compare the convergence performance for three optimization methods including GD, RMSProp, and Adam with mini-batch and full batch in optimizing the proposed P-DL model. The batch size for mini-batch training is 128. Fig. 2 (a) and (b) demonstrates the training convergence of three optimization techniques using mini-batch and full batch, respectively. When mini-batch is engaged in P-DL training (Fig. 2 (a)), GD presents the slowest convergence rate. RMSProp demonstrates the fastest loss dropping rate at beginging and illustrates a similar convergence speed as Adam after 5000 iterations. Adam demonstrates the best convergence performance, whose loss is droped to 0.0026 when the training ends.

When the P-DL model is trained with full batch shown in Fig. 2 (b), for the first 1000 epochs, GD demonstrates the smallest convergence rate and RMSProp provides the fastest loss-reducing speed among three optimization strategies. As the neural network training continues, RMSProp and Adam reach the local minimum and manage to escape from it successively. After 5000 epochs, the losses of the P-DL model optimized by RMSProp and Adam are reduced to 0.0211 and 0.0010, respectively. In contrast, for full batch training, the GD is stuck at a local minimum and fails to further optimize the neural network. Table 1 reveals the relative error (*re*) of pressure head ψ and volumetric content θ generated from the P-DL model optimized by GD, RMSProp, and Adam using mini-batch and full batch.

As shown in Fig. 2(a-b), full-batch training can lead to convergence with a smaller number of times for backpropagation, i.e., iterations compared with mini-batch. Mini-batch training requires about 10000 iterations for RMSProp and Adam to converge, and 32000 iterations for GD to converge. In contrast, Adam and RMSProp only take less than 3000 iterations to converge for full-batch training. If the Adam optimizer is engaged, finishing one backpropagation using mini-batch requires approximate 94 milliseconds (ms), whereas the full batch training needs about 124 ms to update for 1 iteration. Thus, mini-batch with batch size 128 requires 16 min to reach convergence, whereas the full batch only demand about 6.4 min to converge. Furthermore, the predictive dynamics generated from the full batch also outperform that from the mini-batch if Adam is engaged to optimize the P-DL, as shown in Table 1. Therefore, in the paper, Adam optimizer with full batch training is recommended for optimizing the soil moisture P-DL model.

3.2 P-DL performance using full batch training

In the current investigation, we compare the soil moisture dynamics predicted by the P-DL model optimized with GD, RMSProp, and Adam only using full batch. The prediction of volumetric water content $\hat{\theta}$ and the hydraulic conductivity $\hat{K}$ are inferred by plugging the neural network prediction, i.e., the pressure head ψ, into the van Genuchten model detailed in Eq. (2). Fig. 2(c,d) shows the constitutive relationships, i.e., HCF and WRC, estimated by the P-DL model using different optimization methods. The P-DL model optimized by GD generates a tiny range prediction of $\hat{\psi}$, which makes both GD-based estimations of WRC and HCF localized in small magnitude and results in an unsuccessful prediction. Among these three optimization strategies, Adam generates the most similar WRC and HCF curve pattern with the ground truth, which indicates the robust estimation of pressure head $\hat{\psi}$.

Fig. 3 (a) and (b) show the ground truth water content distribution and its predictive counterpart generated from the P-DL model with full-batch Adam optimizer in the soil geometry, respectively. Because the soil moisture dynamic is changing over time, Fig. 3 only depicts the mappings at one specific time step, i.e., $t = 15$. The estimated water

https://doi.org/10.21872/2023IISE_3407

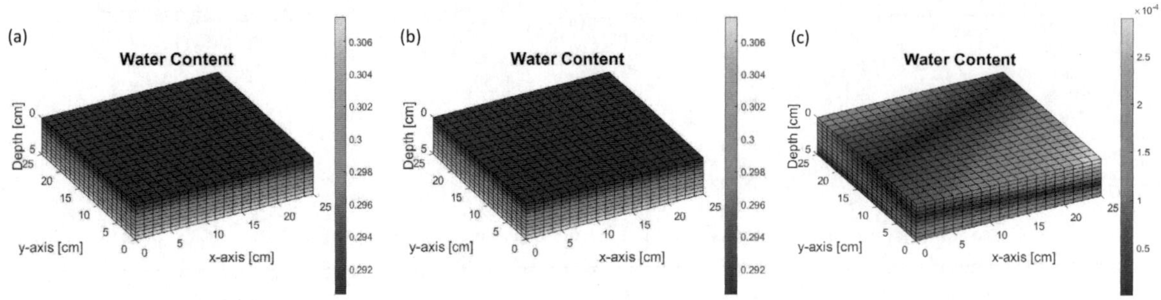

Figure 3: (a)The benchmark water content distribution θ, (b) the predicted $\hat{\theta}$ obtained from P-DL trained by full-batch Adam optimizer, (c) the discrepency mapping between the ground truth and the prediction.

potential pattern is remarkably similar to ground truth mapping, indicating that, the P-DL model, collaborated with proper optimization strategy, is able to effectively predict the spatiotemporal soil moisture dynamics by harnessing the physics-based principles and sensor observation. Fig. 3 (c) displays the mapping of the absolute difference between the prediction and ground truth, which keeps a low value with the magnitude of 10^{-4}.

4. Conclusions

We propose a physics-constrained deep learning (P-DL) framework to solve the Richardson-Richards equation, reconstruct the soil moisture dynamics, and recover the WRCs and HCFs. A feed-forward neural network is engaged to simulate the non-linear relationship between the spatiotemporal instances and the pressure head given the sensor measurements. The DNN is trained not only to satisfactorily match the sensor measurements but also to respect the Richardson-Richards equation with the emprical correlations. Since successful and efficient optimization is of vital importance to obtain the optimal solution to the Richardson-Richards equation, this project aims to investigate different optimizing algorithms for training the neural network. Three different optimizers, Adam, RMSprop, and GD, are engaged to minimize the loss function of P-DL. We compare their performance in minimizing the loss function of the P-DL model. The experimental result shows that the predictive model optimized with Adam using full batch demonstrates the best performance compared to other optimization strategies.

References

[1] L. A. Richards, "Capillary conduction of liquids through porous mediums," *Physics*, vol. 1, no. 5, pp. 318–333, 1931.

[2] E. Buckingham, "Studies on the movement of soil moisture," 1907.

[3] M. Raissi, P. Perdikaris, and G. E. Karniadakis, "Physics-informed neural networks: A deep learning framework for solving forward and inverse problems involving nonlinear partial differential equations," *Journal of Computational physics*, vol. 378, pp. 686–707, 2019.

[4] J. Xie and B. Yao, "Physics-constrained deep learning for robust inverse ecg modeling," *IEEE Transactions on Automation Science and Engineering*, 2022.

[5] T. Bandai and T. A. Ghezzehei, "Physics-informed neural networks with monotonicity constraints for richardson-richards equation: Estimation of constitutive relationships and soil water flux density from volumetric water content measurements," *Water Resources Research*, vol. 57, no. 2, p. e2020WR027642, 2021.

[6] M. T. Van Genuchten, "A closed-form equation for predicting the hydraulic conductivity of unsaturated soils," *Soil science society of America journal*, vol. 44, no. 5, pp. 892–898, 1980.

[7] K.-A. Lie, *An introduction to reservoir simulation using MATLAB/GNU Octave: User guide for the MATLAB Reservoir Simulation Toolbox (MRST)*. Cambridge University Press, 2019.

[8] A. Paszke, S. Gross, S. Chintala, G. Chanan, E. Yang, Z. DeVito, Z. Lin, A. Desmaison, L. Antiga, and A. Lerer, "Automatic differentiation in pytorch," 2017.

[9] D. P. Kingma and J. Ba, "Adam: A method for stochastic optimization," *arXiv preprint arXiv:1412.6980*, 2014.

https://doi.org/10.21872/2023IISE_3426

Proceedings of the IISE Annual Conference & Expo 2023
K. Babski-Reeves, B. Eksioglu, D. Hampton, eds.

Logistics Hub Capacity Deployment in Hyperconnected Transportation Network Under Uncertainty

Abstract ID: 3426

Xiaoyue Liu, Jingze Li, Benoit Montreuil
Physical Internet Center, H. Milton Stewart School of Industrial & Systems Engineering,
Georgia Institute of Technology, Atlanta, GA 30332

Abstract

Modern logistics systems worldwide are facing unprecedented challenges due to the explosive growth of e-commerce, driving the need for resilient systems to tackle problems such as vulnerable supplies, volatile demands, and fragile transportation networks. Motivated by the innovative concept of the Physical Internet, this paper focuses on resilient capacity deployment of open-access logistics hubs in hyperconnected transportation under demand uncertainty and geographical disruptions. We propose a two-stage stochastic optimization model, aiming to smartly deploy the hub capacity to achieve delivery timeliness, high consolidation and network resilience while minimizing hub set-up budget and truck fleet cost. Four optimal hub network configurations are derived by applying scenarios at four stress testing levels into the optimization model, including deterministic demands without hub disruptions, deterministic demands with hub disruptions, stochastic demands without hub disruptions as well as stochastic demands with hub disruptions. To test the performances of different optimal networks, a simulation-based study is then performed over an automotive delivery-to-dealer network and dataset in the Southeast US region. Our results demonstrate the key impacts of various uncertainties on hub capacity deployment in terms of capacity configuration distribution, network resilience, delivery timeliness, and cost-effectiveness. Overall, this study provides a reliable network capacity deployment approach with persistent and sustainable economic and social performances in hyperconnected networks, and the results validate the relationship between capacity deployment and network resilience under different types of uncertainties.

Keywords

Hyperconnected Transportation, Resilient Capacity Deployment, Transportation Hub Network Design, Two-stage Stochastic Optimization Programming, Demand Uncertainty, Network Disruptions, Physical Internet, Stress Testing

1. Introduction

The rapid growth of e-commerce brings challenges to traditional transportation networks, as it boosts the demands for logistics services and introduces complexities in policy and society management. Those issues aggravate temporal and spatial logistic uncertainties as shown in Fig.1. Uncertainties such as unpredictable demand fluctuations and geographical disruptions lead to shipment delays, traffic congestion, economic losses, and resource wastage, further exacerbating the vulnerability of transportation networks. As a result, effective adaptation of transportation networks to cope with uncertainty has risen in importance. Transportation hub networks are strategically designed systems comprised of logistics hubs that serve as focal points for the movement of freight and/or passengers. Their high freight consolidation and flexible origin-destination (OD) connectivity can improve delivery efficiency and reducing transport costs beyond traditional end-to-end transportation [1]. We here focus on enhancing how they cope with uncertainty.

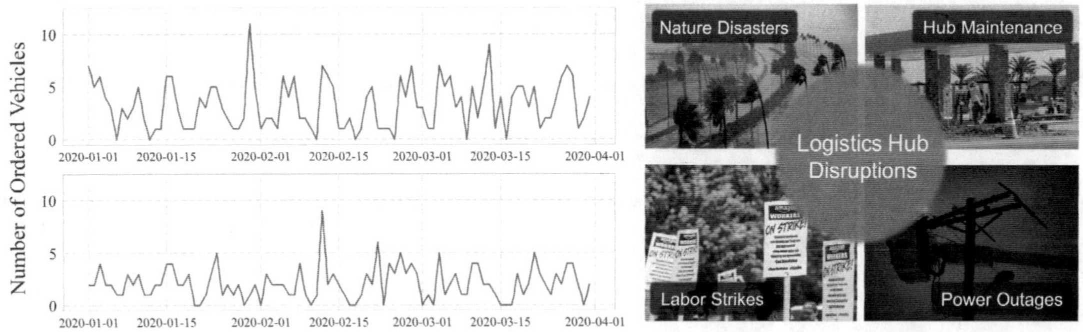

(a) (b)

Figure 1: Example (a) shows fluctuation of demands and (b) displays some primary causes of hub disruptions

https://doi.org/10.21872/2023IISE_3426

Proceedings of the IISE Annual Conference & Expo 2023
K. Babski-Reeves, B. Eksioglu, D. Hampton, eds.

Given an existing network and the potential locations of open-access logistics hubs, the scope of this paper is to strategically allocate the capacity of each hub to enhance the resilience of transportation networks and better serve the logistic needs under uncertainty. This paper primarily delves into the throughput capacity of logistics hubs, with a particular emphasis on crucial factors such as the quantity of open loading docks, the availability of outbound drivers, the equipment in use and so forth. Unlike physical capacity (determined by hub size, layout, and infrastructure), this capacity is flexible and can be adjusted through planning. We further concentrate on the hyperconnected relay transportation network, where long-haul freight shipments are to be shipped via consecutive short-haul route segments through open-access logistics hubs in a relay way. In the paper, we present a two-stage stochastic optimization model to address spatial-temporal uncertainty, which can be efficiently solved using the Sample Average Approximation (SAA) method. To evaluate the performance of our model and analyze the effects of uncertainty types on logistics hub capacity allocation, we test our optimal hub capacity deployment solutions through a large-scale automotive vehicle delivery simulation study under different types of uncertainty scenarios.

The rest of this paper is organized as follows. Section 2 illustratively describes the problem setting and states research assumptions. In Section 3, we summarize literature reviews of the most relevant papers. Section 4 formulates the capacity deployment optimization model based on two-stage stochastic programming. Section 5 presents a simulation-based study and analyzes experimental results. Section 6 concludes our findings and discusses future research avenues.

2. Problem Description

This paper focuses on the resilient capacity deployment problem of open-access logistics hubs for long-haul freight, which require to be transferred by short-haul drivers through multiple hubs in hyperconnected transportation networks. In our general problem setting, given the demand orders, freights are to be shipped from their sources (i.e., origins) to dealer clusters (i.e., destinations) via hubs. In the specific context of vehicle delivery, freight is a car, a utilitarian vehicle, etc.; destination is a dealer or a compact cluster of dealers; source is a plant, port, or railhead used by multiple Original Equipment Manufacturers (OEMs); and open-access hubs in hyperconnected networks enable demand flow through functions such as consolidation, sorting and transfer, providing enhanced hub accessibility and connectivity.

In the simulation study, we use an automotive finished vehicle network in the southeastern USA. Fig. 2 (a) depicts the network nodes consisting of 24 open-access logistics hubs, 13 sources and 50 dealer clusters. Fig. 2 (b) depicts the hyperconnected transportation network with its edges designed to keep the travel time along each edge to less than 5.5 hours, allowing for a short-haul driver round trip to be completed within 11 hours, respecting driving limits imposed by US driving regulations. Demand flow is across the southeast region as shown in Fig. 2 (c) and fluctuates daily according to customer needs. We consider two types of uncertainties: demand uncertainty and disruption uncertainty. Disruptions randomly occur at hubs and cause significant delays in travel time on the arcs connecting to the affected hubs. Given these settings, we aim to determine hub capacity in this hyperconnected transportation network to serve logistics demands while minimizing hub costs and fleet costs under various uncertainty scenarios.

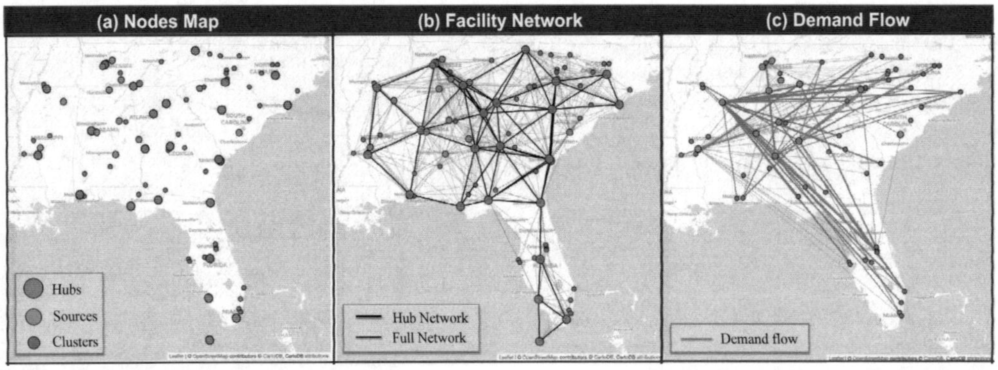

Figure 2: Nodal map (a), facility network (b) and demand flow (c) for multiple automotive OEMs in Southeast USA

3. Related Research

Much work has been done on exploring network configurations for logistic and transportation systems. The earliest and simplest configuration is the direct node-to-node linkages from origins to destinations. However, its discrete commodity routing, handling, and sorting leads to massive network operating costs. To mitigate this problem, the hub-

https://doi.org/10.21872/2023IISE_3426

and-spoke configuration has been introduced to leverage scale economies by consolidating freight flows [2]. The hub-and-spoke network is widely adopted in transportation, and numerous studies have been developed based on this configuration. However, its centralized structure causes inefficient detours, extended delivery time, and potential network crashes from hub disruptions. None of these issues could be solved intrinsically until the hyperconnected network configuration was initially proposed as a key Physical Internet concept [3]. In contrast to hub-and-spoke networks, hyperconnected networks are multi-plane interconnected meshed hub networks that can smoothly support dynamic freight routing and hub-based consolidation and strengthen connectivity between logistics hubs [4]. By taking advantage of modularity, containerization, and hyperconnectivity, these networks can offer faster and more reliable deliveries from origins to destinations, thereby improving efficiency, service capability, resilience, and sustainability.

Various studies have been conducted to enable and leverage resilient network design. [5] develops a bi-level, three-stage stochastic optimization model to tackle network design for disaster management including pre-event mitigation, preparedness, and post-event response. In [6], a trilevel optimization model is proposed for the resilient network design problem to minimize the total travel time under uncertain disruptions. When considering the resilience of the Physical Internet, its hyperconnectivity tends to provide more efficient and resilient networks compared to traditional networks [7]. Thus, studying the full potential of the Physical Internet to enhance network resilience is a valuable research area.

In the field of capacity deployment, numerous papers have tackled the capacity facility location problem (CFLP) for hub networks. [8,9] formulate the hub location problem under capacity constraints and optimize the size of hubs by determining the discrete capacity levels for each hub. In [10], a dynamic deployment optimization model is proposed to address the modular smart lockers relocation problem in hyperconnected logistics systems, taking into account uncertain events. The model comprises a two-stage stochastic optimization approach, and to efficiently solve large-scale instances of this problem, Bender's decomposition, combined with an acceleration method, is developed in [10].

In this paper, we aim to fill the gaps of the resilient capacity hub network design in hyperconnected transportation by optimizing the capacity deployment of logistics hub to enhance network resilience. Compared to the present state of research, we make the following contributions. First, our study explores the potential of hyperconnected network resilience through a concrete methodology and simulation analysis. Second, we validate that the deployment of logistics hub capacities has a significant impact on the transportation network resilience, opening avenues for future research on resilient capacity deployment. Third, the hub networks resulting from our work can become inputs for other studies based on the Physical Internet, such as operating system design [11] and simulation modeling [12].

Table 1: Notations of sets, parameters, and decision variables in the two-stage stochastic model

	Sets		**Parameters**
$\boldsymbol{O}$	set of origins, indexed by o	$q_{od}(\omega)$	demands from origin o to destination d
$\boldsymbol{D}$	set of destinations, indexed by d	$t(a,\omega)$	travel time on arc a in scenario ω
$\boldsymbol{H}$	set of logistics hub, indexed by h	s_h	hub operational cost per capacity h
$\boldsymbol{V}$	set of nonempty demand OD pairs	f_h	fixed cost of setting up a new hub h
$\boldsymbol{A}$	set of arcs, indexed by a	v_a	fleet cost per hour on arc a
$\boldsymbol{W}$	set of scenarios, indexed by ω	m	truckload
$\boldsymbol{\delta^+(i)}$	set of arcs come out from node i	b	maximal capacity based on the budget limit
$\boldsymbol{\delta^-(i)}$	set of arcs come into node i		
	First-stage Decision variables		**Second-stage Decision variables**
X_h	binary variable, whether to activate hub h or not	$F_o^d(a,\omega)$	amount of freight from origin o to destination d on arc a in scenario ω
C_h	capacity to deploy at logistics hub h	$Tr(a,\omega)$	number of trucks on arc a in scenario ω

4. Methodology

We propose a two-stage stochastic optimization model that aims to minimize the hub capacity costs and the fleet costs over uncertainty scenarios. In the framework of the two-stage stochastic programming, the first stage is to make hub capacity decisions before freight routing schedules are established, while the second stage is to evaluate the total fleet cost due to a realized network configuration, considering the impact of network disruptions and delay penalties on the chosen arcs. The model notations are displayed in Table 1 and its objective function and constraints are as follows:

$$\text{Minimize} \quad \textstyle\sum_{h\in \boldsymbol{H}} (f_h \cdot X_h + s_h \cdot C_h) + E_\omega[\sum_{a\in \boldsymbol{A}} v_a \cdot t(a,\omega) \cdot Tr(a,\omega)] \qquad (1)$$

$$\text{subject to} \quad \textstyle\sum_{a\in\delta^+(o)} F_o^d(a,\omega) = q_{od}(\omega) \qquad \forall (o,d) \in \boldsymbol{V}, \forall \omega \in \boldsymbol{W} \quad (2)$$

https://doi.org/10.21872/2023IISE_3426

$$\sum_{a\in\delta^-(d)} F_o^d(a,\omega) = q_{od}(\omega) \qquad \forall(o,d)\in \boldsymbol{V}, \forall\omega\in \boldsymbol{W} \quad (3)$$

$$\sum_{a\in\delta^+(h)} F_o^d(a,\omega) = \sum_{a\in\delta^-(h)} F_o^d(a,\omega) \qquad \forall h\in \boldsymbol{H}, \forall(o,d)\in \boldsymbol{V}, \forall\omega\in \boldsymbol{W} \quad (4)$$

$$\sum_{(o,d)\in V} F_o^d(a,\omega) \le m\cdot Tr(a,\omega) \qquad \forall a\in \boldsymbol{A}, \forall\omega\in \boldsymbol{W} \quad (5)$$

$$\sum_{(o,d)\in V}\sum_{a\in\delta^-(h)} F_o^d(a,\omega) \le C_h \qquad \forall h\in \boldsymbol{H}, \forall\omega\in \boldsymbol{W} \quad (6)$$

$$C_h \le b\cdot X_h \qquad \forall h\in \boldsymbol{H} \quad (7)$$

$$C_h\in \mathrm{N},\ F_o^d(a,\omega)\in \mathrm{N}, Tr(a,\omega)\in \mathrm{N},\ X_h\in\{0,1\}\ \forall h\in \boldsymbol{H}, \forall a\in \boldsymbol{A}, \forall(o,d)\in \boldsymbol{V}, \forall\omega\in \boldsymbol{W} \quad (8)$$

The first term in the objective function (1) represents the first-stage hub costs, including fixed hub setup cost and daily operational cost. The second term is the expected value of the second-stage fleet costs, where $\sum_{a\in A} v_a\cdot t(a,\omega)\cdot Tr(a,\omega)$ refers to the total fleet costs under scenario ω. In the formulation, v_a, s_h and f_h are cost parameters based on realistic factors such as the cost of land, regional average salary, and the level of local economic development. Flow conservation constraints (2) – (4) ensure that each freight demand is routed from its source to its destination. Constraints (5) provide the minimal number of trucks needed on each arc given the maximal loading capacity of the truck, while constraints (6) prevent capacity restriction violation for each logistics hub. Constraints (7) guarantee that logistics hubs are open when they have active capacity. Constraints (8) specifies the domains of the model variables.

The objective function (1) is computationally intractable due to integration over the continuous distribution of uncertainty scenarios. However, it can be approximated by the formula (9) using the SAA method.

$$\text{Minimize}\quad \sum_{h\in H}(f_h\cdot X_h + s_h\cdot C_h) + \frac{1}{|\boldsymbol{W}|}\sum_{\omega\in W}\left(\sum_{a\in A} v_a\cdot t(a,\omega)\cdot Tr(a,\omega)\right) \quad (9)$$

The true distributions of daily demands and random disruption are estimated by the empirical distribution of the $|\boldsymbol{W}|$ independent and identically distributed samples. As $|\boldsymbol{W}|$ approaches infinity, the optimal value and solutions of the model using objective (9) converge to the optimal value and solutions of the original one under mild assumptions [13].

We effectively handle uncertainties by utilizing the two-stage stochastic optimization approach. Regarding demand uncertainty, the demands between each OD pair changes in each scenario ω, as reflected in $q_{od}(\omega)$. The model finds a feasible hub capacity solution to meets demands in all scenarios by satisfying constraints (2) to (4) (network flow constraints). In terms of disruption uncertainty, a hub causes local delays during disruption, leading to the travel time of arcs connecting to this hub increases, as reflected in $t(a,\omega)$. Such increased travel time will result in a rise in fleet costs, which represents a delay penalty for the affected arc. Then the model avoids using those disrupted arcs as we minimize the total fleet costs in objective function (9).

5. Simulation-Based Stress Testing Investigation

In this section, we apply our two-stage stochastic optimization model to the Southeast US finished vehicle delivery network presented in Section 2. The synthetic dataset includes around 35,000 vehicle orders over a 3-month planning horizon. In our Southeast US dataset, we observe that utilizing this hyperconnected transportation network can enable all shipments to be delivered within a single day, without carrying over to the next day's transportation resources. Our model exploits this observation to reduce model complexity by avoiding the introduction of a temporal dimension. To generate uncertainty scenarios from the dataset, we estimate the distribution of daily demands on each OD pair using the Kernel Density Estimation (KDE) approach and then take 50 samples from the estimated distribution as its demand scenarios. As for the disruption scenarios, we randomly generate 50 scenarios by assuming that open-access logistics hubs are independent of each other and the disruption rate for each hub h is given by a parameter p_h.

To examine the separated effects of demand uncertainty and random geographical disruptions on capacity deployment, we stress test our model at four stress testing levels. The first level does not consider any uncertain conditions, the second level assumes there is no disruption in the network but account for uncertain demands, the third level only accounts for random disruptions given certain demand information, and the fourth level considers both types of uncertainties. Then we achieve these four stress testing levels by inputting four types of scenarios into our model as follows. In the first level, we solve our model using the same 50 scenarios with a 70-percentage daily demands and no disruption in any of the scenarios. Being that all scenarios are identical; this level is equivalent to a deterministic optimization model. For the second level, we use our 50 generated demand scenarios with no disruption in scenarios. In the third level, where we only consider random disruptions, we use our 50 generated disruption scenarios with the 70-percentage daily demands. In the final level, which includes both types of uncertainties, we use the 50 demand scenarios combined with the generated disruption scenarios as our input scenario set.

https://doi.org/10.21872/2023IISE_3426

Solving our two-stage stochastic optimization model at these four stress testing levels yields four optimal networks: the baseline deterministic network (BDN), the stochastic demand network (SDN), the stochastic disruption network (SDiN), and the integrated stochastic network (ISN) as presented in Fig. 3. Combining Table 2 and Fig. 3, we can see that the BDN and SDN tend to allocate large capacity to a small group of logistics hubs located in central areas. It reasonably reflects that when there is no disruption in the network, utilizing few central hubs leads to shorter shipment routes and improved consolidation efficiency, resulting in a decrease in both hub costs and fleet costs. In addition, demand uncertainty exacerbates this centralized tendency and drives model results with higher network connectivity when comparing the SDN to BDN. However, as we can see from the network configurations of SDiN and ISN, accounting for the uncertainty of disruptions results in more decentralized capacity allocation, smaller logistics hub designs and more robust network interconnection. This is because irreplaceable high-capacity hubs are more likely to incur greater penalty costs in hub failures. Finally, the ISN balances the tradeoff between centralization from demand uncertainty and decentralization from disruption uncertainty by establishing 11 logistics hubs across the US Southeast region, with an average daily vehicle throughput capacity of 49.6. It uses relatively few hubs to create a highly connected network where each hub is connected to around 4 other hubs, leading to a resilient and stable network.

Table 2: Basic information of the four optimal hub networks

Optimal Networks for Each of Stress Testing Levels	BDN	SDN	SDiN	ISN
Hub Network Throughput Capacity (Vehicles per day)	348	481	485	546
Number of Active Hubs	7	9	13	11
Hub Average Throughput Capacity (Vehicles per day)	49.7	53.4	37.3	49.6
Hub Connectivity (Average degree of hub nodes)	2.0	3.1	4.9	4.2

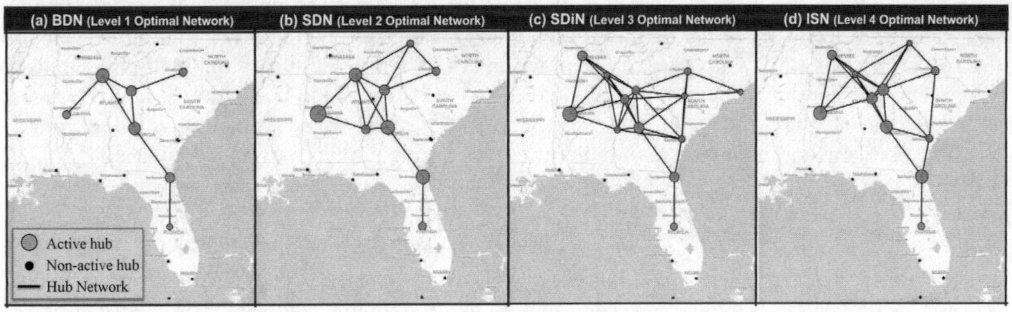

Figure 3: Optimal hub network configuration at each stress testing level

To test the performances of the four hub networks, we perform simulation studies for each optimal network at four stress testing levels. In the simulation, the demand scenarios are the actual daily demands from the dataset, while the disruption scenarios are randomly generated daily. We consider the case instance where the simulation runs for three months, the delivery deadline is one day, and all disruptions last for one day. The simulation's key performance indicators include delivery timeliness and cost-effectiveness. Overall, increasing scenario stress testing levels from 1 to 4 results in decreased on-time delivery rates and increased daily total costs for each network as illustrated in Fig. 4.

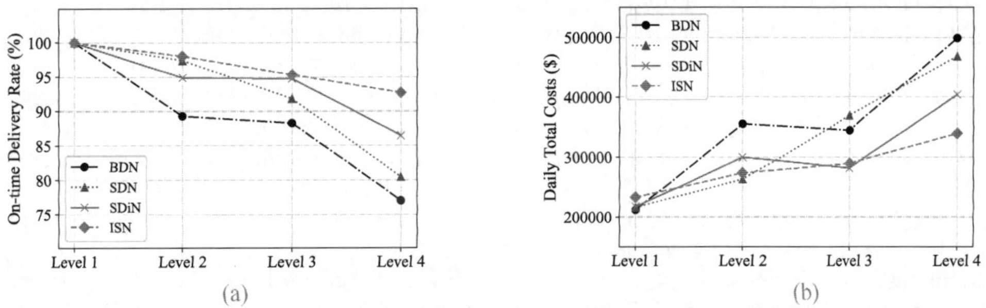

Figure 4: On-time delivery rate and average daily total cost for each network at each stress testing level

Fig. 4 (a) shows that the ISN consistently has the highest on-time delivery rate compared to the other networks and is more resilient to uncertainties since it has a smaller decreasing slope as uncertainty increases. In Fig. 4(b), the optimal hub network at each uncertain level tends to have the lowest cost in its corresponding level, since the goal of our

https://doi.org/10.21872/2023IISE_3426

optimization model is to minimize total cost. However, moving from level 1 to level 4 as uncertainty becomes more complex, the ISN is more resilient and robust to keep daily total costs relatively low under $350,000 through all stress testing levels. Other networks, specifically the BDN from the deterministic optimization model, are more sensitive to uncertainties, and their daily costs increase rapidly as uncertainty becomes complex. Overall, our two-stage stochastic optimization model yields different hub network configurations based on various sets of input scenarios. Incorporating more uncertainties, such as demand fluctuations and geographical disruptions, leads to a more resilient network. The aforementioned illustrates the significant impact of capacity deployment on network resilience.

6. Conclusion

In this paper, we have proposed a two-stage stochastic optimization model to address spatial-temporal uncertainty in hyperconnected transportation networks, which can be solved efficiently utilizing the SAA method. We further analyzed the separate and combined effects of demand uncertainty and geographical disruptions on hub network configurations by solving the model under different stress testing levels. Our stress testing simulation results demonstrate that our approach provides practical , effective, and resilient hub network capacity deployment solution with high on-time delivery rates, and low total cost under multiple demand and disruption uncertainty scenarios.

Overall, we have proposed an adaptive logistics hub capacity deployment approach in hyperconnected transportation settings. Additionally, this model can be extended to multi-day shipping by modifying each scenario in the input scenario set from a one-day scenario to a k-day scenario and applied to a variety of hub networks that require freight transfer via hubs. With prior knowledge of uncertainties, our model can provide a hub network that effectively adapts to uncertainty and deliver high performances in terms of network resilience, cost savings, and delivery timeliness.

We here propose several research avenues for future research. First, advanced computational methods like Bender's decomposition can be studied to speed up the solution approach and make it more efficient for large-scale instances. Second, beyond our current static hub capacity deployment, it is worthwhile to explore the dynamic capacity deployment by accounting for the changes of throughput flow across time. Third, hub capacity can be further divided into multiple categories, including the loading zone capacity, parking zone capacity, driver-switching zone capacity and so on. Such capacity categorization is usually studied together with hub facility and layout design.

References

[1] Campbell, J.F., De Miranda, G., De Camargo, R.S., O'Kelly, M.E. "Hub Location and Network Design with Fixed and Variable Costs," in *48th Hawaii International Conference on System Sciences (HICSS)*, pp. 1059-1067, 2015

[2] O'Kelly, M.E., Miller, H.J. "The hub network design problem: A review and synthesis," *Journal of Transport Geography*, vol. 2, no. 1, pp. 31-40, 1994.

[3] Montreuil, B. "Toward a Physical Internet: Meeting the Global Logistics Sustainability Grand Challenge," *Logistics Research*, vol. 3, no. 2, pp. 71-87, 2011.

[4] Montreuil, B., Buckley, S., Faugere, L., Khir, R., Derhami, S. "Urban parcel logistics hub and network design: The impact of modularity and hyperconnectivity," in *15th IMHRC Proceedings, Savannah, Georgia*, 2018.

[5] Faturechi, R., Miller-Hooks, E. "Travel time resilience of roadway networks under disaster, " *Transportation research part B: methodological*, 70, pp. 47-64, 2014.

[6] Rahdar, M., Wang, L., Dong, J., Hu, G. "Resilient transportation network design under uncertain link capacity using a trilevel optimization model." *Journal of Advanced Transportation*, pp. 1-16, 2022.

[7] Kulkarni, O., Dahan, M., Montreuil, B. "Resilient hyperconnected parcel delivery network design under disruption risks," *International Journal of Production Economics*, vol. 251, pp. 108499, 2022.

[8] Rodriguez, V., Alvarez, M.J., Barcos, L. "Hub Location Under Capacity Constraints," *Transportation Research Part E: Logistics and Transportation Review*, vol. 43, issue 5, pp. 495-505, 2007.

[9] Correia, I., Nickel, S., Saldanha-da Gama, F. "Single-Assignment Hub Location Problems with Multiple Capacity Levels," *Transportation Research Part B: Methodological*, vol. 44, issue 8-9, pp. 1047-1066, 2010.

[10] Faugère, L., Klibi, W., White III, C., Montreuil, B. "Dynamic Pooled Capacity Deployment for Urban Parcel Logistics," *European Journal of Operational Research*, 2022.

[11] Li, J., Montreuil, B., Campos, M. "Trucker-sensitive Hyperconnected Large-Freight Transportation: An Operating System." *In Proceedings of IISE Annual Conference*, 2022.

[12] Furtado, P., Fakhfakh, R., Frayret, J. M., Biard, P. "Simulation of a Physical Internet — Based transportation network." *In proceedings of international conference on Industrial Engineering and Systems Management*, 2013.

[13] Dentcheva, D., Ruszczyński, A., Shapiro, A. "Lectures on Stochastic Programming: Modeling and Theory, " *Society for Industrial and Applied Mathematics*, 2009.

https://doi.org/10.21872/2023IISE_3442

Proceedings of the IISE Annual Conference & Expo 2023
K. Babski-Reeves, B. Eksioglu, D. Hampton, eds.

Route Optimization for Mining Operations

Abstract ID: 3442

Willem Peters
South Dakota School of Mines and Technology
Rapid City, SD

Dr. Saurav Kumar Dubey
South Dakota School of Mines and Technology
Rapid City, SD

Abstract

With the increase in demand for minerals and precious metals, it is increasingly necessary for optimization in the Mining Industry. Autonomous hauling systems allow for continuous mining operations while increasing worker safety and overall productivity. A typical haul truck on a mine site is used to move material from the excavation location to the processing location. The research explores developing a network diagram that attempts to accurately represent how a haul truck can move through a mine site. The network diagram has the nodes as the intersections of the haul roads, the excavation location, and the processing location. The arcs for the network diagram are the haul roads and the flow through the network are the haul trucks. The network diagram will then be used to develop a linear programming model to find the optimal route for a haul truck to travel. A Travelling Salesman Problem (TSP) vs Shortest path comparison will be made to verify model assumptions made to solve the problem. The model developed is then tested using a network diagram of an existing mine site. The network diagram of the existing mine site will be developed by using google earth.

Keywords

Mining , Programming, and Route Optimization

1. Introduction

With the increase in demand for minerals and precious metals, it is increasingly necessary for optimization in the Mining Industry. As excavation of the minerals and precious metals increase to meet this need, the use of haul trucks will increase for above ground operations. The objective of the haul truck is to move material from the excavation location to the processing location. The increase in the use of haul trucks will lead to an increase of carbon emissions on mine sites. In this paper a linear programing model is developed that will help lower carbon emissions. The linear programming model aims to achieve this goal by developing an objective function that will minimize the distance travelled by the haul trucks on a mine site. Data needed for the model was gathered by using Google Earth to determine distances between key points on a mine site. The mine site used for this paper is the Wharf mine site in South Dakota. This mine site is an above ground operation that utilizes heavy duty haul trucks to move excavated material.

2. Problem Description

In order to decrease carbon emissions of haul trucks a linear programming model must be developed that finds the optimal route. The optimal route will be determined by minimizing the total distance travelled by the haul trucks at the Wharf mine site. The route travelled starts at the loading zones and moves to the crusher in order to dump the material being hauled. Additionally the haul trucks must travel back to the loading zones after dumping the material off at the crusher. Google earth was used to develop a network diagram of the mine site which is shown in Figure 1. In Figure 1, the blue arcs are roads that can be traversed either direction, but only one haul truck can fit on that road. This limits those arcs to only allow flow of the haul trucks in one direction, but the model chooses which direction is optimal. For instance, a model could choose to travel from node 5 to node 7 or from node 7 to node 5, but not both directions.

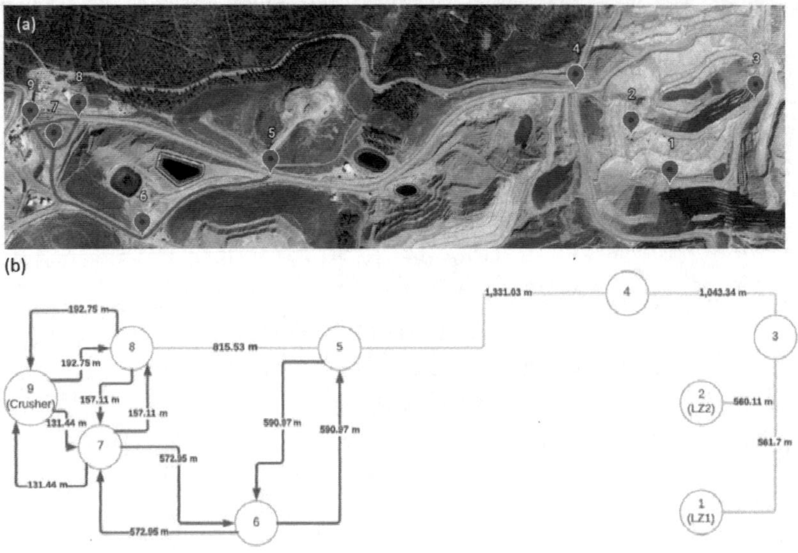

Figure 1, **(a, b)** View of the possible routes that could be travelled on google earth at the Wharf mine site and the network diagram for those routes. Arcs in yellow are two-way roads and arcs in blue are narrow roads with the distance shown in meters.

In the network diagram shown in Figure 1 the haul trucks would start at the loading zones which are nodes 1 and 2. The haul trucks would then travel to node 3. After arriving at node 3, the haul trucks would travel together towards the crusher which is node 9 in the network diagram. When the haul trucks arrive at node 9 they would then travel back to node 3. After arriving at node 3 they would separate and return to their respective starting nodes. The narrow roads identified in Figure 1 are bidirectional roads, but once a direction of travel is chosen no trucks can travel the opposite direction. This creates a limitation that if a narrow road is travelled going towards node 9 that same road cannot be travelled on the way back to nodes 1 and 2. The linear programming model developed would need to capture the limitations described above as well as determine the optimal route based upon distance travelled. In this problem the nodes identified are key areas such as loading zones, the crusher, and intersections in the haul road. In this model there will be two haul trucks with the first haul truck starting at node 1 and the second haul truck starting at node 2. The decisions variables in this problem are the routes travelled by the two trucks. A distance matrix for the distances between each node is shown in Table 1.

Table 1: Distance Matrix for Distances between Nodes at the Wharf Mine Site with All Values in Meters

All Values in meters		Node j								
		Node 1	Node 2	Node 3	Node 4	Node 5	Node 6	Node 7	Node 8	Node 9
Node i	Node 1			561.7						
	Node 2			560.11						
	Node 3	561.7	560.11		1043.34					
	Node 4			1043.34		1331.03				
	Node 5				1331.03		590.97		915.83	
	Node 6					590.97 m		572.95		
	Node 7						572.95		157.11	131.44
	Node 8					915.83		157.11		192.75
	Node 9							131.44	192.75	

In Table 1, the row indicates node $_i$ and the column indicates node $_j$. The intersection points between the row and column are the value for D_{ij}. D_{ij} is the route travelled between node $_i$ to node $_j$. D_{ij} is the same value as D_{ji}.

3. Related Research

Optimization of mine sites focus on decreasing the idle time of loaders in order to improve the total amount of ore transported. This is often achieved by using a fleet scheduling system which determine the best assignment of trucks. A stochastic integer programming model was developed in [1] that makes dispatching decisions while considering the uncertainties in truck travel time. Fleet scheduling models focus on optimizing costs while meeting the production target [1]. This leads to a focus on production optimization rather than route optimization. Route optimization can be used to decrease the distance travelled by the haul trucks which in turn could lower carbon emissions. Route optimization models take on a variety of forms such as the TSP formulation. TSP formulations solve for the optimal route that travels to every node in the network diagram while starting and ending at the same node [2]. Several applications of the TSP problems are explored in reference [2]. In this problem a TSP formulation was unable to be used for the whole network diagram because it is not possible to travel to every node exactly one time. Nodes 3, 4, and 5 have to be travelled to multiple times in order to start and end at the loading zones. Additionally, travelling to every node will lead to excess distance travelled because it is only required to travel to the crusher and then back to the loading zones. A shortest path problem formulation finds the shortest path between two nodes in a network diagram [3]. Multiple applications of the shortest path formulation are explored in reference [3]. The shortest path formulation can be used to find the shortest path from the loading zones to the crusher as well as the shortest path from the crusher to the loading zones. The two separate shortest path formulations could be used as phases of a larger model similar to the two-phase method for solving a hub location problem and a vehicle routing problem in reference [4]. The research problem being explored needs to be solved in two phases in order to find the optimal route. The first phase finds the shortest path from the loading zones to the crusher. The second phase finds the shortest path from the crusher to the loading zones. Both parts needed to be solved simultaneously to find the optimal route.

4. Methodology

The objective function's goal is to minimize the total distance travelled by the haul trucks. The generic formulation for the objective function can be seen in Equation (1). A two-phase method was developed in order to capture the flow of the haul trucks in one linear programming model. The first phase consists of finding the shortest path from the starting nodes to the crusher. The second phase consists of finding the shortest path from the crusher to the starting nodes. The two phases are solved simultaneously in order to find the optimal solution of both phases. The first step of the first phase is to initialize the two haul trucks departing from nodes 1 and 2 (Loading Zones). This was achieved by using departure constraints from a generic TSP formulation which is shown in Equation (2). Next, the flow of the haul trucks from node 3 to the crusher was captured by using flow conservation constraints which is shown in Equation (3). The first step of the second phase captures the flow of haul trucks from the crusher to node 3 using flow conservation constraints which is shown in Equation (4). The first phase was connected to the second phase by setting the flow out of the first phase equal to the flow in of the second phase which is also established in Equation (4). Next the two trucks returned back to the starting nodes using the arrival constraints from a generic TSP formulation which is shown in Equation (5). An additional constraint was established to limit the number of haul trucks that could travel down the narrow roads which is shown in Equation (6). This constraint limits the model to only allowing two haul trucks to travel down the narrow road. Limiting the number of haul trucks that can travel down the narrow roads will constrain the model to only choose one direction of travel for those arcs. The final set of constraints established that the number of haul trucks travelling through an arc is greater than or equal to zero which is shown in Equation (7).

Notation

X_{ij}= Route travelled by a haul truck from node $_i$ to node $_j$
D_{ij}= Distance between node i and node j in meters
Both i and j go from 1-9

$$Minimize\ Z = \sum_{j=1}^{9}(D_{ij}X_{ij} + D_{ji}X_{ji})\ for\ i = 1,2,\ldots,9, i \neq j \tag{1}$$

Subject to:

$$\sum_{j=1}^{9} X_{ij} = 1\ for\ i = 1,2, i \neq j \tag{2}$$

$$\sum_{j=1}^{9}\left(X_{ij} - X_{ji}\right) = 0\ for\ i = 3,4,\ldots,9, i \neq j \tag{3}$$

https://doi.org/10.21872/2023IISE_3442

$$\sum_{j=1}^{9}(X_{ji} - X_{ij}) = 0 \ for \ i = 3,4,\ldots,9, i \neq j \quad (4)$$

$$\sum_{j=1}^{9} X_{ji} = 1 \ for \ i = 1,2, j \neq i \quad (5)$$

$$\sum_{j=5}^{9}(X_{ij} + X_{ji}) = 2 \ for \ i = 5,6,\ldots,9, i \neq j, X_{ij} \neq X_{58}, X_{85} \quad (6)$$

$$X_{ij} \geq 0 \ for \ all \ i \ and \ j \quad (7)$$

During the first implementation of this model it was discovered that some possible routes would not satisfy the constraints of the model proposed. The arcs of X_{78} and X_{87} are the flow in and flow out of nodes 7 and 8 which appear in both phases of the model in the flow conservation constraints. When one of those arcs are travelled the value of that arc appears in both phases which creates a duplicate set of haul trucks. This duplicate set of haul trucks will not satisfy Equation (5) because the model is constrained to having two haul trucks. This issue was alleviated by creating a set of decision variables for each phase of the model for those two arcs. This eliminates the duplication of haul trucks which allowed for any routes including the arcs of X_{78} and X_{87} to be travelled.

5. Results

The model developed found multiple optimal solutions for the optimal route for the Wharf mine site. Both optimal routes determined the minimum distance travelled for the haul truck starting at node 1 to be 7,984.5 meters and the distance travelled by the haul truck starting at node 2 to be 7,981.32 meters. The reason there is a slight difference between the two haul trucks is because the distance from the starting nodes to node 3 is slightly different for each haul truck. The first possible route for the model is shown in Figure 2 with the routes colored in green. Both haul trucks followed the same route for all arcs except for the arcs between node 1 and 3 and the arcs between node 2 and node 3.

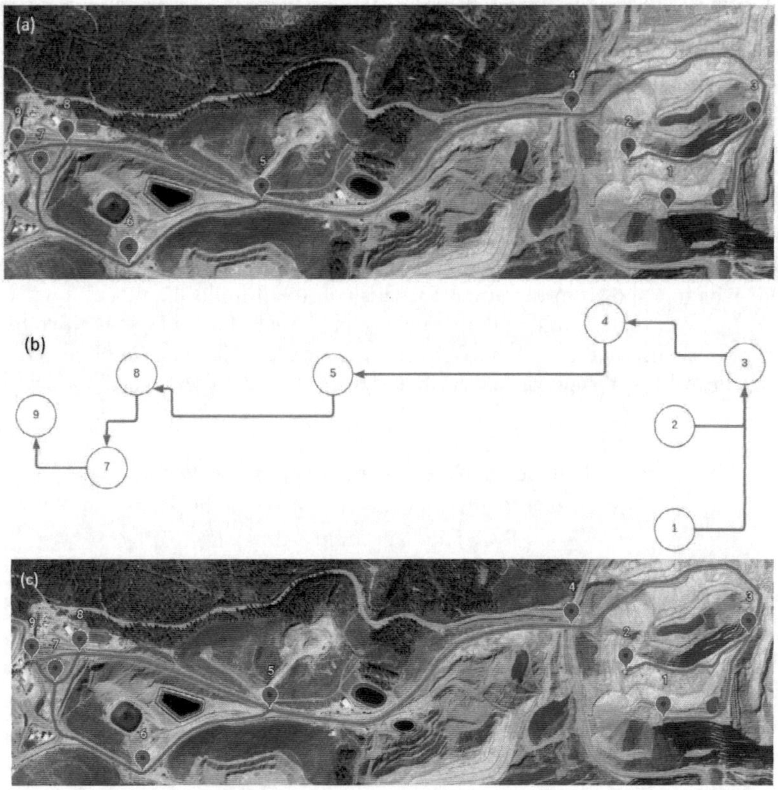

https://doi.org/10.21872/2023IISE_3442

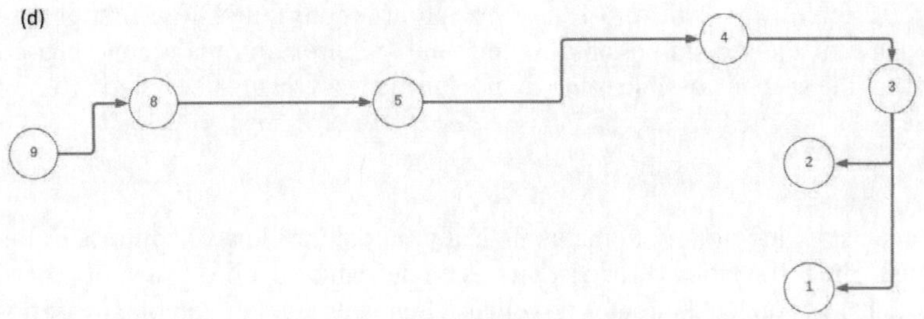

Figure 2: **(a, b)** The first possible route for the haul trucks to travel in the first phase is shown by the green arcs. **(c, d)** The first possible route for the haul trucks to travel in the second phase is shown by the green arcs.

This model kept the haul trucks travelling together where possible because the optimal route for first truck is often the optimal route for the second truck. The second possible route found is shown in Figure 3.

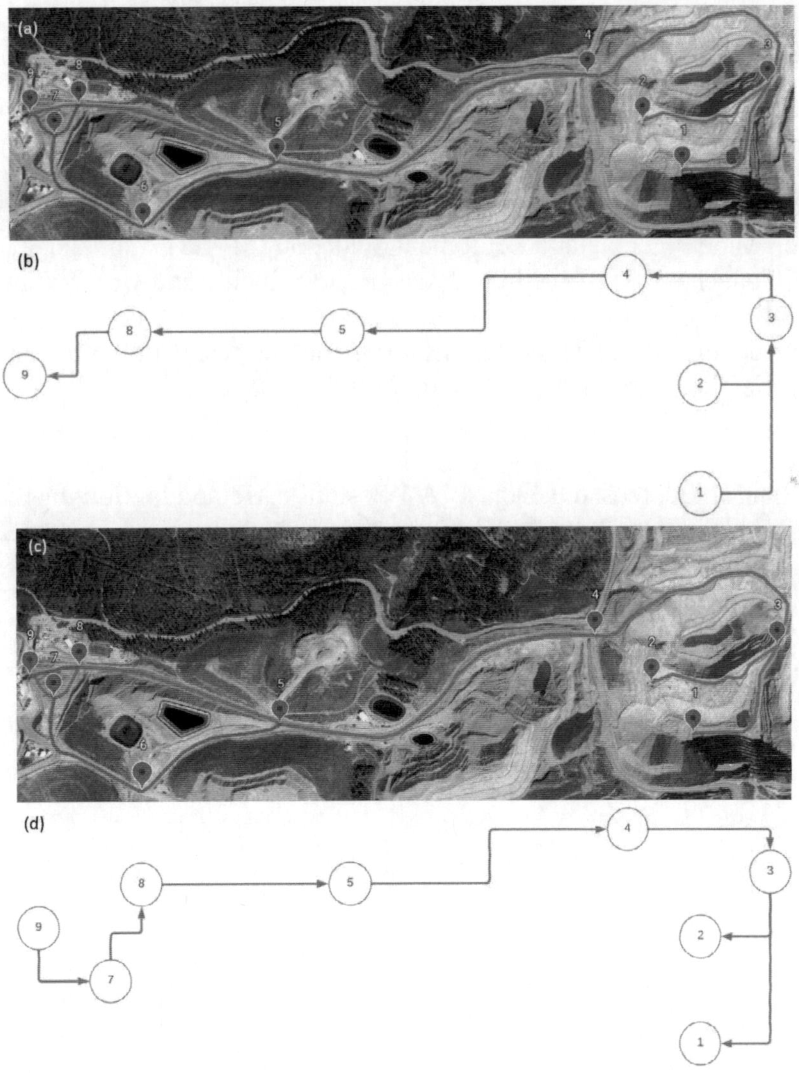

Figure 3: **(a, b)** The second possible route for the haul trucks to travel in the first phase is shown by the green arcs. **(c, d)** The second possible route for the haul trucks to travel in the second phase is shown by the green arcs.

The second possible route found was a mirror of the first optimal route found. The first phase of the first possible route was the second phase of the second possible route found. Additionally the second phase of the first possible route was the first phase of the second possible route found. Further criteria are needed to determine the optimal route between the two found.

6. Conclusions

In this paper a better understanding of how to mathematically model the flow of haul trucks through above ground mine sites was developed. The formulations proposed in this model can be applied to a variety of different applications in the mining industry to help optimize the routes travelled. The lower level of complexity of this model allowed for the output of the model to be validated by visual inspection and hand calculations. This study is a first step towards applying route optimization towards larger mine sites. Additionally modifications to the objective function can be made to find the optimal route based upon carbon emissions while travelling between nodes. Carbon emission on mine trucks would vary travelling each direction due to things such as road conditions and elevation changes. This new level of complexity would not change the constraints for the model proposed because the flow of the haul trucks would remain the same. A new objective function would need to be developed which would have D_{ij} set equal to the carbon emission for that arc. The ability to minimize carbon emissions for haul trucks would help to reduce the carbon footprint of each mine site tremendously. Additionally the model can be used for a sensitivity analysis to determine which haul roads should have more maintenance in order to further reduce the carbon emissions.

References

[1] A. Afrapoli, N. Hooman, "A Transportation Problem-Based Stochastic Integer Programming Model to Dispatch Surface Mining Trucks under Uncertainty," 2018 27th International Symposium on Mine Planning and Equipment Selection (MPES), Santiago, Chile, 2018

[2] G.Gutin, AP.Punnen,"The Travelling Salesman Problem and Its Variations", Kluwer Academic Publishers, 2004.

[3] X. Wang, "The comparison of Three Algorithms in the Shortest Path Issue" Journal of Physics: Conference Series, vol. 1087, no. 2, 2018. [Online Serial]. Available: https://iopscience.iop.org/article/10.1088/1742-6596/1087/2/022011. [Accessed Nov. 28, 2022].

[4] X. Yang, Y. Bian, N. Bostel and P. Dejax, "A Two-Phase Method for Solving the Hub Location and Routing Problem," 2019 4th IEEE International Conference on Cybernetics (Cybconf), Beijing, China, 2019, pp. 1-6, doi: 10.1109/Cybconf47073.2019.9436586.

https://doi.org/10.21872/2023IISE_3452

Proceedings of the IISE Annual Conference & Expo 2023
K. Babski-Reeves, B. Eksioglu, D. Hampton, eds.

Open-Source Optimization of Green Last Mile Delivery Systems

Abstract ID: 3452

John Blake Sooter
Chase Rainwater, Ph.D.
Department of Industrial Engineering
University of Arkansas
Fayetteville, AR 72701

Abstract

New methodologies to estimate energy consumption from road attributes present an opportunity to optimize delivery systems for energy use rather than distance. Last mile, time window constrained delivery services are a carbon intensive application area for the well-studied vehicle routing problem. These deliveries, often carried out by gig economy workers, require various side constraints to accurately model the system. In this research, we propose a model of a time window constrained last mile delivery network and demonstrate the value of energy consumption as the optimization criteria in vehicle routing. Our study shows that using energy as the optimization criteria provides energy savings, even under homogenous fleet conditions. In addition, we develop and provide an environment to utilize real routing and address data as parameters at scale for no cost.

Keywords
Green Optimization, Last Mile, Open Source, Sustainability, Vehicle Routing Problem

1. Introduction

The vehicle routing problem (VRP) [1] is among the most well-studied NP-hard optimization problems. Recent developments have provided a new context for the study of the VRP. First, the rise of gig economy delivery services, such as grocery delivery, has dramatically increased consumer adoption of rapid local delivery networks of independent contractors. Second, research from the National Renewable Energy Lab [2] has made it feasible to accurately estimate the energy consumption of proposed routes from their attributes, enabling optimization based on energy consumption rather than the typical distance or duration objectives. Third, the last decade has seen the rise of credible, performant open-source optimization frameworks and routing engines, allowing for academic research that uses real world routing data without incurring large Application Programming Interface (API) costs. In this paper, we present a version of the Time-Independent Green Vehicle Routing Problem with Time Windows (GVRPTW) [3] representative of rapid, independently contracted local delivery systems, formulated as an integer program. In addition, we justify a metaheuristic algorithm for the problem, identify energy savings from optimizing for energy rather than distance or duration for common customer/vehicle sets, evaluate energy savings from speed governance, and compare the transportation energy consumption of grocery delivery to grocery pickup.

2. Problem Statement and Formulation

2.1 Integer Programming Formulation

Consider a system that is comprised of n nodes in the set $N = \{1, \dots, n\}$, with node 1 and node n representing the delivery depot, and nodes 2 through n-1 representing delivery customers. The arc (i,j) represents travel from node $i \in N$ to node $j \in N$. The nodes are serviced by m delivery vehicles in the set $H = \{1, 2, \dots, m\}$ during p delivery periods in the set $K = \{1, 2, \dots, p\}$. A set of binary decision variables, $x_{hijk} \in X$, represent whether vehicle h traversed arc (i,j) during delivery period k. The length of a period, in minutes, is defined as l. For each vehicle-arc-period combination, there is a corresponding input parameter for energy consumption e_{hijk}, travel time t_{hijk}, and travel distance d_{hijk}. For each customer-period combination, there is a binary parameter a_{jk} representing whether a period is acceptable for delivery. Every vehicle has a discrete order capacity $c_h \in C$. The objective function (1) is comprised of the total energy consumption of the proposed set of routes and a linear penalty M imposed on excess vehicle capacity.

https://doi.org/10.21872/2023IISE_3452

$$\min \sum_{x_{hijk} \in X} e_{hijk} x_{hijk} + \sum_{\{x_{hijk} \in X | h \geq V_{min}\}} M * h x_{hijk} \tag{1}$$

The capacity penalty is imposed on vehicle capacity beyond the theoretical minimum number of vehicles, V_{min}, required to serve all customers. We adopt the uniform distribution for demand, so $V_{min} = \max\left(\frac{n-2}{p*avg(C)}, 1\right)$. The index h is included in the penalty because it scales with the size of violation. If one more vehicle is used than the theoretical minimum, a penalty M is incurred for using that vehicle to service any arc. If 2 more vehicles are used, a penalty of $2M$ is incurred. Vehicles $h \in H$ are homogeneous, so using the index to restrict capacity does not impact the energy objective. To complete all deliveries, every customer is arrived at once (2) and departed from once (3).

$$\sum_{h \in H} \sum_{j \in N \setminus \{i,1,n\}} \sum_{k \in K} x_{hijk} = 1 \ \forall_{i \in N} \tag{2}$$

$$\sum_{h \in H} \sum_{i \in N \setminus \{j,1,n\}} \sum_{k \in K} x_{hijk} = 1 \ \forall_{j \in N} \tag{3}$$

Every route must depart from (constraint 4), and ultimately arrive at (constraint 5), the depot node.

$$\sum_{j \in N \setminus \{1\}} x_{h1jk} = 1 \ \forall_{h \in H} \ \forall_{k \in K} \tag{4}$$

$$\sum_{i \in N \setminus \{n\}} x_{hink} = 1 \ \forall_{h \in H} \ \forall_{k \in K} \tag{5}$$

If some arc (i,j) is used to visit a customer node i during period k, there must also be an arc used that departs from customer node i using the same vehicle h during the same period k.

$$x_{hijk} - \sum_{\bar{j} \in N \setminus \{n\}} x_{hi\bar{j}k} = 0 \ \forall_{h \in H} \ \forall_{i \in N \setminus \{1,n\}} \ \forall_{j \in N \setminus \{1,n\}} \forall_{k \in K} \tag{6}$$

If some arc (i,j) is used to depart from a customer node j during period k, there must also be an arc used to arrive at customer node j using the same vehicle h during the same period k.

$$x_{hijk} - \sum_{\bar{i} \in N \setminus \{n\}} x_{h\bar{i}jk} = 0 \ \forall_{h \in H} \ \forall_{i \in N \setminus \{1,n\}} \ \forall_{j \in N \{1 \setminus n\}} \forall_{k \in K} \tag{7}$$

Any delivery must take place during a customer approved period, a_{jk}.

$$x_{hijk} \leq a_{ik} a_{jk} \ \forall_{h \in H} \ \forall_{i \in N} \ \forall_{j \in N} \forall_{k \in K} \tag{8}$$

No vehicle may exceed the available time in a period (9) or its own order capacity (10).

$$\sum_{i \in N} \sum_{j \in N} t_{hijk} x_{hijk} \leq l \ \forall_{h \in H} \ \forall_{k \in K} \tag{9}$$

$$\sum_{i \in N \setminus \{n\}} \sum_{j \in N \setminus \{1\}} x_{hijk} \leq c_{hk} + 1 \ \ \forall_{h \in H} \ \forall_{k \in K} \tag{10}$$

No vehicle may travel to the depot "exit door" (11), no arc can travel from the depot "entry door" (12), and no arc can travel from a node to itself (13).

$$x_{hi0k} = 0 \ \forall_{h \in H} \ \forall_{i \in N} \ \forall_{k \in K} \tag{11}$$

$$x_{hnjk} = 0 \ \forall_{h \in H} \ \forall_{j \in N} \ \forall_{k \in K} \tag{12}$$

$$\sum_{i \in N} x_{hiik} = 0 \ \forall_{h \in H} \ \forall_{k \in K} \tag{13}$$

Lastly, a subtour elimination constraint is required. Our formulation uses the classical Miller-Tucker-Zemlin (MTZ) formulation, first proposed for the travelling salesperson problem [4]. A set of integer decision variables $u_{hik} \in U$, can be used to track the order in which nodes are visited when the MTZ constraint is implemented (14).

$$u_{hik} - u_{hjk} + (n-1) x_{hijk} \leq n - 2 \ \forall_{h \in H} \ \forall_{i \in N} \ \forall_{j \in N} \ \forall_{k \in K} \tag{14}$$

2.2 Pseudo-random Elements and Parameter Assumptions

To define the depot and the set of customer nodes $N \setminus \{1,n\}$, assumptions must be made to define the location of the depot, delivery region of the depot, selection of *n-2* nodes (customer nodes) from the delivery region, calculation of arc costs ($e_{hijk}, t_{hijk}, d_{hijk}$), and delivery period acceptability, a_{jk}, for the selected customers. Walmart is a leader in gig-economy based grocery delivery, with its past partnership with DoorDash and current rollout of the in-house Spark Driver app [5]. As of August 2022, the Spark Driver app accounted for 75% of Walmart deliveries, with 84% household penetration [5]. To build out the set of nodes, N, address data is queried from a local instance of

https://doi.org/10.21872/2023IISE_3452

OpenStreetMap. Given their market leading position, a Walmart supercenter is used as the delivery depot in our model. The delivery region for a supercenter location is defined as the set of residential addresses sharing the same 5-digit postal code. To define the set of customers, $N\backslash\{1,n\}$, every residential address within the delivery region is queried and *n-2* addresses are randomly selected. To generate arcs and their corresponding attributes, a local instance of the *Openrouteservice* routing engine is employed. *Openrouteservice* allows for the selection of many parameters, three of which influence the costs of arc *(i,j)* [6].

Table 1: *Openrouteservice* shortest path algorithm parameters that influence arc costs [6]

	Parameter	Definition
Shortest path parameters	share_factor	Maximum fraction of the route that alternatives may share with the optimal route
	target_count	Target number of alternative routes to compute
	weight_factor	Maximum factor by which route weight may diverge from the optimal route

Since *Openrouteservice* recommends routes primarily on time and distance, and *not* on energy, for our purpose it is important that *Openrouteservice* return numerous possible routes so that the best route for the minimization objective can be selected. Table 2 shows the impact on an energy consumption objective value for a small instance solved to optimality under varying constraints on the shortest path algorithm.

Table 2: Objective value for (n = 20, m = 2, p =12, zip_code = 72701, node_seed = 1) problem instance solved to optimality with CP-SAT under least and most constrained shortest path algorithms

		share_factor	target_count	weight_factor	objective value (GGE)
Shortest path constraints	Most Constrained	0.01	2	1.01	3.62
	Default Constraints	0.6	2	1.4	3.61
	Least Constrained	.99	10	1.99	3.60

A 0.61% gap in the objective value is seen when comparing the least constrained arc generation algorithm to the most constrained arc generation algorithm. Generating several reasonable routes for each arc *(i,j)* and selecting the best based on the objective of the current model has an observable impact on model results. We adopt the least constrained shortest-path algorithm parameters to minimize arc costs. Relative performance of the optimization model defined in section 2.1 can be evaluated by solving a model instance three times: once with energy consumption as the objective, once with distance, and once with duration. To provide an accurate comparison, all arc costs $(e_{hijk}, t_{hijk}, d_{hijk})$ must be recomputed prior to the solving of each model with the best route for arc *(i,j)* selected based on the objective criteria. Selecting a route is effectively an initial optimization step. Delivery period acceptability, a_{jk}, is assigned to each customer node according to the uniform random distribution $X{\sim}U(1,p)$. For each customer, a random variate is drawn from $X{\sim}U(1,p)$, and both the period equal to the random variate and the one immediately following it are considered acceptable to that customer. If the period equal to the random variate is *p*, the period immediately before it is considered acceptable to the customer. This reflects the desire of the retailer to levelize demand across the day and the willingness of the consumer to provide modest flexibility in delivery time.

3. Metaheuristic and Exact Solution Methods

The classical VRP is NP-hard, thus making exact solutions to large problem instances inherently intractable. However, when establishing the energy savings derived solely from using energy as the optimization criteria, it is important to guarantee that the true energy savings and trade-off time and distance costs are not obscured by a metaheuristic. Google's open-source CP-SAT solver, a fully featured integer programming solver [7], is used. CP-SAT has won numerous gold medals in the MiniZinc Challenge against proprietary and open-source competitors and performed very well on VRP benchmarks [8]. To select a metaheuristic, the problem formulated in section 2.1 was implemented using OR-Tools' Routing library, which allows for the selection of a first solution heuristic and a metaheuristic algorithm [9]. To select the best set of strategies, a grid search was employed to test all combinations of first solution and metaheuristic algorithms for instances of size $n = [200,400]$ under maximum run times $t = [5,15,25,35]$. Computational experiments were conducted on an 11th Gen Intel i9-11900 2.5 GHz CPU with 64 GB RAM.

Table 3: Best Combination of First Solution and Metaheuristic algorithms, for various *n* and time limit

https://doi.org/10.21872/2023IISE_3452

n	Time (min)	Objective	First Solution	Metaheuristic	Performance vs Worse Case
200	5	27.35	local cheapest insertion	simulated annealing	2.97%
	15	27.17	local cheapest insertion	guided local search	3.61%
	25	27.17	local cheapest insertion	guided local search	3.61%
	35	27.17	local cheapest insertion	guided local search	3.61%
300	5	40.05	parallel cheapest insertion	guided local search	2.34%
	15	39.73	parallel cheapest insertion	greedy descent	4.39%
	25	39.73	parallel cheapest insertion	greedy descent	4.39%
	35	39.66	parallel cheapest insertion	tabu search	4.56%

Cheapest insertion strategies always provided the best first solutions, while guided local search (GLS) provided good solutions quickly, regardless of *n.* In cases where GLS did not provide the best solution, it provided a solution that was near parity with the best solution. In instances with a longer time limit and large *n,* Tabu Search provided the best solution. Based on the results in Table 3, we adopted local cheapest insertion, guided local search, and a time limit of 15 minutes to achieve low computation time and generalizable performance for various *n*.

4. Computational Results

To verify that there is, in fact, a savings in energy from optimizing for energy rather than related factors distance or duration, small instances were run to optimality on the CP-SAT solver for various problem sizes and maximum vehicle speeds. Speed is a relevant consideration because it impacts energy and duration. Energy consumption in gasoline gallon equivalent (GGE) under an energy objective is benchmarked against energy consumption under distance and duration objectives, tradeoff costs of optimizing energy consumption are measured, and the energy usage of the energy-optimal delivery network is benchmarked against individual customer trips to the store ("Pickup") in an equivalent vehicle. The cost of Pickup is defined as arc (i,n) + arc $(1,i)$ for every customer *i*.

Table 4: Delivery Energy Savings, Tradeoff Costs, and Energy Savings vs. Pickup (Exact Solution)
(m = 1, c = 5, p = 12, zip code = 72701, node_seed = 1, 2016 Toyota Camry)

		Energy Consumption (GGE) for Delivery			**Energy Objective Tradeoff Costs**		**Energy Consumption (GGE) for Pickup**	
n	max speed (mph)	Energy Objective	vs Distance Objective	vs Duration Objective	Distance vs Optimal Distance	Duration vs Optimal Duration	Energy Objective	Energy Savings, Delivery vs. Pickup
20	30	3.72	-1.36%	-0.84%	1.31%	1.33%	7.40	-49.7%
	35	3.67	-1.47%	-0.91%	0.92%	1.13%	7.36	-50.0%
	40	3.67	-1.35%	-0.97%	0.92%	1.13%	7.33	-50.0%
	45	3.66	-1.40%	-1.17%	0.91%	1.34%	7.33	-50.0%
	50	3.67	-1.56%	-1.05%	0.93%	1.16%	7.33	-50.0%
	55	3.67	-1.41%	-1.13%	0.93%	1.35%	7.33	-50.0%
30	30	5.45	-0.96%	-0.51%	1.66%	0.93%	12.00	-54.6%
	35	5.35	-1.17%	-0.38%	1.53%	0.66%	11.87	-55.0%
	40	5.31	-0.65%	-0.61%	0.88%	1.30%	11.82	-55.0%
	45	5.31	-0.56%	-0.66%	0.88%	1.25%	11.81	-55.0%
	50	5.31	-0.68%	2.15%*	0.72%	3.92%	11.82	-55.1%
	55	5.31	-0.68%	2.06%*	0.72%	3.94%	11.82	-55.1%

https://doi.org/10.21872/2023IISE_3452

In every case, optimizing for energy led to a reduction in energy consumption versus a distance objective, with an average reduction of 1.10% achieved solely from using energy as the optimization criteria and not from heavier utilization of more energy-efficient vehicles, dropping of costly customers, or other "green" operational changes. Energy savings from heavier utilization of fuel-efficient vehicles are not quantified here because it obscures the added value of energy as the optimization criteria. Savings from energy-efficient vehicles could be derived without this information by prioritizing the utilization of fleet vehicles with the highest energy efficiency. Energy savings versus a duration objective was also achieved, with two notable exceptions (*). Intuitively, it may seem impossible that optimizing for duration could ever cause *worse* energy consumption. This possibility is the result of our arc cost optimization (section 2.2) procedure that selects the best arc *(i,j)* based on the objective criteria. For the two instances marked by (*), selection of the least energy intensive arcs generated enough tradeoff duration costs that more delivery routes were required to serve all customers while still adhering to time window constraints.

Table 5: Delivery Energy Savings, Tradeoff Costs, and Energy Savings vs. Pickup (GLS metaheuristic)
(m=8, c=5, p=12, zip code = 72701, node_seed=1, 2016 Toyota Camry)

		Energy Consumption (GGE) for Delivery			**Energy Objective Tradeoff Costs**		**Energy Consumption (GGE) for Pickup**	
n	max speed (mph)	Energy Objective	vs Distance Objective	vs Duration Objective	Distance vs Optimal Distance	Duration vs Optimal Duration	Energy Objective	Energy Savings, Delivery vs. Pickup
200	35	27.04	0.21%	1.04%	2.88%	2.59%	82.52	-67.2%
	45	27.03	-0.54%	0.48%	4.19%	2.57%	81.81	-67.0%
	55	27.07	-0.02%	-0.11%	4.41%	0.44%	81.75	-66.9%
300	35	40.57	-0.50%	-0.29%	1.96%	1.44%	125.24	-67.6%
	45	40.53	-0.03%	1.13%	5.54%	3.33%	123.89	-67.3%
	55	40.90	1.66%	-0.36%	6.89%	-0.12%	123.79	-67.0%

Importantly, we have identified that in all cases, grocery delivery provides a substantial savings in energy consumption, with an average savings of ~67% across the large-scale instances in Table 5. This confirms that the findings of Wygonik and Goodchild [10], who found substantial emissions savings with grocery delivery on heavy duty (35 order-capacity) delivery vehicles, hold when applied to the small passenger car delivery network that has developed as grocery delivery has become mainstream. Even if you assume that delivery services generate demand for store "trips", delivery would have to increase trip demand 3-fold to approach the breakeven point with pickup. Previous models at optimality identified that the energy savings vs a distance objective is ~1% (Table 4). Results from large model instances solved via the GLS metaheuristic show both small increases and small decreases in energy consumption relative to a distance objective. This suggests that the true savings are obscured by the absolute gap between the best-found solution and the true optimal. Given that most all large-scale routing optimization is performed using metaheuristic algorithms, this presents a practical impediment to realizing the benefits of using energy as the optimization criteria at scale. In our experiments, we have assumed that a customer is willing to accept delivery in two adjacent, customer selected time windows, and the optimization model considers a full day of orders. While the exact degree of flexibility is debatable, it is certain that the customer must be able to pick a specific window of time for the delivery of perishable goods. Therefore, with small modifications, it is possible to segment a day's orders into discrete optimization problems. Maintaining our assumption that the customer chooses two adjacent windows, if we were willing to provide the option to the customer to choose 1 of {8 AM-10 AM, 10 AM-12 AM,…6 PM-8PM}, rather than choose 2 of {8 AM-9 AM,9 AM-10 AM,…,7 PM-8 PM}, our optimization problem would be comprised of nonoverlapping periods and could be segmented into individual models, each with two one hour periods and a set of delivery customers. By doing so, we drastically reduce the scale of each optimization model, potentially reducing the gap between the best-found solution and the true optimal.

Table 6: Metaheuristic Performance for various "subproblems"
(m=6, c=5, p=2, zip code = 72701, node_seed=1, 2016 Toyota Camry)

Energy Consumption (GGE) for Delivery	**Energy Objective Tradeoff Costs**

n	max speed (mph)	For Energy Objective	vs Distance Objective	vs Duration Objective	Distance vs Optimal Distance	Duration vs Optimal Duration
20	45	2.58	-1.20%	-1.08%	1.03%	1.32%
30	45	4.10	-1.13%	-0.77%	1.01%	0.71%
40	45	4.92	-2.50%	-1.05%	2.18%	0.78%

The energy savings seen in these subproblems exhibit more variation, as expected with an inexact solution approach, but are consistent with the energy savings seen in the exact solution instances. Subproblems of size n=40 represent a 12-period day of 240 customers. This is in line with the size of instances in Table 5 and shows that the subproblem approach offers much improved performance for an overall problem of equivalent size.

5. Conclusion and Future Directions

In this paper, we proposed an integer program modeling a last mile grocery delivery system optimized for energy consumption. We identified the need to generate arc costs from *Openrouteservice* under the least constrained parameters, then selecting the best arc based on the optimization criteria of the problem instance. We evaluated various initial solution and metaheuristic approaches to solving large instances of a day's orders, showing that local cheapest insertion and guided local search provided the best balance between solution quality and computation time. We found that optimization for energy consumption provided ~1% energy savings compared to a distance objective, while any benefit from imposing a maximum driving speed was unclear. It is also seen that grocery delivery represents a ~67% reduction in energy consumption relative to grocery pickup. It proved difficult to realize the ~1% energy savings identified in small, optimal instances when the GLS metaheuristic was applied to large model, presenting a practical challenge to realizing the identified energy savings. To remedy this, an alternative approach was proposed to segment the model into discrete subproblems by forcing the customer to select contiguous, nonoverlapping delivery periods. This sacrifices a very small amount of customer flexibility and provides a large reduction in size of each problem, allowing the GLS approach to find solutions closer to the true optimal and affording a great degree of scalability. In testing, the average subproblem returned an energy savings of 1.61% over a distance objective. The subproblem approach allows optimization to occur closer to the time of deliveries and can be highly parallelized. In the future, the model performance for a traditional gasoline vehicle can be compared to an all-electric vehicle, as differences in electric and internal combustion motors could alter model results. Speed optimization merits further consideration.

Acknowledgments

This research was supported by an Honors College Research Grant from the University of Arkansas Honors College. Assistance was provided by Jake Holden and NREL in receiving access to the RouteE energy consumption model.

References

[1] G. B. Dantzig and J. H. Ramser, "The Truck Dispatching Problem," Management Science, vol. 6, no. 1, pp. 80–91, Oct. 1959, doi: 10.1287/mnsc.6.1.80.
[2] J. Holden, N. Reinicke, and J. Cappellucci, "Routee: A vehicle energy consumption prediction engine," SAE Technical Paper Series, 2020.
[3] P. Toth and D. Vigo, "Chapter 15: Green Vehicle Routing" in *Vehicle routing: Problems, methods, and applications*. Philadelphia, Pennsylvania: Society for Industrial and Applied Mathematics, 2015.
[4] W. Winston and J. Goldberg, "9. Integer Programming" in *Operations Research: applications and algorithms*. Pacific Grove, California: Thomson/Brooks/Cole, 2004.
[5] K. Souza, "Walmart and DoorDash end their delivery partnership," Talk Business & Politics, Aug. 22, 2022. https://talkbusiness.net/2022/08/walmart-and-doordash-end-their-delivery-partnership/ (accessed Dec. 21, 2022).
[6] "ORS," openrouteservice.org. https://openrouteservice.org/dev/#/api-docs/v2/ (accessed Dec. 22, 2022).
[7] "or tools - CP-SAT over CBC for MIP, is it worthwhile?," Operations Research Stack Exchange. https://or.stackexchange.com/questions/4119/cp-sat-over-cbc-for-mip-is-it-worthwhile (accessed Dec. 22, 2022).
[8] "MiniZinc," www.minizinc.org. https://www.minizinc.org (accessed Dec. 22, 2022).
[9] "Routing Options | OR-Tools," Google Developers. https://developers.google.com/optimization/routing
[10] E. Wygonik and A. Goodchild, "Evaluating the Efficacy of Shared-use Vehicles for Reducing Greenhouse Gas Emissions: A U.S. Case Study of Grocery Delivery," Journal of the Transportation Research Forum, vol. 51, no. 2, Sep. 2012, doi: 10.5399/osu/jtrf.51.2.2926.

https://doi.org/10.21872/2023IISE_3535

Modeling Data Uncertainty and Non-Homogeneity in Data Envelopment Analysis: A Methodological Perspective

Abstract ID: 3535

S. Mohsenirad[1], K. Triantis[2], T. Topcu[3]

1, 2, 3: Grado Department of Industrial and System Engineering, Virginia Tech

Abstract

This paper investigates the challenges of using Data Envelopment Analysis (DEA), a method used to evaluate relative efficiency performance of several decision-making units (DMUs). DEA model selection can be challenging due to the various analytical formulations and underlying assumptions in modeling. The paper highlights two principal data related conditions, namely data uncertainty and non-homogeneity, where basic DEA assumptions are not held. Under these circumstances the single efficiency frontier of the basic DEA is not accurate and instead multiple frontiers are required to compute the DMUs' performance. The contribution of the paper is to outline the limitations of DEA under these circumstances, which can lead to inaccurate evaluations and poor decision making. The paper proposes opportunities for DEA research by suggesting that DEA modeling should be combined with data pre-processing and post-analytical techniques such as clustering and robustness analysis to properly address these problems.

Keywords: Data Envelopment Analysis (DEA), Data Uncertainty, Measurement Imprecision, Non-homogeneity

1. Introduction

Data Envelopment Analysis (DEA) is a linear programming method for evaluating a group of decision-making units (DMU) based on their technical efficiency performance [1]. In this context, a DMU is a production unit, such as an enterprise, that transforms some inputs into outputs. The concept of efficiency performance originates from microeconomic production theory, where the production function is defined as the maximum amount of output, given the input and existing technological capabilities [2]. The production function then indicates the most efficient production plans and serves as a benchmark for how much efficiency can be improved [3]. When conducting empirical analysis, it can be challenging to ascertain the absolute maximum or minimum values of outputs/ inputs due to limited data. Therefore, relative efficiency performance is typically measured in empirical analysis, which involves evaluating performance in relation to other data that have been observed [28]. In order to accomplish this, a methodology is necessary to establish a connection between production theory and data analysis. One of the seminal works that proposed a solution to this problem is Farrell's (1957) that provides a methodology to approximate the relative production frontier based on data [4]. Building on Farrell's approach, Charnes, Cooper, and Rhodes (CCR) developed DEA to estimate the *production possibility frontier,* extending earlier works to the multiple output case [5]. DEA has two key advantages over alternative methods such as stochastic frontier analysis (SFA) for efficiency analysis [6]. First, because it accounts for multiple outputs and multiple inputs, it provides a framework to evaluate multifaceted performance. In particular, when DMUs undertake different activities, the method is still applicable and gives a comprehensive evaluation. Second, because DEA is a non-parametric method, prior knowledge about the functional forms of the technology is not required. Oftentimes, this knowledge is unavailable for the researcher, and DEA thereby eliminates a significant obstacle. Due to these advantages, the method has been widely applied in a variety of contexts since its inception [7].

The widespread applications of the DEA have resulted in different innovations in DEA modeling. Specifically, to adapt the basic model to various conditions, researchers have introduced many extensions in the literature. In this regard, Imprecise DEA [15], Fuzzy DEA [16], Network DEA [30], Dynamic DEA [31], Weight-restricted DEA [32] are a few examples. In addition, the modeling evolution pushed the basic formulations into contexts that are seemingly unrelated to where the model's creators intended. Namely, Charnes, Cooper, and Rhodes (CCR) (1978) developed the model to evaluate *public programs* that are *not-for-profit entities* where inputs and outputs are not priced based on a market system [5]. This means that the most significant difference between DMUs is their level of *technical efficiency*, not *price/cost efficiency*, which makes it a reasonable context for using the DEA methodology in technological settings. However, in the current practice, the applications include a wide range of performance measurements that are not restricted to either the public sector or non-profit activities [9]. Furthermore, selecting the DEA model that is most appropriate for a given application is challenging due to the model's various versions [10]. In particular, DEA model selection requires to account for data irregularities such as data uncertainty, mixed data scales, non-homogeneity, etc. [11]. The underlying assumptions of basic DEA are that data is perfectly precise, homogenous, and without any uncertainty. However, in real-world scenarios, data are often subject to varying degrees of uncertainty, imprecision, and non-homogeneity. The unique properties of the data being used can impose specific constraints on the optimization problem that DEA aims to solve. Failure to account for these factors in DEA modeling can lead to biased or inaccurate results. However, in many instances, the analysis tends to use the original models (CCR, BCC [5], [12]), that are less disputable but do not necessarily address the data characteristics. In this regard, there are some major DEA application pitfalls discussed in [13]. In order to perform an accurate performance measurement, the data characteristics are required to be identified and accounted properly in DEA modeling.

This paper explores two principal data-related conditions where the basic DEA models may not be appropriate. Specifically, DEA is tasked with computing a unique convex production frontier. The identification of a production frontier is a critical step in DEA as it helps to identify efficient DMUs and evaluate their relative efficiency. However, when data are uncertain due to measurement imprecision or missing data, they can lead to errors in the measurement of DMUs' performance. These errors can cause variations in the DMUs' efficiency scores, which can lead to inaccurate results. In addition, when dealing with non-homogeneous DMUs, a DMU performs differently relative to similar and dissimilar units. Therefore, a single benchmark may not be appropriate to evaluate all DMUs. In these cases, it is may be essential to compute multiple frontiers to capture the different data characteristics accurately. We explore these circumstances in sections 2 and 3. In addition, we look into complications that may arise due to these data issues in DEA modeling.

The contribution of this paper is twofold. First, it discusses how, despite the fact that DEA is a non-parametric method, it relies on specific assumptions that may not be met in empirical conditions, resulting in inaccurate evaluations of efficiency performance that can cause practitioners to reach erroneous decisions. Second, it introduces new opportunities for DEA research by demonstrating that DEA modeling must be coupled with data pre-processing and post-analytical techniques, such as clustering and robustness analysis.

2. Data Uncertainty

2.1. Efficiency measurement under data uncertainty

Data uncertainty refers to the lack of accuracy of the data used in the analysis [14]. This can arise primarily from measurement error, missing data, and qualitative scales of data measurement [15], [16]. Measurement error arises when the exact value of an input or output is unknown and instead, an estimate of these values is available that is subject to estimation error. For example, the number of employees of a bank's branches may be reported as an interval estimation (e.g. [3, 5] instead of 3) [17]. Another source of data uncertainty is missing data, which occurs when data on inputs/outputs for at least one of the DMUs are missing. In this case, the DMU with missing data is either excluded from the sample or the missing values are estimated [18]. The latter option is frequently more reasonable, yet it results in measurement error [19]. Qualitative scales of data measurement are another source of data uncertainty in DEA [20] and refer to non-numeric measurements that are used to describe attributes or characteristics of DMUs. For example, a Likert scale may be used to rate the quality of customer service in a hotel [21]. However, since such scales are subjective and dependent on the individual making the rating, they can introduce variability into the results of a DEA analysis. These sources of uncertainty can lead to biased results and impact the validity of DEA efficiency scores [11].

To address data uncertainty in DEA, several approaches have been proposed. Namely, Fuzzy DEA and Imprecise DEA (IDEA) are two major solutions in this regard [15, 16]. While these methods are developed based on different theoretical frameworks about data uncertainty, they are common in imposing extra constraints in the DEA optimization problem that characterize the variability of inputs/outputs data. To keep the discussion concise, we only look into IDEA formulation. In Problem (1), the DMUs are represented by $j = 1, \dots, n$, while their inputs and outputs are denoted by x_{ij}, where $i \in \{1, \dots, m\}$, and y_{rj}, where $r \in \{1, \dots, s\}$, respectively. Furthermore, $\underline{y}_{rj} \leq \boldsymbol{y_{rj}} \leq \bar{y}_{rj}$ and $\underline{x}_{ij} \leq \boldsymbol{x_{ij}} \leq \bar{x}_{ij}$ are extra constraints added to the original DEA that indicate the inputs and outputs with data imprecision. The impreciseness or uncertainty in the input and output data is represented by ranges instead of precise values. The ranges are specified using lower and upper bounds that define the impreciseness of the data. Since the measurement of inputs/outputs is not precise under data uncertainty, the resulting efficiency score and production frontier are not unique. To illustrate, three $DMUs$ are compared in Figure (1) based on one input and one output. $DMUs$ of B, D, and F have uncertain inputs and outputs shown by horizontal and vertical segments respectively. Given

https://doi.org/10.21872/2023IISE_3535

the data uncertainty of the input and output, the boxes B, D, and F in figure (1) represent possible positions of a DMU with respect to its input-output relationships. Then the production frontiers can envelope different set of points in these boxes, and the figure shows that multiple frontiers can be identified based on the available data. The figure presents the frontiers for a single input and output, and as the number of DMUs and dimensions of inputs and outputs increase, the "exact" location of the frontier will be increasingly more imprecise [35].

$$\max_{u_r, v_i} \sum_{r=1}^{s} u_r y_{r0} \quad (1)$$

$s.t.$

$$\sum_{i=1}^{m} v_i x_{i0} = 1 \quad (2)$$

$$\sum_{r=1}^{s} u_r y_{rj} - \sum_{i=1}^{m} v_i x_{ij} \leq 0, \quad j = 1, \ldots, n, \quad (3)$$

$$\underline{y}_{rj} \leq \boldsymbol{y_{\hat{r}j}} \leq \bar{y}_{rj}, \quad r \in \{1, \ldots, s\}, \quad j = 1, \ldots, n, \quad (4)$$

$$\underline{x}_{ij} \leq \boldsymbol{x_{\hat{i}j}} \leq \bar{x}_{ij}, \quad i \in \{1, \ldots, m\}, \quad j = 1, \ldots, n, \quad (5)$$

$$u_r, v_i \geq \varepsilon, \quad r = 1, \ldots, s, \quad i = 1, \ldots, m \quad (6)$$

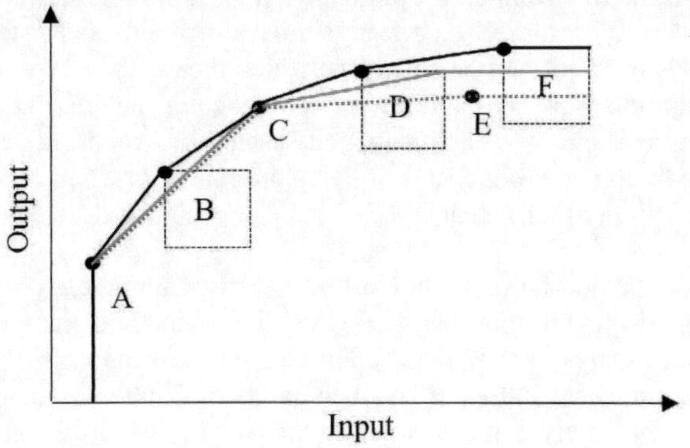

Figure 1 Production Possibility Frontiers with Uncertain Data. We look at six DMUs, with DMUs B, D, and C having imprecise input and output and the remainder having precise data. Because of data uncertainty, multiple frontiers can be identified, indicating varying performance of a single DMU. For example, depending on which frontier is selected, DMU B, with imprecise data, and DMU E, with precise data, are labeled as efficient (on the frontier) or inefficient (below the frontier).

Depending on which frontier is selected for evaluation, a DMU may be positioned on or below the frontier, indicating that it can be rated as either efficient or inefficient. This raises the question of whether a single frontier may be selected among others as a benchmark. If so, which frontier should be selected? and what are the selection's underlying assumptions? Section 2.2 provides more details on these questions.

2.2. Modeling complications

As we discussed in section 2.1 data uncertainty can result in significant variations in the efficiency scores obtained from DEA, making it difficult to interpret and apply in practice. Data uncertainty leads to the existence of multiple frontiers, and consequently, the performance of a DMU may be evaluated as being efficient or inefficient depending on which frontier is benchmarked. This can raise questions regarding the fairness of the evaluation, as the researcher's standards can influence the analysis. Some research recommends using the most optimistic scenario for computing the efficiency score [15]. To achieve this objective, the lowest input and maximum output for the DMU under evaluation are considered, as well as the maximum input and minimum output for other DMUs. Consequently, the analysis yields the maximum possible efficiency score. However, this is merely the upper bound of the possible efficiency scores and under less optimistic scenarios the results would be different. To overcome this problem, some studies propose to find an interval for the efficiency score rather than a single value [19]. In this situation, the most optimistic and pessimistic (i.e., maximum inputs and minimum outputs) efficiency scores are computed. Thus, this reveals the lower and upper bounds of the performance, which is a more comprehensive evaluation of the possible performance. The issue with the latter is that unlike a single value, the comparison of interval efficiency scores is not straightforward, and it complicates the ranking of DMUs.

In addition, fuzzy set theory can be applied to model data uncertainty in DEA [16]. According to fuzzy set theory when data are uncertain, different values of a variable are not equivalent in terms of their accuracy. As a result, a DMU's multi efficiency scores presumably differ in terms of the analysis's accuracy rather than the analysis outlook. Based on this concept, different approaches are developed to incorporate fuzzy set theory in DEA. While these approaches differ in several aspects, they share the use of fuzzy numbers and membership functions to account for varied levels of precision in DEA modeling. Fuzzy DEA computes multiple frontiers that represent differing degrees

https://doi.org/10.21872/2023IISE_3535

of precision. However, the models include extra parameters related to the membership function in the analysis, which can make the analysis more complicated [26].

3. Non-homogeneity

3.1. Efficiency measurement under non-homogeneity

Recalling that the objective of DEA is to evaluate the relative efficiency of DMUs by comparing their performance to the production frontier, the fundamental assumption of the basic DEA is that DMUs are homogeneous. In other words, DMUs are presumed to perform similar operations and face similar constraints. Nonetheless, in practical situations, DMUs may not be homogeneous, which means they may encounter various technological, environmental, or other conditions. This non-homogeneity complicates the DEA analysis and requires specific methods to handle it [22]. In an extreme case of non-homogeneity each DMU is considered to have its own set of inputs and outputs, and therefore, its own unique production frontier. Consequently, when non-homogeneity is present among DMUs, the fundamental assumption of DEA that DMUs can be benchmarked based on a single frontier is not accurate.

To handle non-homogeneity in DEA, various approaches have been developed with the objective of making relative comparisons feasible [23], [24], [25]. These methods may rely on a single frontier, multiple frontiers, and/or meta-frontier analysis [1], [33], [34]. The single frontier assumes the non-homogeneity is negligible, and relatively speaking, is the simplest solution. When the number of frontiers is increased, the analysis can identify non-homogeneous groups more accurately, but this comes at the cost of increased complexity. The reason is that it requires more data and can burden computational resources. Additionally, incorporating insights from multiple frontiers can become more challenging, making it harder to interpret the results. DEA modeling must therefore account for a compromise between evaluation simplicity and accuracy. Figure (2) shows the two frontiers and the meta frontiers for two groups of DMUs that are presumably non-homogenous. The gap between the group frontier and the meta-frontier shows that a DMU may be identified efficient with respect to its group but inefficient relative to non-homogeneous groups.

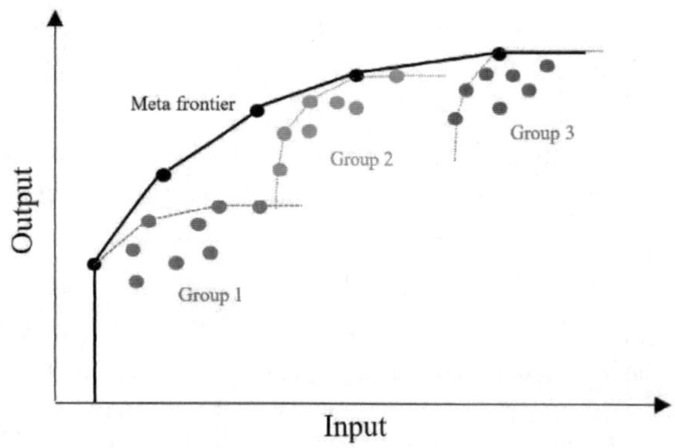

Figure 2: Non-homogeneous Groups Frontiers and the Meta-Frontier. The difference between in-group performance and the meta frontier implies that a DMU can be characterized as efficient with respect to homogeneous peers but inefficient with respect to non-homogeneous DMUs.

Non-homogeneity among DMUs can pose analytical challenges in DEA. First, the non-homogeneity can result in biased efficiency scores, leading to incorrect conclusions about the relative efficiency of DMUs. This can occur when the DEA model does not fully capture the differences among DMUs, leading to an inaccurate representation of their true efficiency. Second, there is no universally accepted method for identifying and addressing non-homogeneity in DEA, and consequently, model selection is a challenging task. In section 3.2 we discuss in more detail these challenges.

3.2. Modeling complications

As we mentioned in section 3.1 non-homogeneity is another challenge of computing an accurate frontier in DEA. In comparison to other problems [27], accounting for non-homogeneity is more obscure. Because, first, different researchers may have different interpretations of what constitutes non-homogeneity. Previous studies attributed the problem to a variety of factors, including environmental differences (contextual variations), discrepancies in technology, differences in production scale, etc. The diverseness of these factors results in a lack of consensus on the identification of non-homogeneity. The second problem is measurement of non-homogeneity. DMUs are assumed to

https://doi.org/10.21872/2023IISE_3535

be distinct from one another, and hence a level of dissimilarities is required to carry out the analysis. In this regard, non-homogeneity is often an issue of intensity rather than existence. This necessitates an index of non-homogeneity to measure the magnitude of the problem. While some research has offered solutions to this objective [24], a universally accepted indicator is yet to be established in the field. Third, in order to provide an overall performance analysis of DMUs, non-homogenous frontiers need to be combined (see Figure 2). To this aim, meta-frontier analysis provides an overall performance measurement; nevertheless, it is debatable if combining different production frontiers that presumably result from different underlying assumptions and production processes is informative.

4. Discussion & Conclusions

In this paper, we discussed DEA modeling data issues regarding uncertainty and non-homogeneity. Under these circumstances, multiple frontiers may be required to measure efficiency performance, which is distinct from the basic DEA model that has a single frontier. The multiple frontiers result in a variety of performance measurements for a DMU that may be difficult to comprehend. Consequently, this leads to questions about the reliability of the results. To overcome this challenge, DEA could be combined with other methods to carry out pre- and post-analysis. The pre-analysis is necessary to identify data uncertainty and non-homogeneity in a specific dataset and scenario. The magnitude of these problems should be established to instruct the model selection given the trade-off between the simplicity and accuracy of the DEA model. The post-analysis is also necessary to classify, rank, and/or combine multiple efficiency scores obtained from multiple frontiers analysis. Additionally, it is essential to examine the robustness of the results when inaccuracies in performance measurement occur. As such, the analysis should not just focus on determining "What is the performance measurement of a DMU?" but also consider "Under what conditions the performance measurement remain consistent?". In addition, it's important to note that while data uncertainty and non-homogeneity are often treated as separate issues in DEA modeling, they can coexist in practice, adding further complexity. To overcome this, a more advanced methodology that consistently integrates these two modeling issues is required. Namely, conventional DEA can be integrated with fuzzy set theory and statistical classifiers to model concurrent data uncertainty and non-homogeneity. In conclusion, it is critical for practitioners to pay attention to the data characteristics in DEA modeling described in this paper. Applying the model blindly and ignoring data attributes can result in "data in garbage out," where the output is flawed even if the input data are adequate. Instead, a thoughtful pre- and post-analysis is necessary to identify potential issues with data and examine the consequences of these issues on performance measurement, ensuring that the results are both accurate and robust.

5. References

[1] Emrouznejad, A. and Yang, G.L., 2018. A survey and analysis of the first 40 years of scholarly literature in DEA: 1978–2016. Socio-economic planning sciences, 61, pp.4-8.
[2] Ribeiro, V.M., Varum, C. and Daniel, A.D., 2020. Introducing microeconomic foundation in DEA: the average-based approach. Spatial Economic Analysis, 15(2), pp.145-164.
[3] Tone, K. and Sahoo, B.K., 2003. Scale, indivisibilities and production function in data envelopment analysis. International Journal of Production Economics, 84(2), pp.165-192.
[4] Farrell, M.J., 1957. The measurement of productive efficiency. Journal of the royal statistical society: series A (General), 120(3), pp.253-281.
[5] Charnes, A., Cooper, W.W. and Rhodes, E., 1978. Measuring the efficiency of decision making units. European journal of operational research, 2(6), pp.429-444.
[6] Mortimer, D., 2002. Competing methods for efficiency measurement: a systematic review of direct DEA vs SFA/DFA comparisons.
[7] Liu, J.S., Lu, L.Y., Lu, W.M. and Lin, B.J., 2013. A survey of DEA applications. Omega, 41(5), pp.893-902.
[8] Ahn, T., Charnes, A. and Cooper, W.W., 1988. Efficiency characterizations in different DEA models. socio-economic planning sciences, 22(6), pp.253-257.
[9] Liu, J.S., Lu, L.Y. and Lu, W.M., 2016. Research fronts in data envelopment analysis. Omega, 58, pp.33-45.
[10] Panwar, A., Olfati, M., Pant, M. and Snasel, V., 2022. A Review on the 40 Years of Existence of Data Envelopment Analysis Models: Historic Development and Current Trends. Archives of Computational Methods in Engineering, pp.1-30.
[11] Zhu, J. and Cook, W.D. eds., 2007. Modeling data irregularities and structural complexities in data envelopment analysis. Springer Science & Business Media.
[12] Banker, R.D., Charnes, A. and Cooper, W.W., 1984. Some models for estimating technical and scale inefficiencies in data envelopment analysis. Management science, 30(9), pp.1078-1092.

https://doi.org/10.21872/2023IISE_3535

[13] Dyson, R.G., Allen, R., Camanho, A.S., Podinovski, V.V., Sarrico, C.S. and Shale, E.A., 2001. Pitfalls and protocols in DEA. European Journal of operational research, 132(2), pp.245-259.
[14] Wen, M., 2015. Uncertain data envelopment analysis. Springer.
[15] Zhu, J., 2003. Imprecise data envelopment analysis (IDEA): A review and improvement with an application. European Journal of Operational Research, 144(3), pp.513-529.
[16] Hatami-Marbini, A., Emrouznejad, A. and Tavana, M., 2011. A taxonomy and review of the fuzzy data envelopment analysis literature: two decades in the making. European journal of operational research, 214(3), pp.457-472.
[17] Jablonsky, J., Fiala, P., Smirlis, Y. and Despotis, D.K., 2004. DEA with Interval Data: An Illustration Using the Evaluation of Branches of a Czech Bank. Central European Journal of Operations Research, 12(4).
[18] Chen, Y., Li, Y., Xie, Q., An, Q. and Liang, L., 2014. Data envelopment analysis with missing data: a multiple imputation approach. International Journal of Information and Decision Sciences, 6(4), pp.315-337.
[19] Kao, C. and Liu, S.T., 2000. Data envelopment analysis with missing data: an application to university libraries in Taiwan. Journal of the Operational Research Society, 51, pp.897-905.
[20] Cook, W.D., 2011. Qualitative data in DEA. Handbook on data envelopment analysis, pp.151-172.
[21] Shirouyehzad, H., Lotfi, F.H., Shahin, A., Aryanezhad, M.B. and Dabestani, R., 2012. Performance evaluation of hotels by data envelopment analysis based on customers' perception and gap analysis. International Journal of Services and Operations Management, 12(4), pp.447-467.
[22] Cook, W.D., Harrison, J., Imanirad, R., Rouse, P. and Zhu, J., 2015. Data envelopment analysis with non-homogeneous DMUs. Data Envelopment Analysis: A Handbook of Models and Methods, pp.309-340.
[23] Banker, R.D. and Morey, R.C., 1986. Efficiency analysis for exogenously fixed inputs and outputs. Operations research, 34(4), pp.513-521.
[24] Triantis, K., Sarayia, D. and Seaver, B., 2010. Using multivariate methods to incorporate environmental variables for local and global efficiency performance analysis. INFOR: Information Systems and Operational Research, 48(1), pp.39-52.
[25] Topcu, T.G., Triantis, K. and Roets, B., 2019. Estimation of the workload boundary in socio-technical infrastructure management systems: The case of Belgian railroads. European Journal of Operational Research, 278(1), pp.314-329.
[26] Emrouznejad, A. and Mustafa, A., 2010. Fuzzy assessment of performance of a decision making units using DEA: A non-radial approach. Expert Systems with Applications, 37(7), pp.5153-5157.
[27] Chen, L. and Wang, Y.M., 2022. Data envelopment analysis cross-efficiency method of non-homogeneous decision-making units. Journal of the Operational Research Society, pp.1-15.
[28] Kalirajan, K.P. and Shand, R.T., 1999. Frontier production functions and technical efficiency measures. Journal of Economic surveys, 13(2), pp.149-172.
[29] Li, Y., Chen, Y., Liang, L. and Xie, J., 2012. DEA models for extended two-stage network structures. Omega, 40(5), pp.611-618.
[30] Kao, C., 2014. Network data envelopment analysis: A review. European journal of operational research, 239(1), pp.1-16.
[31] Mariz, F.B., Almeida, M.R. and Aloise, D., 2018. A review of dynamic data envelopment analysis: state of the art and applications. International Transactions in Operational Research, 25(2), pp.469-505.
[32] Joro, T. and Viitala, E.J., 2004. Weight-restricted DEA in action: from expert opinions to mathematical models. Journal of the Operational Research Society, 55(8), pp.814-821.
[33] Färe, R., & Grosskopf, S., 2000. Nonparametric efficiency and divergence measures for heterogeneous technologies. Journal of Productivity Analysis, 13(2), 125-141.
[34] Kao, C., & Hwang, S. N. (2008). Efficiency decomposition in two-stage data envelopment analysis: An application to non-life insurance companies in Taiwan. European Journal of Operational Research, 185(1), 418-429.
[35] Azizi, H., 2013. A note on data envelopment analysis with missing values: an interval DEA approach. The International Journal of Advanced Manufacturing Technology, 66(9-12), pp.1817-1823.

https://doi.org/10.21872/2023IISE_3587

Proceedings of the IISE Annual Conference & Expo 2023
K. Babski-Reeves, B. Eksioglu, D. Hampton, eds.

Cost Analysis of Mass Finishing Additively Manufactured Ti6Al4V Parts

Abstract ID: 3587

Mr. Foxian Fan, Mr. Kazi Safowan Shahed and Dr. Guha Manogharan, Pennsylvania State University, University Park, PA

Abstract

Postprocessing for surface improvement has become one of the major cost factors in Additive Manufacturing (AM). Mass Finishing (MF) technologies, such as Centrifugal Disc Finishing (CDF), can batch-processing and fixture-free-processing to achieve the desired surface roughness and characteristics. Therefore, applications of MF technologies for metal AM surface improvement are potentially more cost-effective and offer higher design freedom. In this study, comprehensive numerical models for unit part cost are developed to understand the economic benefits of AM+MF hybrid manufacturing system. A case study on AM Ti6Al4V with CDF surface improvement was conducted for cost estimation, followed by a sensitivity analysis to validate the model. It was shown that the batch size is the key influencer in the AM+MF cost model (53% cost reduction for batch size n=10), where the largest contributing cost component varies according to processing batch size. Finally, the unit cost of AM+MF is compared with sand blasting, and the result showed that an optimal batch size exists for MF to become the more cost-effective choice. In addition, this developed model can potentially benefit the AM community by serving as a decision-making tool when considering using MF for surface improvement.

Keywords
Cost Modelling, Additive Manufacturing, Mass Finishing, Ti6Al4V, Centrifugal Disc Finishing

1. Introduction

Additive Manufacturing (AM) is developing at a rapid pace, with diverse applications in various industries such as aerospace, automobile, and biomedical. However, postprocessing are frequently conducted for as-built AM parts to improve surface finish, since surface finish is related to mechanical properties, corrosion resistance, and general applicability. A 3D reconstructed as-built AM surface example of Laser Powder Bed Fusion (L-PBF) fabricated Ti6Al4V parts is shown in Figure 1, where powder bed based as-built metal AM parts typically contains many surface defects such as porosity and partially melted powder. Varieties of postprocessing techniques for AM surface improvement are utilized. These methods can be categorized as mechanical, thermal, chemical, and coating-based treatments [1]. Despite high abundancy of postprocessing techniques, fixture requirements and low processing capacity in many techniques highly limit the design freedom of AM and frequently results in high postprocessing cost. Therefore, batch-processing, fixtureless Mass Finishing (MF) techniques, such as Centrifugal Disc Finishing (CDF), are gaining popularities in r

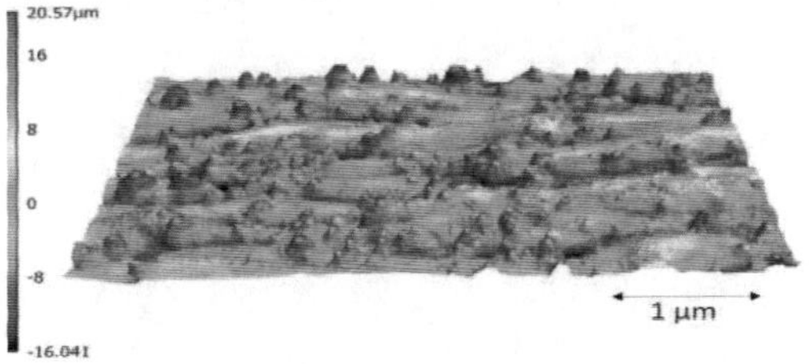

Figure 1. 3D microscopy of as-built L-PBF AM surface showed high amount of surface defects such as partially melted powder, porosity, and crevice.

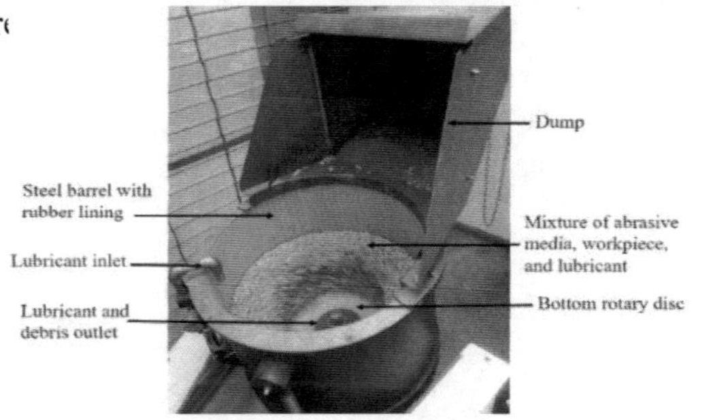

Figure 2. Illustration of Walther Trowal TT-45 Centrifugal Disc Finisher. Reprinted from [2] with author permission.

https://doi.org/10.21872/2023IISE_3587

2. Problem Description

MF is a deburring technology which uses abrasive media to batch process workpieces (see Figure 2. for an illustration of CDF). The surface finish is improved through mechanical interactions of ploughing, cutting, and impinging between abrasive media and workpiece surfaces[6], [7]. While many economic analyses have been conducted on AM [8]– [10], however, there is a need to understand the overall financial feasibility and applicability of AM+MF hybrid manufacturing system. In addition, it is important to understand if the cost reduction can be achieved and optimized through MF postprocessing. Therefore, this study aims to provide an economical insight through cost analysis of MF applications in metal AM postprocessing.

3. Methodology

A unit cost model is developed for L-PBF AM+MF hybrid manufacturing system in this study. In this model, the unit total cost ($C_{total,unit}$) has two components: 1). the fabrication cost of a single metal AM part ($C_{manufacturing,unit}$), and 2). the unit cost of postprocessing the AM part for surface finish improvements ($C_{finishing,unit}$). This relationship is shown in Eqn. (1):

$$C_{total,unit} = C_{manufacturing,unit} + C_{finishing,unit} \tag{1}$$

A generic cost model is developed for the L-PBF process based on the cost model by Rickenbacher et al. [11]. Furthermore, comprehensive cost analysis on powder feedstock and reusability was conducted by Barclift et al. [12] This AM+MF cost model combines modified cost models from AM literatures and MF literatures to understand cost effectiveness of the hybrid manufacturing system. The cost model in this study assumes only one type of part is fabricated per build plate. The parts are also printed in batches where the batch size (n) can be one or larger. The total number of parts are constrained by the build volume of the AM printer.

3.1 AM Cost Analysis

This model considers that the manufacturing process starts after the digital design CAD model (Computer Aided Design) of the part is received. It is assumed that the CAD models are watertight and error-free. There are commercially available software options (e.g., Netfabb) that can take the design and prepare the parts for printing. The orientation of the build, support structures, stacking multiple parts on the build plate and other geometry sensitive parameters are assumed to be validated by an operator before printing. During the preparation time ($T_{prepare}$) , cost is incurred for the operator ($C_{operator}$), and computing facility with software and PC (C_{soft}). The total unit cost incurred during AM preparation stage ($C_{preparation,\ AM}$) of n ($n \geq 1$) parts is expressed in Eqn. (2):

$$C_{preparation,AM} = \frac{(C_{operator} + C_{soft}) \times T_{prepare}}{n} \tag{2}$$

After digitally preparing the parts, the layout of the build plate with parts are uploaded to the printer. The total unit cost of printer setup cost ($C_{machine-setup(AM)}$) of setup duration T_{setup} of n ($n \geq 1$) parts is expressed in Eqn. (3):

$$C_{machine-setup,AM} = \frac{\left(C_{operator} + C_{machine}\right) \times T_{setup}}{n} \tag{3}$$

AM machine usage and its consumed energy cost ($C_{machine}$), and inert gas usage cost (C_{gas}) are incurred during fabrication. Thus, the total machine cost ($C_{build-machine}$) and fabrication time (T_{build}) of n ($n \geq 1$) parts is expressed in Eqn. (4):

$$C_{build-machine} = \frac{(\, C_{machine} + C_{gas}) \times T_{build}}{n} \tag{4}$$

During fabrication, it is assumed only virgin powder is consumed, and the material cost is incurred ($C_{build-material}$). The volume of the part and support structure decides the amount of material needed for the print. Furthermore, degradation of unmelt powder from process inefficiency prevents reuse of the powder and requires more powder than

https://doi.org/10.21872/2023IISE_3587

the volume of the part. In addition, powder feedstock can get trapped inside the support structures which cannot be recovered and adds to the cost, as found in [12]. In Eq. (5), $C_{build-material}$ is the material cost for a unit product. M is the mass of the powder required for a single part. From Eqn. (6), V_p and V_s are the volume of the part and volume of the support structure, respectively. Process inefficiency is denoted by η and ω percentage of powder that gets trapped in the support structure. ρ_w and ρ_t are the wrought density and tap density of the powder feedstock. After the parts are printed, the substrate is removed from the machine. Thus, the unit cost of bulk material is expressed in Eqn. (5) & Eqn. (6):

$$C_{build-material} = M \times C_{material} \tag{5}$$

$$\text{Where, } M = (1+\eta) \times (V_p + V_s) \times \rho_w + (\omega \times V_s \times \rho_t) \tag{6}$$

In addition, the substrate is removed, and machine is cleaned. This time ($T_{removal}$) is independent of the parts and batch size. The total unit cost of the removal $C_{removal-machine}$ of n ($n \geq 1$) parts is shown in Eqn. (7) below:

$$C_{removal-machine} = \frac{(C_{operator} + C_{machine}) \times T_{removal}}{n} \tag{7}$$

In this model, we considered typical practice of using wire Electric Discharge Machining (w-EDM) to remove the as-built parts from the substrate. The w-EDM process is charged per substrate, and it is independent of the number of parts on the substrate. Thus, the cost of w-EDM can be distributed over the number of parts. Additionally, tools are used for removing the support structures which incurs tool cost (C_{tools}) and tooling time (T_{tools}) for each part. The cost of removing the part from the substrate and removing support structures is given below in Eqn. (8):

$$C_{removal-part} = \frac{C_{EDM}}{n} + C_{tools} \times T_{tools} \tag{8}$$

In summary, the total unit cost during AM stage $C_{manufacturing(unit)}$ in summation form is shown in Eqn. (9):

$$C_{manufacturing,unit} = C_{preparation,AM} + C_{machine-setup,AM} + C_{build-machine} + C_{build-material} + C_{removal-machine} + C_{removal-part} \tag{9}$$

3.2 Cost of Mass Finishing

Comprehensive cost analysis for several types of MF processes was studied by Gillespie [6]. However, the CDF machine is used as the primary example in this study for postprocessing cost analysis because of its fast processing, batch processing, and fixtureless features. Operating CDF machine results in machine depreciation, which is described as an hourly cost ($C_{machine,MF}$). In addition, the machine requires periodic maintenance. The maintenance cost ($C_{maintenance,MF}$) can be divided over the operating hours. Furthermore, CDF machine operation incurs energy cost ($C_{energy,MF}$) from energy consumption. Due to the nature of high capacity in MF and for the sake of simplicity, it can be assumed that the required finishing time for postprocess AM parts (T_{finish}) is invariable with batch size. Thus, the total unit cost of using MF machine to postprocess n ($n \geq 1$) parts ($C_{machine,finishing}$) is expressed in Eqn. (10):

$$C_{machine,finishing} = \frac{(C_{machine,MF} + C_{maintenance,MF} + C_{energy,MF})}{n} \times T_{finish} \tag{10}$$

In MF, the abrasive media worn out over processing time (T_{finish}). It is assumed that the media wear rate is constant, and the same number of media are used for every batch. Furthermore, lubricant and water mixture flow into the working container as shown in Figure 2, and the usage volume depends on the processing time. Thus, the cost of media depreciation ($C_{media,MF}$) and the cost of lubricant ($C_{water,MF}$) are described as an hourly rate. The material cost of MF operation ($C_{material,finishing}$) is shown in Eqn. (11):

$$C_{material,finishing} = (C_{media,MF} + C_{water,MF}) \times T_{finish} \tag{11}$$

Manual labor is involved in loading, unloading, cleaning the machine and parts in MF processes. These procedures incur labor cost ($C_{labor,finishing}$) and cleaning-material cost ($C_{cleaning,finishing}$) at an hourly rate. Additionally, there

https://doi.org/10.21872/2023IISE_3587

is an hourly overhead cost ($C_{overhead,finishing}$) associated with facility operation. Assume the time required loading and unloading is $T_{handling}$, and the time required for cleaning is $T_{cleaning}$, these costs is expressed in Eqn. (12) as:

$$C_{handling,finishing} + C_{cleaning,finishing} + C_{overhead,finishing} = \frac{C_{labor,MF}}{n} \times T_{handling} + \frac{C_{cleaning,MF}}{n} \times T_{cleaning} + \frac{C_{overhead,MF}}{n} \times T_{finish} \quad (12)$$

Thus, the total cost of postprocessing batch size n ($n \geq 1$) is summed in Eqn. (13):

$$C_{finishing,unit} = C_{machine,finishing} + C_{material,finishing} + C_{handling,finishing} + C_{cleaning,finishing} + C_{overhead,finishing} \quad (13)$$

There, substitute the $C_{finishing,unit}$ and $C_{manufacturing,unit}$ in Eqn. (1) with Eqn. (9) & Eqn. (13) provides the total unit cost of AM+MF ($C_{total,unit}$) hybrid manufacturing system, as show in Eqn. (14):

$$C_{total,unit} = C_{preparation,AM} + C_{machine-setup,AM} + C_{machine-setup,AM} + C_{build-material} + C_{removal-machine} + C_{removal-part} + C_{machine,finishing} + C_{material,finishing} + C_{handling,finishing} + C_{cleaning,finishing} + C_{overhead,finishing} \quad (14)$$

4. Case Study

A case study was conducted to validate this model. Ten Ti6Al4V tensile bars were designed based on ASTM-E8b and fabricated via 3D Systems ProX 320 L-PBF system. A cost analysis up to batch size of ten was conducted. The detailed value of each cost component is based on literature [6], [11], [12] and adjusted accordingly to fit the current inflation rate, or experimental finding at current state. The AM cost and MF cost decreases as the batch size increases as shown in Figure 3. This can be explained due to fixed cost components are diluted over the batch size. The cost decreases rapidly initially but diminishes as batch size increases. This phenomenon is due to the high material cost for L-PBF, as batch size n increase, the material cost reduces rapidly initially but almost become a fixed cost once reached maximum number of allowable builds per plate. The batch size (n=1&10) greatly affects AM & MF cost components.

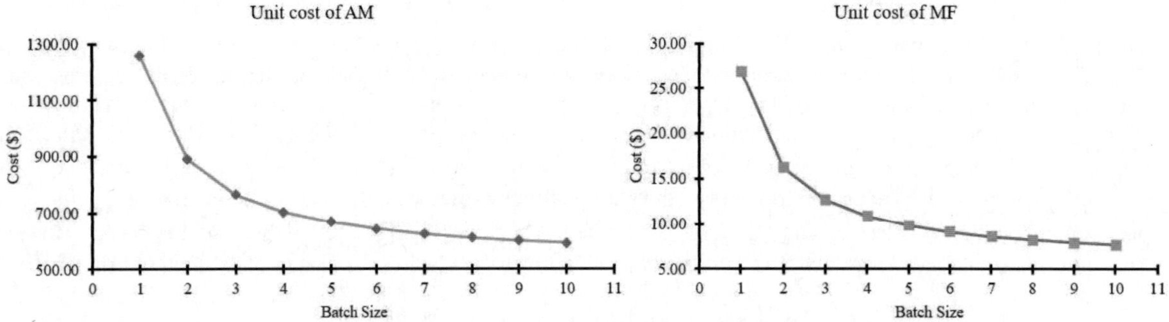

Figure 3. Batch size effects on AM fabrication cost, and MF postprocessing cost.

5. Sensitivity Analysis

Base on the case study finding, a sensitivity analysis was conducted to identify critical cost components for a batch size n=10. It is evidential that material cost ($C_{build-material}$) in Eqn. (4) and machine cost ($C_{build-machine}$) in Eqn. (5) are the leading contributor for AM, while the media cost ($C_{media,MF}$) and lubricant cost ($C_{water,MF}$) in Eqn. (11) are most critical for MF. Thus, these four components are selected for further sensitivity analysis at optimistic level (50% reduction) and most-optimistic level (90% reduction). The sensitivity analysis of these cost components is shown in Figure 5. While $C_{build-material}$, $C_{media,MF}$, and $C_{water,MF}$ are independent variables, the machine operational cost ($C_{build-machine}$) is largely affected by the multiplier build time (T_{build}). Thus, the sensitivity analysis was conducted on T_{build} instead of $C_{build-machine}$. By reducing the build time (T_{build}) by 50% and 90%, total unit fabrication cost

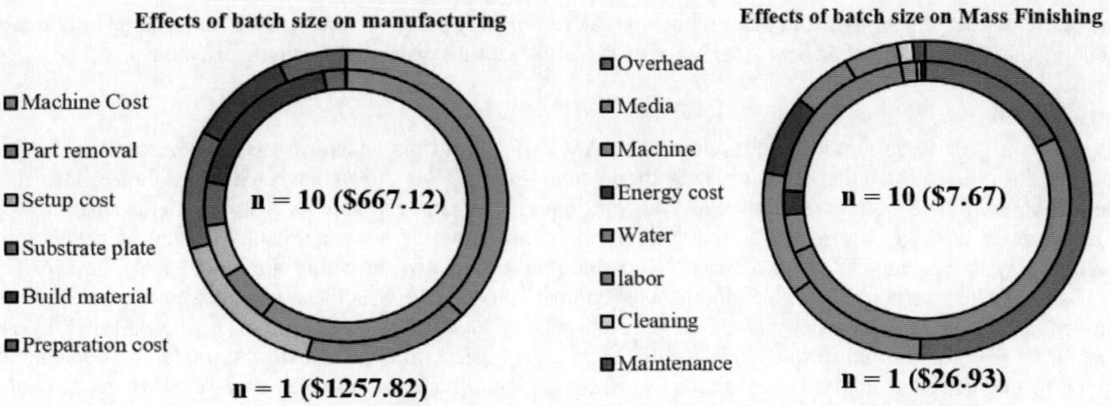

Figure 4: Batch size (n =1 for outer circle, and n=10 for inner circle) effects on AM + MF cost components.

reduces by 33.1% and 60%, respectively. Whereas, reducing the material cost ($C_{build-material}$) by 50% and 90% reduces the unit fabrication cost reduced by 11% and 19.8%. The media depreciation cost ($C_{media,MF}$) and lubricant cost ($C_{water,MF}$) were found to be the critical components from the MF process. The analysis shows that a 25.6% reduction and a 45.9% reduction of total unit postprocessing cost ($C_{finishing,unit}$) can be expected by reducing the media cost ($C_{media,MF}$) by 50% and 90%, respectively. A 10.6% and a 19% reduction on total postprocessing unit cost can be expected from lubricant and water cost ($C_{water,MF}$) at optimistic and most-optimistic level, respectively.

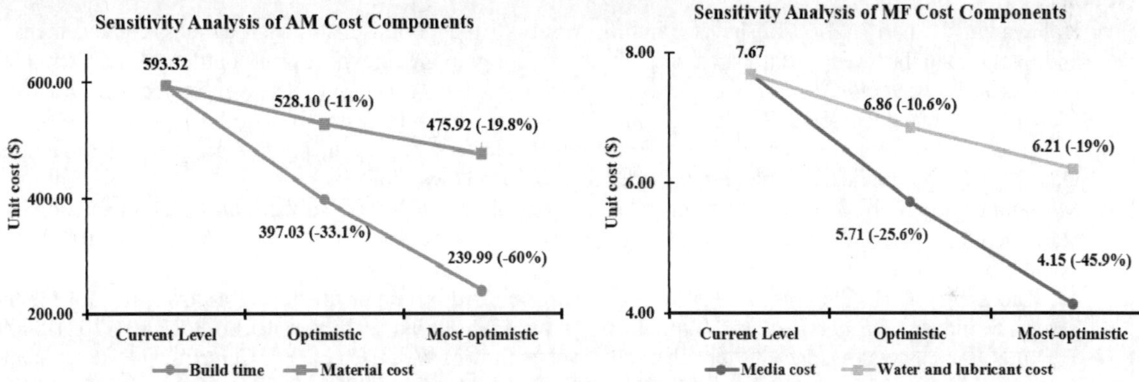

Figure 5. Sensitivity analysis of critical cost components in AM+MF cost model.

6. Discussion & Comparison

Based on the case study of the developed AM+MF cost model, an optimal batch size exists for this hybrid manufacturing system. To further investigate this optimal batch size, the postprocessing cost is compared with sandblasting. Both sandblasting and MF can achieve similar surface finish, however, sandblasting typical can only process one part at a time. Thus, the batch size n is considered invariable to total sandblasting cost. The sandblasting cost information is collected from Penn State Engineer Service. For the same fabricated part, the sandblasting cost $15/unit. Thus, for a batch size of ten, the total postprocessing costs $150, almost twice the amount compared to $76.7 of MF. However, for a batch size of n=1, sandblasting is more economical. It can be argued that batch size is critical when selecting postprocessing methods. To understand the effects of batch size, Eqn. (15) was developed to determine when MF is more beneficial for the given batch size:

$$C_{MF(n)} - n \times C_{AF(finish)} \leq 0 \tag{15}$$

$C_{MF(n)}$ is the cost of MF (equivalent to $C_{finishing,unit} \times n$) for a batch size of n. $C_{AF(finishing)}$ is the cost of alternative surface finish improvement method. To validate this equation, an estimation using sandblasting unit cost and MF unit cost was conducted, which resulted in a cutoff value of batch size $n = 3$. It is important to understand that this model does not consider the geometry complexity of AM parts. Thus, part geometry can be a deciding factor which may render the MF applicability in AM and make other alternative finishing process more suitable. This model only serves as a generic guideline of when and how AM+MF hybrid manufacturing system is economic effective.

7. Conclusion

In this study, a generic cost model is developed for AM+MF hybrid manufacturing system. A case study was conducted using both adjusted literature data and experimental findings. It is shown that batch size is the key influence of whether AM+MF is economical effective. AM machine operation cost, bulk material cost, MF media attrition cost, and MF lubricant cost are determined as the critical cost contributors at AM stage and MF stage. Furthermore, sensitivity analysis was conducted on the critical cost components at two levels as summarized in Table 1. The AM+MF cost model was validated with additional cost comparison of AM + Sandblasting. An optimal batch size was determined. This generic cost model serves as a decision-making tool and economical tool for the AM+MF hybrid manufacturing system. The future work includes developing AM orientation-based MF cost model, and potentially incorporating geometry factor based on AM topology optimization to further consummate the AM+MF cost model.

Table 1: Effects of critical components at Optimistic Level (OPT) & Most Optimistic Level (M-OPT)

Components	$C_{build\text{-}material}$		T_{build}		$C_{media,\ MF}$		$C_{water,\ MF}$	
	OPT	M-OPT	OPT	M-OPT	OPT	M-OPT	OPT	M-OPT
$C_{manufacturing,\ unit}$	-33.1%	-60%	-11%	-19.8%	N/A		N/A	
$C_{finishing,\ unit}$	N/A		N/A		-25.6%	-45.9%	-10.6%	-19%

References

[1] H. Fayazfar, J. Sharifi, M. K. Keshavarz, and M. Ansari, "An overview of surface roughness enhancement of additively manufactured metal parts: a path towards removing the post-print bottleneck for complex geometries," *International Journal of Advanced Manufacturing Technology*. Springer Science and Business Media Deutschland GmbH, 2023. doi: 10.1007/s00170-023-10814-6.

[2] F. Fan *et al.*, "Effects of Centrifugal Disc Finishing for Surface Improvements in Additively Manufactured Gears," 2021. [Online]. Available: http://dx.doi.org/10.26153/tsw/17678

[3] M. Jamal and M. N. Morgan, "Design process control for improved surface finish of metal additive manufactured parts of complex build geometry," *Inventions*, vol. 2, no. 4, Dec. 2017, doi: 10.3390/inventions2040036.

[4] M. Khorasani, A. H. Ghasemi, B. Rolfe, and I. Gibson, "Additive manufacturing a powerful tool for the aerospace industry," *Rapid Prototyp J*, vol. 28, no. 1, pp. 87–100, Jan. 2022, doi: 10.1108/RPJ-01-2021-0009.

[5] F. Fan, S. Jalui, N. Lajoie, and G. Manogharan, "Investigation of Wear Behavior of Centrifugal Disc Finishing on Additively Manufactured Ti6Al4V Samples," Austin, TX, 2022. [Online]. Available: http://dx.doi.org/10.26153/tsw/44335

[6] L. Gillespie, *Mass Finishing Handbook*, First Edition. New York: Industrial Press, 2007.

[7] M. Kopp, E. Uhlmann, and C. Kneider, "Experimental investigations of the workpiece-media-interaction and the surface topography formation in centrifugal disc finishing," *Procedia CIRP*, vol. 115, pp. 24–29, 2022, doi: 10.1016/j.procir.2022.10.044.

[8] J. Ding, M. Baumers, E. A. Clark, and R. D. Wildman, "The economics of additive manufacturing: Towards a general cost model including process failure," *Int J Prod Econ*, vol. 237, Jul. 2021, doi: 10.1016/j.ijpe.2021.108087.

[9] M. Schröder, B. Falk, and R. Schmitt, "Evaluation of cost structures of additive manufacturing processes using a new business model," in *Procedia CIRP*, 2015, vol. 30, pp. 311–316. doi: 10.1016/j.procir.2015.02.144.

[10] G. Costabile, M. Fera, F. Fruggiero, A. Lambiase, and D. Pham, "Cost models of additive manufacturing: A literature review," *International Journal of Industrial Engineering Computations*, vol. 8, no. 2, pp. 263–282, Apr. 2016, doi: 10.5267/j.ijiec.2016.9.001.

[11] L. Rickenbacher, A. Spierings, and K. Wegener, "An integrated cost-model for selective laser melting (SLM)," *Rapid Prototyp J*, vol. 19, no. 3, pp. 208–214, Apr. 2013, doi: 10.1108/13552541311312201.

[12] M. Barclift, S. Joshi, T. Simpson, and C. Dickman, "Cost Modeling and Depreciation for Reused Powder Feedstocks in Powder Bed Fusion Additive Manufacturing," in *International Solid Freeform Fabrication Symposium*, 2016.

https://doi.org/10.21872/2023IISE_3627

Proceedings of the IISE Annual Conference & Expo 2023
K. Babski-Reeves, B. Eksioglu, D. Hampton, eds.

Learning with the Pump: Freshmen to Seniors

Abstract ID: 3627

Jun-Ing Ker and Mary Fendley
Industrial Engineering Program
College of Engineering and Science
Louisiana Tech University
Ruston, LA 71272

Abstract

At Louisiana Tech University, all freshmen engineering students are required to take two fundamental engineering courses, Engineering Problem Solving I and II (ENGR 120 and ENGR 121), in their first two quarters of study. Funded by a NSF grant, this curriculum has been the core of the Integrated Engineering Curriculum at Louisiana Tech. These courses require student teams to fabricate a small centrifugal pump driven by a DC motor, to test the pump performance, and to control the temperature and salinity of a small volume of water through a fish tank project. Although designed to benefit all engineering majors, industrial engineering seniors felt differently, that the project-based experience was not directly related to the knowledge learned in their upper-level industrial engineering courses. In the IISE Annual Conference and Expo 2020, the lead author of this paper presented a plan to integrate the centrifugal pump freshmen make in ENGR 120 through several industrial engineering courses in the curriculum. In this paper, the authors present the implementation of this teaching plan in two upper-level industrial engineering courses: Work Design, and Lean Manufacturing and Management Systems. Students in these two courses were asked to determine the standard time for assembling the pumps and set up a Kanban production line to fabricate pumps. Results show that the students felt using a product they were familiar with, helps them connect the concepts learned in classes to real world applications.

Keywords

Industrial Engineering Education Innovation, Engineering Curriculum Integration, Project-Based Learning, Time Study Methods, Kanban Production System

1. Introduction

At Louisiana Tech University, all freshmen engineering students are required to take two fundamental engineering courses, Engineering Problem Solving I and II (ENGR 120 and ENGR 121) in their first two quarters of study. Funded by a NSF grant, this curriculum has been the core of the Integrated Engineering Curriculum at Louisiana Tech. These courses require student teams to fabricate a small centrifugal pump driven by a DC motor, to test the pump performance, and to control the temperature and salinity of a small volume of water through a fish tank project. Figure 1 shows some pump fabrication and assembly operations.

Although designed to benefit all engineering majors, industrial engineering seniors felt differently. They felt that the project-based experience was not directly related to the knowledge learned in their upper-level industrial engineering courses. To bridge the gap between the upper-level industrial engineering courses and the freshman engineering courses, the lead author of this paper perceived the value of integrating the freshman pump into upper-level industrial engineering courses in 2014. This integration started with asking the juniors who had completed INEN 401 Engineering Statistics II, to analyze root causes for low pump efficiency (normally 1% to 3%) and present them in the form of a fishbone diagram. Consequently, they developed a new YouTube video to provide better fabrication and assembly instructions. This fishbone development activity then became an in-class exercise when INEN 401 was taught the following year. Figure 2 shows the fishbone diagram developed by a group of students who took INEN 401 in 2015.

https://doi.org/10.21872/2023IISE_3627

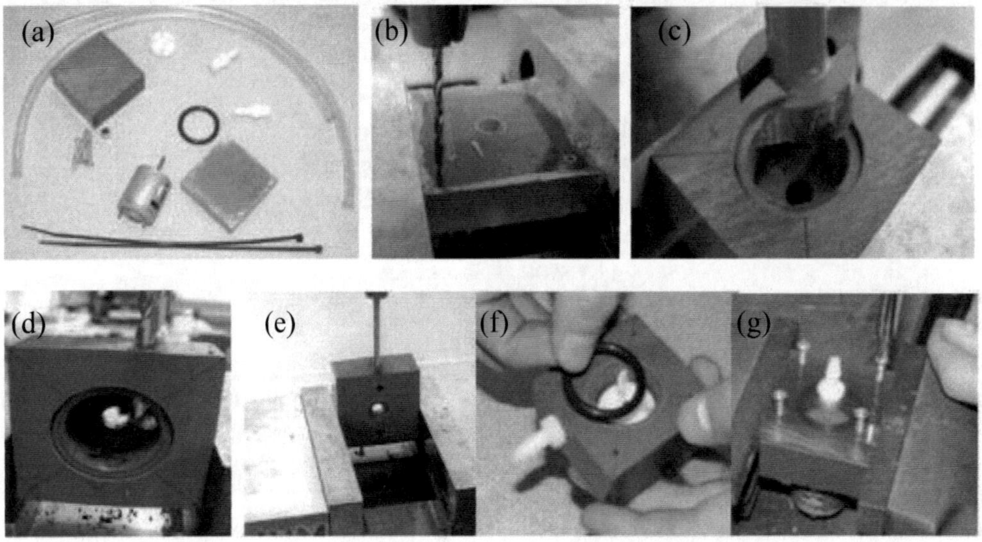

Figure 1: Pump part fabrication and pump assembly **(a)** pump parts **(b)** drilling faceplates **(c)** milling center holes **(d)** drilling side hole for barbed fitting **(e)** drilling side hole for zip tie **(f)** assembly of O-ring and impeller **(g)** assembly of faceplate to pump body

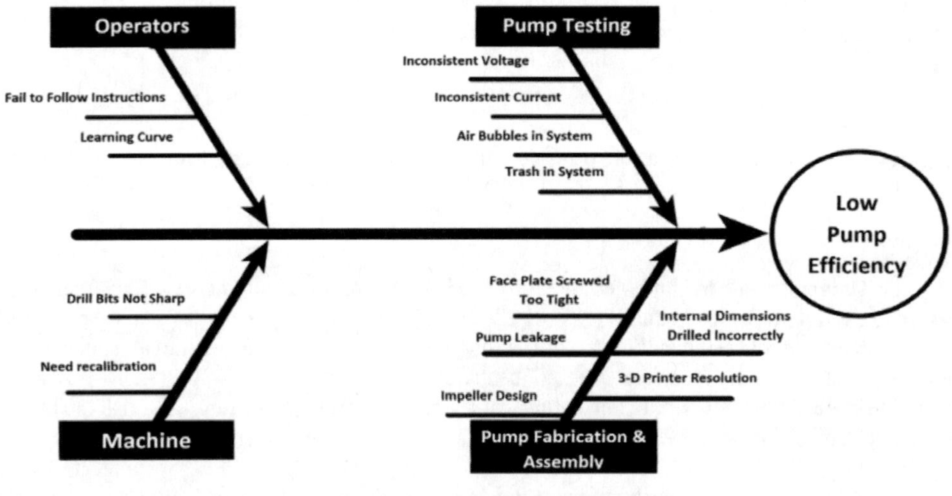

Figure 2: Fishbone diagram demonstrating possible root causes for low pump efficiency

Furthermore, in 2018 a Capstone Design team, comprised of two seniors, was assigned to validate the root causes that appeared on the fishbone diagram and to find effective ways to improve the pump efficiency. The results were reported in the IISE Engineering Lean and Six Sigma Conference 2019, that pump efficiency could be significantly improved by increasing the inner diameters of discharge hose and suction hose from 3/16" to 3/8", as shown in Figure 3.

https://doi.org/10.21872/2023IISE_3627

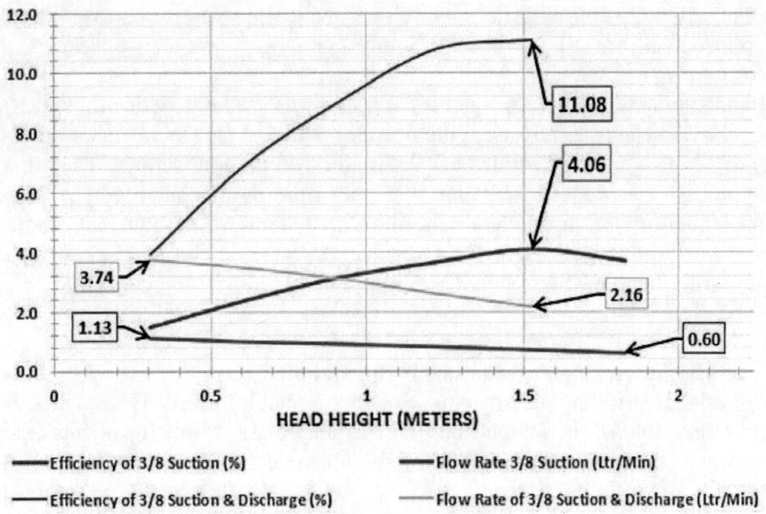

Figure 3: Efficiency and flow rate at different head height

Additionally, in the IISE Annual Conference and Expo 2020, the lead author of this paper presented a plan to integrate the centrifugal pump that freshmen make in ENGR 120 across several industrial engineering courses in the curriculum. In this paper, the authors present the implementation of this teaching plan in two upper-level industrial engineering courses: INEN 409 Work Design and INEN 408/511 Lean Manufacturing and Management Systems in the 2022-2023 academic year.

1. Related Research

In an effort to improve engineering education, recommendations for engineering curriculum design include (among others) establishing relevance of course material, balancing concrete and abstract information, integrating practical elements, and promoting active learning [1, 2]. Integration of project-based learning can address these points. Project-based learning is a method which engages students in learning and acquiring skills while completing a carefully designed project with intentional tasks [3, 4]. Projects in the freshman engineering class are designed to aid in retention while teaching foundational engineering principles, teamwork and communication skills, and connection education and work [5, 6, 7].

Maher, et al. [8] recommended that hands-on projects include additional exercises beyond the scope of the original course work based on feedback from students wishing to continue learning from the projects. They additionally introduced a freshman robot project to two interdisciplinary senior design teams, with positive feedback from the students [8]. Richmond, et al. [9] integrated a project into the curriculum across three courses, allowing retention of the original course design while increasing the time span of the project to help students link concepts with a familiar design. To bridge the gap between freshman engineering courses and upper-level industrial engineering courses, Ker and Khan [10] showed that industrial engineering juniors, after applying quality tools learned in classes, were able to produce pumps with higher efficiency compared to those made by freshmen. In a broader effort, Cline, et al. [11] proposed a model that incorporates projects across an undergraduate mathematics curriculum.

2. Methods

2.1. INEN 409 Work Design

INEN 409, Work Design, is a junior-level industrial engineering course. Topics taught in this class include methods engineering, work measurement, production standards, and workplace analysis and design. It is a three-hour class that includes two 75-minute lectures and a three-hour lab session each week. Prior to the introduction of the pump assembly lab to this class, the lab videos used for direct time studies were out-of-date, and were short processes lasting only a few seconds with each repetition. The pump project was integrated into the INEN 409 course taken by upper-level industrial engineering students starting in the Fall 2021 Quarter. The pump assembly video was made by students who took this class in the Fall 2020 Quarter as a class exercise in which the pump was assembled while recording the process several times (video available at https://www.youtube.com/watch?v=u4czv4oh45w).

This video was used in practicing two different concepts. First, it was used to practice directed time study. There are two elements, in which the first element ends when the operator uses two zip ties to attach the DC motor to the pump body. The second element ends when the operator has fastened down the four screws. The pump assembly provided a real-world assembly process that students were familiar with from the freshman course. Second, the recordings were used in the course laboratory to illustrate the concept of predetermined motion time studies. The students were asked, in small groups, to determine work elements and calculate cycle time in time measurement units (TMU) and seconds using the Maynard Operation Sequence Technique (MOST) and Methods Time Measurement (MTM-1) systems.

2.2 INEN 408/511 Lean Manufacturing and Management Systems

INEN 408 is an upper-level industrial engineering course taken by seniors. INEN 511 is a graduate level course cross listed with INEN 408. INEN 408/511 is a three-hour, lecture only class in which students learn the basic concepts and methods of Lean Manufacturing, such as voice of the customer, quality function deployment, value stream mapping, 6S, 9 forms of wastes, Kanban System, etc. In addition, the course covers many traditional industrial engineering methods aimed at improving productivity such as line balancing, machine coupling, materials handling system design, and plant and office layout.

In the Fall 2022 Quarter, students in INEN 408/511 were asked to conduct an in-class simulation exercise to compare the performance of a push system with a pull Kanban system. Students were divided into two groups of six people (stations) to produce two types of pumps: red pumps (with a red impeller and one water outlet) and blue pumps (with a blue impeller and two water outlets). Production activities occurred at each station were as follows: 1) Station 1: drilling holes on the face plate, 2) Station 2: milling holes into the red pump body and drilling holes for a Barbed fitting and a zip tie, 3) Station 3: milling holes into the blue pump body and drilling holes for two Barbed fittings and a zip tie, 4) Station 4: assembling the bushing with the pump, 5) Station 5: assembling the pump with a red impeller and one Barbed fitting, and 6) Station 6: assembling the pump with blue impellers and two Barbed fittings.

Prior to the simulation exercise, four undergraduate students from each group who have taken ENGR 120 were asked to fabricate three parts and obtain the cycle time per unit for Stations 1 to 4, respectively. Time for Station 5 was obtained from direct time studies of the video used in the INEN 409 lab. Time for Station 6 was obtained by adding time elements for assembling one more Barbed fitting to the blue pump body based on the MTM-1 system learned in INEN 409. During the simulation exercise, the members of each station were to set a timer according to the data that was previously determined. This data showed length of time at each station, so parts could be flowed through stations at correct time intervals according to the production orders given. Table 1 shows cycle times and manufacturing cell times produced by the two teams. In the simulation exercise, everyone was assigned to a station (S). Note that red pumps were produced by following the sequence of S1-S2-S4-S5 while blue pumps were made following the sequence of S1-S3-S4-S6. Figure 4 shows the structure of this pump production system. Figure 5 shows students using the lab equipment to manufacture pump parts for this project.

Table 1: Station times and manufacturing cell times used in the pump simulation exercise

Station	Team 1 (s)	Team 2 (s)
S1	167	175
S2	209	251
S3	338	345
S4	22	27
S5	245	245
S6	260	260
Red Pump Mfg. Cell	643	698
Blue Pump Mfg. Cell	787	807

https://doi.org/10.21872/2023IISE_3627

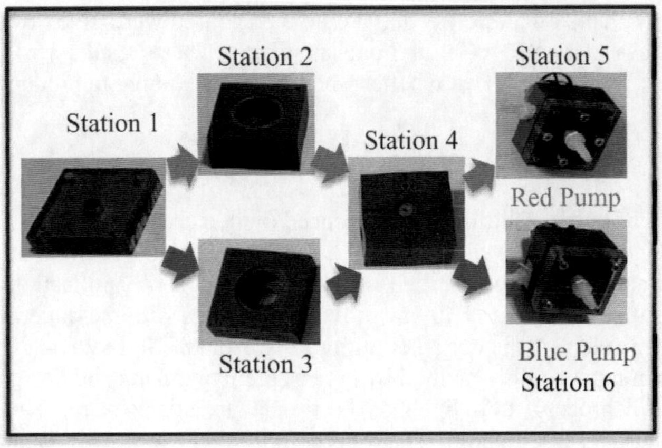

Figure 4: Structure of the pump manufacturing cell

Figure 5: Students fabricating pump parts

The pump production simulation exercise started with producing red pumps and blue pumps, alternatively, based on the push system for twenty minutes. Afterwards, pumps were produced based on the pull system for another twenty minutes. In the pull system, production orders were triggered by Kanban cards periodically issued by the course instructor. Table 2 shows the number of pumps produced and the WIP level reported by the two teams with the push system and the pull system. Note that the WIP inventory included only parts in transition between machines.

Table 2: Number of pumps produced and WIP level by the push system and the pull system

	Push System Team 1	**Push System** Team 2	**Pull System** Team 1	**Pull System** Team 2
Pump produced	1	1	1	1
WIP level	5	6	0	0

As illustrated in Table 2, the superiority of the Pull System in reducing the WIP inventory over the Push System can be clearly seen.

3. Outcomes Assessment Results

3.1. Assessment of INEN 409 Time Study Lab

After implementing the pump videos into the course in the Fall Quarter of 2022, we surveyed the students to get input on whether the pump assembly was helpful in the labs. Nine students were surveyed, and six of the seven respondents stated that, yes, the pump assembly was helpful. One student responded not applicable, as she was a transfer student and did not participate in the freshman pump project. One student commented that this time study project worked well in achieving course outcomes. In addition, comments were positive and included, from a student:

> *"Being able to use what we learned and then practically apply it, rather than it being theoretical. Plus, it was nice to see some stuff we talked about we may have seen in our life already..."*

We also elicited feedback on how the use of the pump assembly for the labs could be improved. Several students commented that the camera angle could be better adjusted. One student asked for more guidance on an acceptable range of calculated time as their answers were not always correct when they were asked to determine the pump assembly times based on MTM-1 and MOST.

3.2 Assessment of INEN 408/511 Pump Assembly Simulation Exercise

A course outcomes survey was conducted at the end of Fall Quarter 2022 to assess the learning effectiveness of this course. Nine out of eleven students returned the surveys. One of the questions asked students if the pump production simulation exercise helped them understand the difference between a push system and a pull system. Sixty-six percent of respondents (six out of nine) responded with "Strongly Agree" or "Agree", one chose "Neutral", and two selected

"Disagree". A more in-depth analysis of the results revealed that the two unfavorable responses were given by two graduate students who did not take ENGR 120 class and INEN 409 at Louisiana Tech. These results were not surprising though. Rather, they support the hypothesis that students learn better when they are exposed to products or processes with which they are familiar.

4. Conclusions

In this paper, the authors present the implementation of pumps that freshmen produced in their first engineering class in two upper-level industrial engineering courses. The assessment results show that undergraduate students felt that these course integration efforts helped them assimilate the concepts learned in class into real world applications. In the future, we plan to further improve the learning outcomes of INEN 409 by providing students with the dimensions of the workstation setup so that they can estimate time elements better when using MOST or MTM-1. We also plan to let students propose ways to improve the current pump assembly method by better organizing parts and designing fixtures for holding the pump body during the assembly process. In INEN 408/511, we plan to establish a real Kanban supermarket in the lab where students fabricate their freshmen pumps to further expose them to a manufacturing environment which closely resembles a real industrial setting. This would also allow graduate students who received their undergraduate education elsewhere to observe the real fabrication processes of pump parts to improve their learning outcomes. The effectiveness of these new implementations will be assessed using data collected over multiple years to increase the sample size. In addition, we plan to develop more laboratory exercises and examples that can be implemented in other industrial engineering courses using the freshmen pump as the platform.

Acknowledgements

The lead author would like to acknowledge the support of the James Emmett Smith Endowed professorship that made this study possible. The James Emmett Smith Endowed Professorship is made available through the State of Louisiana Board of Regents Support Funds.

References

[1] R. M. Felder, D. R. Woods, J. E. Stice, A. Rugarcia, "The Future of Engineering Education: Part 2. Teaching Methods that Work," *Chemical Engineering Education*, vol. 34, no. 1, pp. 26-39, Winter 2000.

[2] J. E. Froyd, P. C. Wankat, K. A. Smith, "Five Major Shifts in 100 Years of Engineering Education," in *Proceedings of the IEEE*, vol. 100, no. Special Centennial Issue, pp. 1344-1360, 13 May 2012, doi: 10.1109/JPROC.2012.2190167.

[3] T. Markham, J. Larmer, J. Ravits, "*Project Based Learning Handbook: A Guide to Standards-focused Project-Based Learning*, 2nd Ed. Novato, CA: Buck Institute for Education, 2003.

[4] J. S. Lee, S. Blackwell, J. Drake, K.A. Moran, "Taking a Leap of Faith: Redefining Teaching and Learning in Higher Education Through Project-Based Learning. *Interdisciplinary Journal of Problem-Based Learning*, vol. 8, no. 2, 2014, doi.org/10.7771/1541-5015.1426.

[5] M. E. Besterfield-Sacre, C.J. Atman, L.J. Shuman, "Characteristics of Freshman Engineer Students: Models for Determining Student Attrition in Engineering," *Engineering Education*, vol. 86, no.1, pp. 139-150, January 2013.

[6] R. M. Abdulaal, A. M. Al-Bahi, A. Y. Soliman, F. I. Iskanderani, "Design and Implementation of a Project-Based Active/cooperative Engineering Design Course for Freshmen", *European Journal of Engineering Education*, vol. 35, no. 4, pp. 391-402, July 2011, doi: 10.1080/03043797.2011.598498.

[7] A. Kolmos, J. E. Holgaard, N. R. Clausen, "Progression of Student Self-assessed Learning Outcomes in Systemic PBL", *European Journal of Engineering Education*, vol. 46, no.2, pp. 67-89, 2021. doi: 10.1080/03043797.2020.1789070.

[8] R. C. Maher, J. Becker, T. Sharpe, J. Peterson, B.A. Towle, "Development and Implementation of a Robot-based Freshman Engineering Course," in *Proceedings of the 2005 American Society for Engineering Education Annual Conference & Exposition, June 2005.*

[9] W. Richmond, D. Banerjee, B.J.White, "Integrating Curriculum Across Courses in the Same Semester and Across Semesters Using a Service Learning Project," *Decision Sciences Journal of Innovative Education*, vol. 6, no. 2, pp. 509-513, July 2008.

[10] J. Ker, M. Khan, "Vertical Integration of Engineering Courses for Effective Learning of Continuous Quality Improvement Concepts," in *Proceedings of the 2017 ASEE Gulf Southwest Section Annual Conference*, Dallas, TX, U.S.A., March 12-14, 2017, doi: 10.18260/1-2-1153-33831.

[11] K. Cline, J. Fasteen, A. Francis, E. Sullivan, T. Wendt, "A Vision for Projects Across the Mathematics curriculum," *PRIMUS*, vol. 30, no. 4, pp. 379-399, March 2020, doi: 10.1080/10511970.2019.1600176.

https://doi.org/10.21872/2023IISE_3687

Proceedings of the IISE Annual Conference & Expo 2023
K. Babski-Reeves, B. Eksioglu, D. Hampton, eds.

A Column Generation-Based Heuristic Algorithm for Solving a Cooperative Two-Echelon Truck and Drone Routing Problem in Humanitarian Relief Operations

Abstract ID: 3687

Md Mahbubar Rahman[1*], Arup Dey[2], Nita Yodo[1] and David Grewell[1]

[1]Department of Industrial and Manufacturing Engineering, North Dakota State University, Fargo, ND 58102, USA
[2]School of Engineering, Math and Technology, Navajo Technical University, Crownpoint, NM 87313, USA

Abstract

In post-disaster relief operations, humanitarian crews deliver emergency supplies to disaster-affected people to alleviate their suffering and casualties. Humanitarian teams find this task challenging because the damage in critical infrastructures often makes it impossible for delivery vehicles, such as trucks, to reach disaster-affected people. Aerial vehicles, such as helicopters and drones, can deliver emergency supplies to trapped people, as they do not need to use ground-based transportation networks. Thus, affected people on the undisrupted and disrupted sides of the transportation network can be served using cooperative trucks and drones. This paper proposes a cooperative two-echelon truck and drone routing problem for the emergency supply delivery operation and a column generation (CG)-based heuristic algorithm for solving the problem. In the problem setup, disaster-affected people on the undisrupted side of the transportation network receive truck delivery, and those on the disrupted side receive drone delivery. A mixed-integer linear programming (MILP) model is formulated for this problem. The MILP model is solvable for small problems containing less than ten nodes within five minutes. For solving practical size problems containing 20 or more nodes, the MILP model is decomposed into a path-based master problem and truck-and-drone subproblems to allow a column generation framework to solve the problem. Each subproblem is solved using a novel hybrid algorithm based on a genetic algorithm, tabu search, and traveling salesman heuristics. The CG-based heuristic finds a feasible solution to a practical size problem with an optimality gap of less than five percent within five minutes.

Keywords

Column generation, cooperative truck and drone routing, humanitarian logistics, last-mile delivery

1. Introduction

Affected people need emergency supplies during and after a disaster. Timely delivery of these supplies helps reduce suffering and casualties. Humanitarian crews use aerial and ground vehicles to deliver the supplies. Typically, trucks and helicopters are used as ground and aerial vehicles, respectively. Helicopters are expensive and often become inadequate at the onset of a catastrophic disaster, such as a hurricane or tsunami. A cheap and widely available alternative to a helicopter is a drone. In this paper, an emergency supply delivery problem is defined using cooperative trucks and drones, and a column generation (CG)-based heuristic algorithm is proposed to solve the problem. The rest of the paper is organized as follows. The emergency supply delivery problem is defined in Section 2. The related research in the field of study is presented in Section 3. Section 4 describes the methodology for the CG-based heuristic algorithm. The result of a case study problem solved using the CG-based algorithm is demonstrated in Section 5. Finally, the concluding remarks of this paper are given in Section 6.

2. Problem Description

For post-disaster emergency supply delivery, a last-mile distribution problem is defined in this section as a cooperative two-echelon truck and drone routing problem. In the problem setting, drones work in tandem with trucks to deliver emergency supplies. Truck-accessible people are served by trucks. On the other hand, truck-inaccessible people are served by drones. Each truck route starts from and ends at the depot. Every drone is deployed from and landed on the

https://doi.org/10.21872/2023IISE_3687

carrier truck while the truck is parked at a truck-accessible location (rendezvous/satellite node). Once all the deployed drones are back to the carrier truck, the truck moves to the next truck-accessible location. Truck routes and drone routes are called first-echelon and second-echelon routes, respectively. Assuming the set C contains all the customer nodes, and they are partitioned into truck nodes, rendezvous nodes (which are truck nodes but allowed to deploy and land drones), and drone nodes as C^t, S, and C^d, respectively, $C = C^t \cup C^d$, and $S \subset C^t$. The set of the depot node is denoted as O, nodes of the ground transportation network $V = O \cup C^t$, and nodes of the drone network $V' = S \cup C^d$. Therefore, trucks and drones traverse through the edges of the undirected graph $G(V, E_1)$ and complete graph $H(V', E_2)$, respectively, where E_1 is the set of functioning edges of the transportation network containing V nodes, and E_2 is the set of all edges among the nodes in V'. The objective is to minimize the total routing cost of trucks and drones. The assumptions of this problem are as follows:

- Multiple number of trucks are available at the depot
- Each truck can carry multiple drones
- Drones are subject to payload and flying range limitations
- A drone can serve multiple people in a single flight
- Truck customers are served by trucks and drone customers are served by drones
- Every deployed drone must return to its carrier truck at the end of its flight.

For developing the mathematical formulations of the proposed last-mile distribution system, the necessary parameters are presented in Table 1, and the variables are listed in Table 2.

Table 1: List of parameters

Symbol	Parameter Name
K_1	Number of first-echelon trucks at the depot
K_2	Number of drones carried by each truck
Q_2	Payload capacity of each drone
d_i	Demand of customer $i \in C$
c_{ij}	Road distance between nodes i and j, where $(i,j) \in E_1$
d_{ij}	Aerial distance between nodes i and j, where $(i,j) \in E_2$
t_{ij}	Drone travel time between nodes i and j, where $(i,j) \in E_2$
T	Flying range of a drone

Table 2: List of variables

Symbol	Variable Name
f_{ij}	Binary variable and its value equal 1 if a truck visits node j from node i, where $i, j \in V$.
x_{ij}^{sk}	Binary variable and its value equal 1 if the arc $(i,j) \in E_2$ is traversed by a drone deployed from satellite $s \in S$.
n_i	Continuous variable and takes the potential value of node i, where $i \in C^t$.
l_i	Continuous variable and takes the potential value of node $i \in S \cup C^d$.
t_s^k	Flight time of drone $k \in K_2$ deployed from satellite node $s \in S$.

The mixed-integer linear programming (MILP) formulation for the cooperative two-echelon truck and drone routing problem is

$$\min \sum_{(i,j)\in E_1} c_{ij} f_{ij} + \sum_{k\in K_2} \sum_{s\in S} \sum_{(i,j)\in E_2} d_{ij} x_{ij}^{sk} \tag{1a}$$

$$\sum_{i:(i,j)\in E_1} f_{ij} = 1 \qquad \forall j \in C^t \tag{1b}$$

$$\sum_{j:(i,j)\in E_1} f_{ij} = \sum_{j:(j,i)\in E_1} f_{ji} \qquad \forall i \in C^t \tag{1c}$$

$$n_i - n_j + f_{ij} \le |C^t| - 1 \qquad \forall i, j \in C^t, i \ne j \tag{1d}$$

$$\sum_{j:(0,j)\in E_1} f_{0j} = |K_1| \tag{1e}$$

https://doi.org/10.21872/2023IISE_3687

$$\sum_{s\in S}\sum_{k\in K_2}\sum_{j:(c,j)\in E_2} x_{cj}^{sk} = 1 \qquad \forall c \in C^d, \exists t_{cs}: t_{cs} \leq \frac{T}{2} \tag{1f}$$

$$\sum_{j:(i,j)\in E_2} x_{ij}^{sk} = \sum_{j:(j,i)\in E_2} x_{ji}^{sk} \qquad \forall i \in S \cup C^d, \forall s \in S, \forall k \in K_2 \tag{1g}$$

$$l_i - l_j + Q_2 \sum_{k\in K_2}\sum_{s\in S} x_{ij}^{sk} \leq Q_2 - d_j \qquad \forall i,j \in C^d, i \neq j \tag{1h}$$

$$\sum_{k\in K_2}\sum_{j:(s,j)\in E_2} x_{sj}^{sk} \leq |K_2| \qquad \forall s \in S \tag{1i}$$

$$\sum_{k\in K_2}\sum_{(i,j)\in E_2, j\in S, j\neq s} x_{ij}^{sk} = 0 \qquad \forall s \in S \tag{1j}$$

$$t_s^k = \sum_{(i,j)\in E_2} x_{ij}^{sk} t_{ij} \qquad \forall s \in S, \forall k \in K_2 \tag{1k}$$

$$t_s^k \leq T, \qquad \forall s \in S, \forall k \in K_2 \tag{1l}$$

$$x_{ij}^{sk} \in \{0,1\} \qquad \forall (i,j) \in E_2, \forall s \in S, \forall k \in K_2 \tag{1m}$$

$$f_{ij} \in \{0,1\} \qquad \forall (i,j) \in E_1 \tag{1n}$$

$$n_i \in R \qquad \forall i \in C^t \tag{1o}$$

$$l_i \in R \qquad \forall i \in C^d \tag{1p}$$

$$t_s^k \in R_+, \qquad \forall s \in S, \forall k \in K_2 \tag{1q}$$

The objective function (1a) minimizes the global routing cost of the trucks and drones. Constraint (1b) ensures that each truck customer receives emergency supplies from a truck. Constraints (1c) ensure the flow balance of truck routes. Constraints (1d) are the Miller-Tucker-Zemlin (MTZ) subtour elimination constraints of truck routes. Constraint (1e) ensures that $|K_1|$ number of trucks must be deployed from the depot node. Constraints (1f) ensure that a drone customer is satisfied if it is within the flying range limitation of a drone. Constraints (1g) are the flow balance constraints of the drone routes. Constraints (1h) are the capacitated MTZ constraints for the drone routes. Constraints (1i) ensure that maximum $|K_2|$ number of drones can be deployed from a satellite node. Constraints (1j) ensure that no drone should have different launching and landing rendezvous nodes in a single flight. Constraints (1k) and (1l) ensure that each drone path must satisfy the maximum flying time limitation. Constraints (1m) to (1q) impose restrictions on the binary and continuous variables of the model.

3. Related Research

The research for cooperative truck and drone routing (CTDR) has gained much attention since the seminal work of Murray and Chu [1]. Murray and Raj [2] propose a multi-drone flying sidekick traveling salesman problem (TSP). Ham [3] proposes parallel drone scheduling TSP with drop off and pick up. Agatz et al. [4] propose TSP with drone considering the concept of drone operation. Kitjacharoenchai et al. [5] propose multiple TSP with drone models and solution heuristics. Although many contributions are made in CTDR research, very few of them appear to be practical in post-disaster emergency supply delivery operations because of the presence of transportation network disruption. Rabta et al. [6] propose a MILP model to deliver relief using drones. The drones are deployed from a central depot, and several battery charging stations are made available for the drones if they need to be charged while performing the deliveries. Rahman [7] proposes three MILP models for delivering emergency supplies using trucks and drones in a post-disaster situation. Rahman et al. [8] propose a greedy randomized adaptive search procedure with a path relinking metaheuristic for solving cooperative ground vehicles and unmanned aerial vehicles routing problem in humanitarian operations. Faiz and Vogiatzis [9] propose a two-echelon VRP framework to identify trapped people and deliver emergency aid in a post-disaster situation, and propose a column generation-based heuristic algorithm to solve the problem. Zhang et al. [10] propose a MILP formulation for humanitarian relief network assessment and propose a column generation-based algorithm to solve the problem.

In the context of a post-disaster situation, finding a solution for truck and drone routes for last-mile distribution of emergency supplies is a challenging task because of the disruption in the transportation network and the limitations of delivery drones. To overcome the challenge, in this study, a cooperative two-echelon truck and drone routing problem is developed, and a CG-based heuristic algorithm is proposed to solve the problem in a timely manner.

https://doi.org/10.21872/2023IISE_3687

4. Methodology

4.1. Path-based Formulation

The MILP model developed in Section 2 is NP-hard and unsuitable for decomposition because of the symmetric structure of the model variables. To decompose the problem, a path-based reformulation is developed. The path-based model requires feasible truck and drone paths. Assuming the set P_t contains feasible truck paths, the set P_d contains feasible drone paths, and the set P_d^s, where $s \in S$ contains all the feasible drone paths from satellite node s, then $P_d = \cup_{s \in S} P_d^s$. The variables and parameters of the reformulation are described in Tables 3, and 4, respectively.

Table 3: Variable of the path-based model

Symbol	Variable Name
π_t	Binary variable and its value is 1 if truck path $t \in P_t$ is selected, zero otherwise.
θ_d	Binary variable and its value is 1 if drone path $d \in P_d$ is selected, zero otherwise.
a_{ij}	Binary variable and its value is 1 if the arc $(i,j) \in E_1$ is visited by a truck, zero otherwise.
b_{ij}	Binary variable and its value is 1 if the arc $(i,j) \in E_2$ is visited by a drone, zero otherwise.

Table 4: Parameters of the path-based model

Symbol	Parameter Name
e_d^{ij}	If a drone path $d \in P_d$ contains the arc $(i,j) \in E_2$, its value equals 1, zero otherwise.
e_d^i	If a drone path $d \in P_d$ contains the node $i \in C^d$, its value equals 1, zero otherwise. Mathematically, $e_d^i = \sum_{j \in C^d \cup \{s\}} e_d^{ij}$, where $s \in S$.
g_d^s	If a drone path $d \in P_d$ starts from and stops at node $s \in S$, its value equals 1, zero otherwise.
h_t^{ij}	If a truck path $t \in P_t$ contains the arc $(i,j) \in E_1$, its value equals 1, zero otherwise.
h_t^i	If a truck path $t \in P_t$ contains the node $i \in S \cup C^t$, its value equals 1, zero otherwise.
f_d	Flight duration of drone path $d \in P_d$, and equals $\sum_{i \in C^d \cup \{s\}} \sum_{j \in C^d \cup \{s\}} e_d^{ij} t_{ij}$, where $s \in S$.
o_d	Payload of a drone traveling through the drone path $d \in P_d$. Mathematically, o_d is equal to $\sum_{i \in C^d} e_d^i d_i$.

The path-based reformulation of the proposed MILP model is given as follows.

$$\min \sum_{(i,j) \in E_1} c_{ij} a_{ij} + \sum_{(i,j) \in E_2} d_{ij} b_{ij} \tag{2a}$$

$$\sum_{t \in P_t} \pi_t = |K_1| \tag{2b}$$

$$\sum_{t \in P_t} \pi_t h_t^{ij} = a_{ij} \quad \forall (i,j) \in E_1 \tag{2c}$$

$$\sum_{t \in P_t} \pi_t h_t^i = 1 \quad \forall i \in C^t \tag{2d}$$

$$\sum_{d \in P_d} \theta_d g_d^s \leq |K_2| \quad \forall s \in S \tag{2e}$$

$$\sum_{d \in P_d} \theta_d e_d^{ij} = b_{ij} \quad \forall (i,j) \in E_2 \tag{2f}$$

$$\sum_{d \in P_d} \theta_d e_d^i = 1 \quad \forall i \in C^d, \ \exists (i,s): t_{is} \leq \frac{T}{2} \tag{2g}$$

$$\pi_t \in \{0,1\} \quad \forall t \in P_t \tag{2h}$$

$$\theta_d \in \{0,1\} \quad \forall d \in P_d \tag{2i}$$

$$a_{ij} \in \{0,1\} \quad \forall (i,j) \in E_1 \tag{2j}$$

$$b_{ij} \in \{0,1\} \quad \forall (i,j) \in E_2 \tag{2k}$$

The objective function (2a) minimizes the total cost of truck and drone paths. Constraint (2b) makes sure that $|K_1|$ number of truck paths are selected. Constraints (2c) ensure that if a truck path traverses through an arc $(i,j) \in E_1$, then the variable a_{ij} gets value 1, and zero otherwise. Constraints (2d) ensure that each truck customer is served by a

https://doi.org/10.21872/2023IISE_3687

truck path. Constraints (2e) ensure that maximum $|K_2|$ number of drone paths are generated from a rendezvous node $s \in S$. Constraints (2f) ensure that if a drone path traverses through $(i,j) \in E_2$, then the variable b_{ij} gets value 1, and zero otherwise. Constraints (2g) ensure that each drone customer within the drone flying range must be served by a drone. Constraints (2h) to (2k) impose binary restrictions on the variables of this model.

4.2. Column Generation-based Heuristic

For developing a CG-based heuristic, the path-based formulation developed in Section 4.1 is relaxed into a restricted master problem (RMP). Pricing functions for truck and drone paths are developed from the dual of the RMP. Then, a CG-based heuristic algorithm is developed using the RMP and pricing functions.

Restricted master problem. The master problem is generated by relaxing the integrality constraints of the variables in the path-based formulation, constraints (2a) to (2k). The master problem is named restricted master problem if the sets P_t *and* P_d contain a subset of all possible truck and drone paths, respectively.

Truck path pricing function. The pricing function of a truck path is

$$-\alpha - \sum_{(i,j)\in E_1, j\notin o} f_{ij}(\beta_{ij} + \gamma_j) - \sum_{(i,j)\in E_1, j\in o} f_{ij}\beta_{ij} \tag{3}$$

where $\alpha, \beta, and\ \gamma$ are dual values of constraints (2b), (2c), and (2d), respectively. The parameter f_{ij} equals 1 if the truck path traverses the arc $(i,j) \in E_1$; otherwise equals 0.

Drone path pricing function. The pricing function of a drone path starting from satellite node $s \in S$ is

$$-\lambda_s - \sum_{(i,j)\in E_2, j\neq s} x_{ij}(\sigma_{ij} + \rho_j) - \sum_{(i,j)\in E_2, j=s} \sigma_{ij} x_{ij} \tag{4}$$

where λ, σ, and ρ are dual values of constraints (2e), (2f), and (2g), respectively. The parameter x_{ij} equals 1 if a drone path traverses the arc $(i,j) \in E_2$; otherwise equals 0.

Truck and drone paths generation. The truck and drone paths are generated using a novel hybrid genetic algorithm. The hybrid algorithm is developed using genetic algorithm, tabu search, and Lin-Kernighan's 2-opt heuristic.

Column generation-based algorithm. The algorithm can be formally stated as follows:

Algorithm 1: Column generation-based algorithm for cooperative truck and drone routing problem

```
Initialize a set of truck and drone paths.
Solve the restricted master problem (RMP) and obtain the values of dual variables.
While True do
    Generate truck and drone paths using the hybrid algorithm.
    Calculate the reduced cost of truck and drone paths using their respective pricing functions.
    If all the new paths have positive reduced costs, then
        Break while loop
    End if
    Add the new paths having negative reduced cost to RMP and solve the problem for dual solution.
End while
Solve the RMP as an integer programming problem to find an integer solution.
```

5. Results

A case study is conducted considering the Fargo-Moorhead area in North Dakota as a disaster-affected area. The cooperative truck and drone routing problem is populated with 44 randomly generated affected people locations, which are categorized as 24 truck-accessible nodes (TCus) and 20 drone-accessible nodes (DCus). Among the truck nodes, 13 nodes are categorized as satellite nodes, and Euclidian distance is used to populate the arc length between each pair of nodes. Figure 1 shows: (1) the locations of the depot, truck customers, and drone customers are shown as red, green, and yellow placeholders, respectively, and (2) the problem parameters, truck and drone route solution, and results found using the CG-based heuristic. The CG-based heuristic was implemented with a desktop computer with configurations of: Processor - Intel(R) Core (TM) i7-4770 CPU @ 3.40GHz 3.40 GHz; Ram - 32.0 GB; System type - 64-bit operating system, x64-based processor.

Route Solution

Experimental Parameters & Results

No. Depot: 1	No. Truck Customers: 24
No. Drone Customers: 20	No. Rendezvous Nodes: 13
Max. Truck Traveling Range (meters): 75000	No. Trucks: 3
Max. Drone Flying Range (meters): 15000	Total Cost (meters): 131012
No. Drones Carried by a Truck: 2	Lower Bound (meters): 125579.74
Max. Drone Capacity: 3	Optimality Gap (%): 4.3 (<5)

Figure 1: Route solution and other results obtained from the column generation-based heuristic

The solution finds three routes for trucks: Route 1 (Depot – TCus 3 – TCus 1 – TCus 23 – TCus 24 – TCus 21 – TCus 20 – TCus 22 – TCus 2 – Depot); Route 2 (Depot – TCus 5 – TCus 11 – TCus 7 – TCus 9 – TCus 13 – TCus 12 – TCus 8 – TCus 10 – Depot); and Route 3 (Depot – TCus 6 – TCus 4 – TCus 17 – TCus 15 – TCus 14 – TCus 16 – TCus 18 – TCus 19 – Depot). The truck Routes 1, 2, and 3 are colored in black, blue, and red polygons, respectively, and the drone routes are presented as yellow polygons in Figure 1. Due to space limitations, all the drone routes are not explicitly presented here. The CG-based heuristic provides a near-optimal solution to the study problem with an optimality gap of less than five percent within five minutes, whereas exact solvers take multiple hours to provide a similar solution. Thus, in post-disaster humanitarian operations, the CG-based heuristic will be of critical importance as the delivery operations can be performed immediately without waiting for hours or days for the exact solution.

6. Conclusions

In this work, a novel cooperative two-echelon truck and drone routing problem is proposed to address post-disaster delivery of emergency supplies. A CG-based heuristic algorithm is developed to solve the problem. A case study is conducted in the Fargo-Moorhead area to demonstrate the efficiency of the proposed method. The CG-based procedure has advantages over exact solution methods because it provides quality solutions within a tractable amount of time. Future works would be directed to improve the performance of the CG-based heuristic algorithm by incorporating the truck and drone path pricing functions in the hybrid genetic algorithm.

References

[1] C. C. Murray and A. G. Chu, "The flying sidekick traveling salesman problem: Optimization of drone-assisted parcel delivery," *Transportation Research Part C: Emerging Technologies,* vol. 54, pp. 86-109, 2015.

[2] C. C. Murray and R. Raj, "The multiple flying sidekicks traveling salesman problem: Parcel delivery with multiple drones," *Transportation Research Part C: Emerging Technologies,* vol. 110, pp. 368-398, 2020.

[3] A. M. Ham, "Integrated scheduling of m-truck, m-drone, and m-depot constrained by time-window, drop-pickup, and m-visit using constraint programming," *Transportation Research Part C: Emerging Technologies,* vol. 91, pp. 1-14, 2018.

[4] N. Agatz, P. Bouman, and M. Schmidt, "Optimization approaches for the traveling salesman problem with drone," *Transportation Science,* vol. 52, no. 4, pp. 965-981, 2018.

[5] P. Kitjacharoenchai, M. Ventresca, M. Moshref-Javadi, S. Lee, J. M. A. Tanchoco, and P. A. Brunese, "Multiple traveling salesman problem with drones: Mathematical model and heuristic approach," *Computers & Industrial Engineering,* vol. 129, pp. 14-30, 2019.

[6] B. Rabta, C. Wankmüller, and G. Reiner, "A drone fleet model for last-mile distribution in disaster relief operations," *International Journal of Disaster Risk Reduction,* vol. 28, pp. 107-112, 2018.

[7] M. M. Rahman, "Two-Echelon Vehicle Routing Problems Using Unmanned Autonomous Vehicles," 2017.

[8] M. M. Rahman, T. Afrina, E. Scully, and N. Yodo, "Two-Echelon Cooperative Vehicle and UAV Routing for Last Mile Humanitarian Relief Operations," 2022: Institute of Industrial and Systems Engineers (IISE), pp. 1-6.

[9] T. I. Faiz and C. Vogiatzis, "Computational Approaches for Solving Two-Echelon Vehicle and UAV Routing Problems for Post-Disaster Humanitarian Operations," *arXiv preprint arXiv:2001.06456,* 2020.

[10] G. Zhang, N. Zhu, S. Ma, and J. Xia, "Humanitarian relief network assessment using collaborative truck-and-drone system," *Transportation Research Part E: Logistics and Transportation Review,* vol. 152, p. 102417, 2021.

https://doi.org/10.21872/2023IISE_3735

Proceedings of the IISE Annual Conference & Expo 2023
K. Babski-Reeves, B. Eksioglu, D. Hampton, eds.

A Mathematical Programming Model for Bi-objective Covid-19 Vaccine Distribution

Labiba Noshin Asha[1,*] and Harun Pirim[1]
[1]North Dakota State University, Fargo, ND
[*]Corresponding author: labiba.asha@ndsu.edu

Abstract

Countries are applying different strategies to prevent the spread of the disease since the Covid 19 outbreak. Vaccination is one of the best ways to mitigate the impact of the disease on human health. However, allocation and distribution of Covid 19 vaccines are not easy, especially in developing countries with a very high vaccine demand. The paper presents a bi-objective mixed integer linear programming model for Covid-19 vaccine distribution. The proposed MILP model ensures the optimal allocation of vaccines among the vulnerable population with the minimum cost and travel time. A simple supply chain network is considered in this case with the vaccine manufacturers, distribution centers, vaccine storage warehouses, and customer groups. The model demonstrates a multi-product, multi-period scenario with different types of Covid 19 vaccines. The vaccines require certain refrigeration systems like ultra-cold, very cold, and cold refrigerators in the storage facilities. As a result, some of the practical assumptions are considered during the model formulation such as time-dependent capacity, budget limitation, refrigeration system of the storage facility, grouping of potential customers, and vaccine shipping options like standard shipping and express shipping. To validate the effectiveness of the model a case study has been demonstrated. The results highlight the model's performance in dealing with multiple objectives simultaneously.

Keywords
Covid 19 Vaccine Distribution, Multi-objective, Mathematical Model, Mixed Integer Linear Programming

1. Introduction

Covid-19 vaccine distribution has presented several challenges, including limited supply, logistical challenges, and the need to prioritize certain groups for vaccination. One of the main challenges with the distribution of Covid-19 vaccines is the limited supply. There are currently several vaccines that have been approved for use, but their production has not been able to keep up with the demand. This has led to shortages in some areas and the need to prioritize certain groups for vaccination.

Duijzer et al. examined the literature on the vaccine supply chain in order to better understand the challenges and opportunities in this field. In their study, they categorized the existing literature into four main groups: product modeling, production modeling, allocation modeling, and distribution modeling [1]. Mathematical modeling has played an important role in the distribution of Covid-19 vaccines. Mathematical models can be used to demonstrate different scenarios and evaluate the potential impacts of different strategies for vaccine distribution. This can help decision makers to understand the tradeoff involved in different approaches and to identify the best solutions to the challenges posed by the pandemic. Tsadikovich [2], Saif and Elhedhli [3], Zandkarimkhani et al. [4] provided multiple examples of the application of mathematical modeling in vaccine distribution.

Multi-objective optimization is used to find the best solution to a problem that has multiple conflicting objectives. This type of optimization considers multiple criteria simultaneously and attempts to find a solution that balances or tradeoff between them. The current study proposes a multi-objective MILP model for optimizing the total cost and total lead time simultaneously for Covid 19 vaccine distribution considering a multi period multi product scenario. The motivation of the model was taken from Tavana et al. [5] where the proposed a multi-product mathematical model considered potential waiting time between ordering and delivery time. We extended the model in multi-objective scenario to evaluate the performance of the model considering optimizing the overall cost and lead time.
The remainder of this paper is organized as follows. Section 2 presents the literature review. Section 3 provides the problem statement. The proposed mathematical model is presented in Section 4. Section 5 highlights the case Study. The results of the proposed model are narrated in Section 6. Finally, conclusions are drawn at Section 7.

https://doi.org/10.21872/2023IISE_3735

2. Literature Review

Applications of multi-objective optimization are noticeable in Covid 19 vaccine distribution. Tang et al. [6] looks into a novel multi-period vaccination planning problem that simultaneously aims to reduce the operational cost and the overall travel distance of vaccine recipients. The bi-objective MILP model was initially solved by weighted-sum and an ε-constraint methods, later a tailored genetic algorithm was developed to obtain a better solution result. A bi-objective MILP model is developed by Govindan et al. [7] to minimize the total cost and risks during the Covid vaccines waste handling simultaneously. Fuzzy Goal Programming approaches are considered for solving the proposed model. Mohammadi et al. [8] describes a MINLP model to minimize both the expected number of deaths and distribution costs of the vaccine campaigns. They investigated the resilience of the distribution network through a scenario-based robust-stochastic optimization approach. Our contribution is on developing a multi-product, multi period, multi objective MILP model that simultaneously minimizes the total cost and travel time during vaccine distribution depending on the vaccine shipping delivery option considered for transportation.

3. Problem Statement

In this paper, a multi-objective model for COVID-19 vaccine distribution is presented. The model ensures the optimal allocation of COVID-19 vaccines by minimizing the overall cost and travel time simultaneously. Some general assumptions are considered for this model formulation.

- The model is considered as multi-period and multi-product.
- Time-dependent capacities are considered for the vaccine manufacturers.
- One order can be placed with each manufacturer in each period.
- The model determines the location of the DC and the capacities of the DC are known.
- Air transport is considered when shipping the vaccines from the manufacturers to the DC. Truck transportation is considered while shipping the vaccine from DC to the state warehouses.
- Both Standard and Express delivery services can be used to ship any type of vaccine from manufacturer to DC and from DC to warehouse.
- Possible future storage is considered in this model.
- Ultra cold refrigeration facilities can be added to DC if it already has a very cold refrigeration system.

4. Mathematical Model

4.1. Indices

Indices of the model are mentioned below.

Notation	Description
V	$Set\ of\ vaccine\ types\ (v = 1,2,\ldots,k)$
D	$Set\ of\ Distribution\ Centers\ (DC)\ (d = 1,2,\ldots,l)$
W	$Set\ of\ State's\ Warehouses\ (w = 1,2,\ldots,m)$
G	$Set\ of\ Population\ Groups\ required\ vaccines\ (g = 1,2,\ldots,n)$
O	$Set\ of\ Ordering\ period\ (o = 1,2,\ldots,r)$
P	$Set\ of\ Delivery\ period\ (p = 1,2,\ldots,s)$
T	$Set\ of\ Transportation\ Option\ (t = 1,2,\ldots,q)$

4.2. Parameters

Dem_{gw}	Demand of Group g in State Warehouse w
OC_{vp}	Ordering cost for vaccine v for the delivery period p
FC_d	Fixed cost of establishing DC d with cold refrigeration
FCC_d	Fixed cost of establishing DC d with very cold refrigeration
PC_v	Purchase Cost of Vaccine v
TC_{vdt}	Unit Ttransportation cost of shipping vaccine v to DC d using shipping option t
TCC_{vdwt}	Unit Tansportation cost of shipping vaccine v from DC d to state warehouse w using shipping option t
HC_{vw}	Holding cost of storing v vaccines in w warehouse
Cap_{vop}	Maximum capacity of producing vaccine v if the ordering period is o and delivery period is p
$CapC_d$	Capacity of distribution center d for cold refrigeration vaccine
$CapVC_d$	Capacity of distribution center d for very cold refrigeration vaccine

$CapUC_d$	Capacity of distribution center d for ultra-cold refrigeration vaccine
CR_g	Minimum percentage coverage rate for group g
RV_v	$\begin{cases} 1 & if\ vaccine\ v\ requires\ very\ cold\ or\ ultra\ cold\ refrigerator \\ 0 & Otherwise\ vaccine\ v\ requires\ cold\ refrigeration \end{cases}$
RT_v	$\begin{cases} 1 & if\ vaccine\ v\ requires\ ultra\ cold\ refrigerator \\ 0 & Otherwise\ vaccine\ v\ requires\ very\ cold\ refrigeration \end{cases}$
$big\ M$	A very big number
$TTmnd_{vdt}$	Unit travel time of shipping vaccine v to DC d using shipping option t by air freight on a single trip
$TTdw_{vdwt}$	Unit travel time of shipping vaccine v from DC d to state warehouse w using shipping option t by truck on a single trip
EC	Extra cost to convert very cold refrigerator to ultra-cold refrigerator

4.3. Decision Variables

X_{vop}	$\begin{cases} 1\ if\ vaccine\ v\ is\ ordered\ at\ period\ o \\ to\ deliver\ at\ period\ p \\ 0\ Otherwise \end{cases}$
Y_d	$\begin{cases} if\ DC\ d\ is\ capable\ of\ handling\ vaccines \\ requireing\ cold\ refrigerator \\ 0\ Otherwise \end{cases}$
YO_d	$\begin{cases} 1\ if\ DC\ d\ is\ capable\ of\ handling\ vaccines \\ requireing\ very\ cold\ refrigerator \\ 0\ Otherwise \end{cases}$
YOO_d	$\begin{cases} 1\ if\ ultra\ cold\ refrigeration\ is\ added\ to\ DC\ d \\ 0\ Otherwise \end{cases}$
XO_{vdop}	Integer doses of vaccine v ordered at period o by DC d to be delivered at period p
DD_{vdp}	Integer doses of vaccine v shipped to DC d at period p
CD_{vgwp}	Integer doses of vaccine v allocated to group g in warehouse w at period p
GD_{vwp}	Integer doses of vaccine v stored in warehouse w at period p
WD_{vdwp}	Integer doses of vaccine v shipped from DC d to warehouse w at period p

4.4. Objective Functions

$$Min\ F_1 = \sum_v \sum_o \sum_p OC_{vp} * X_{vop} + \sum_d FC_d * Y_d + \sum_d FCC_d * YO_d + \sum_d \mathrm{EC} * YOO_d + \sum_v \sum_d \sum_o \sum_p PC_v * XO_{vdop} + \sum_v \sum_w \sum_p HC_{vw} * GD_{vwp} + \sum_v \sum_d \sum_t \sum_p TC_{vdt} * DD_{vdp} + \sum_v \sum_d \sum_w \sum_p \sum_t TCC_{vdwt} * WD_{vdwp} \quad (1)$$

$$Min\ F_2 = \sum_v \sum_d \sum_t \sum_p TTmnd_{vdt} * DD_{vdp} + \sum_v \sum_d \sum_w \sum_t \sum_p TTdw_{vdwt} * WD_{vdwp} \quad (2)$$

Equation 1 and 2 define the objective functions related to the total cost and total travel time respectively.

4.5. Constraints

$$\sum_v \sum_p CD_{vgwp} \geq CR_g * Dem_{gw} \qquad \forall g, w \quad (3)$$

$$GD_{vwp} = \sum_d WD_{vdwp} - \sum_g CD_{vgwp} \qquad \forall v, w, p = 1 \quad (4)$$

https://doi.org/10.21872/2023IISE_3735

$$GD_{vwp} = GD_{vw(p-1)} + \sum_{d} WD_{vdwp} - \sum_{g} CD_{vgwp} \quad \forall v, w, p > 1 \quad (5)$$

$$\sum_{d} XO_{vdop} \leq Cap_{vop} \quad \forall v, o, p \quad (6)$$

$$\sum_{v} DD_{vdp} * (1 - RV_v) \leq CapC_d * Y_d \quad \forall d, p \quad (7)$$

$$\sum_{v} DD_{vdp} * RV_v * (1 - RT_v) \leq CapVC_d - (CapUC_d * YOO_d) \quad \forall d, p \quad (8)$$

$$\sum_{v} DD_{vdp} * RV_v * RT_v \leq CapUC_d * YOO_d \quad \forall d, p \quad (9)$$

$$XO_{vdop} \leq bigM * X_{vop} \quad \forall v, d, o, p \quad (10)$$

$$XO_{vdop} \leq bigM * Y_d \quad \forall v, d, o, p \quad (11)$$

$$XO_{vdop} \leq bigM * YO_d \quad \forall v, d, o, p \quad (12)$$

$$\sum_{o} X_{vop} \leq 1 \quad \forall v, p \quad (13)$$

$$WD_{vdwp} * (1 - RV_v) \leq bigM * Y_d \quad \forall v, d, w, p \quad (14)$$

$$WD_{vdwp} * RV_v * (1 - RT_v) \leq bigM * YO_d \quad \forall v, d, w, p \quad (15)$$

$$WD_{vdwp} * RV_v * RT_v \leq bigM * YOO_d \quad \forall v, d, w, p \quad (16)$$

$$YOO_d \leq bigM * YO_d \quad \forall d \quad (17)$$

$$DD_{vdp} \geq \sum_{w} WD_{vdwp} \quad \forall v, d, p \quad (18)$$

$$DD_{vdp} = \sum_{o} XO_{vdop} \quad \forall v, d, p \quad (19)$$

$$\sum_{v} X_{vop} = 0 \quad \forall o > p \quad (20)$$

$$\sum_{\hat{p}>o}^{\hat{p}<p} \sum_{\hat{o}\leq\hat{p}}^{c} X_{v\hat{o}\hat{p}} \leq bigM * (1 - X_{vop}) \quad \forall v, o, p \quad (21)$$

Equation 3 represents the demand constraints. Inventory balance constraints of the state warehouse are highlighted by equations 4 and 5. Equations 6 to 9 narrate the capacity constraints of the problem. Order allocation constraints are shown in equations 10 to 13. Equation 14 to 19 represents the quantity allocation constraints of the distribution centers. Equation 20 and 21 are the order processing constraints of the problem.

4.6. Multi-objective Model Formulation

The bi-objective model is converted into a single objective model using ε-Constraint Method. In ε-Constraint Method One of the objective functions is selected to be optimized, the other objectives are converted into constraints of the problem.

$$minimize\ F_1$$

Subject to,

$$F_2 \leq F_{2_{Min}} + x \times \frac{F_{2_{Max}} - F_{2_{Min}}}{i} \quad (22)$$

Here, $i = number\ of\ intervals, x = 0, 1, 2, \ldots, i$
and all other constraints from Equation [3-21]

5. The Case Study

The data for the case study is mostly gathered from Tavana et al. [5]. The paper collected the data from Ministry of Health and Family Welfare (MOHFW) in India. The information related to demand for each group in each state, holding cost for the vaccines, and coverage rates have been extracted from historical data proposed by the MOHFW [5]. To transport vaccines from manufacturers to distribution centers, air cargo is assumed to be utilized, with a

https://doi.org/10.21872/2023IISE_3735

shipping capacity of 500,000 doses vaccines in a single trip. Similarly, from DC to warehouse shipment, truck transportation is considered with the capacity of 50000 doses of vaccines. Here two types of shipping options are assumed. the average transportation cost for air freight express shipping is considered between $2 to $8 per dose with average lead time 48-72 hours and standard shipping is considered from $1 to $5 with average lead time 120-168 hours for a single trip. Similarly for truck transportation express shipping is considered between $3 to $5 per dose with average lead time 48-72 hours and standard shipping is considered from $1 to $2 with average lead time 120-168 hours. Besides, the information regarding eight types heterogenous population is collected from the study done by Rastegar et al. [9]. The number of potential DC is considered in the case study is 4. Demand for 36 warehouses is gathered here. Five different types of vaccines are considered which requires different storage facilities. Vaccine types 1 requires ultra-cold refrigeration, vaccine 2, 3 and 4 require very cold refrigeration and vaccine 5 requires cold refrigeration.

6. Results

The proposed model is coded in AMPL software using Cplex solver. The result highlighted a set of Pareto optimal solutions in seven different scenarios which is mentioned in Table 1. The total cost of distribution varied between 4.4 and 5.5 billion USD, and the total travel time between 432 and 1096 thousand hours. In scenario 1, the total cost was minimized but the total travel time was maximized, while in scenario 7, the total travel time was minimized but the total cost was maximized. Scenario 1 to 7 demonstrated the conflicting nature of the two objectives, where improving one would likely result in a deterioration of the other. The outcomes of the scenarios highlighted the changes in the objective functions in each of the cases including the related fixed and transportation costs. The table also included the information of the type of vaccines and DC locations, as well as the information whether the DC required ultra-cold refrigeration facilities. Now, depending on the urgency of the vaccine availability, the decision-makers (DM) can analyze the scenarios and make the decision-making activities by balancing both the total cost and the total travel time. The status of each of the objective functions is demonstrated by Figure 1 (a-b). The tradeoff relationships between the objective functions in each of the scenarios are expressed in Figures 2. Operational aspects of the distribution details from DC 4 to Warehouse 5 in Scenario 4 are presented in Figure 3.

Table 1: Impact in the objective functions (total cost and total travel time) depending on different scenarios.

Scenario	Total Cost, Million USD	Total Travel Time, Thousand Hours	DC Location Required to Open	Transportation Cost, Million USD	Fixed Cost, Million USD	Vaccine Type Selected	DCs Requiring Ultra-Cold Refrigeration
1	4484.871	1096.6	1	1912.61	2572.26	1,2,3,5	1
2	4557.382	930.6	1,4	1915.43	2641.96	1,2,3,5	1,4
3	4592.464	764.6	1,4	1950.51	2641.96	1,2,3,5	1,4
4	4659.010	598.6	1,4	2019.55	2639.46	1,2,3,5	4
5	4679.837	515.6	4	2104.78	2575.06	1,2,3,5	4
6	4783.192	451.2	4	2224.75	2558.45	1,2,3,5	4
7	5523.195	432.6	3,4	2561.9	2961.29	1,2,3,4,5	3,4

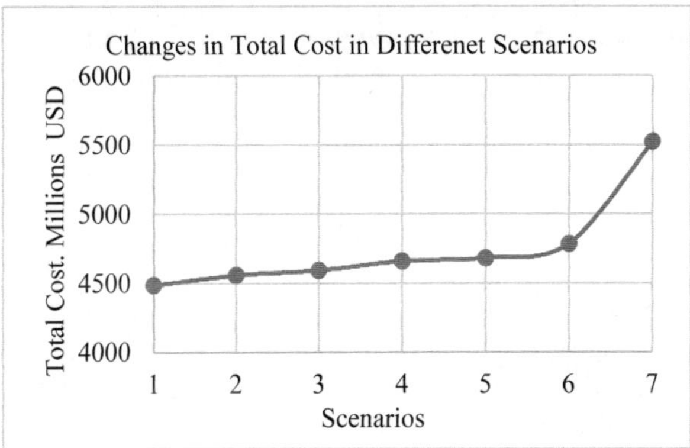

(a)

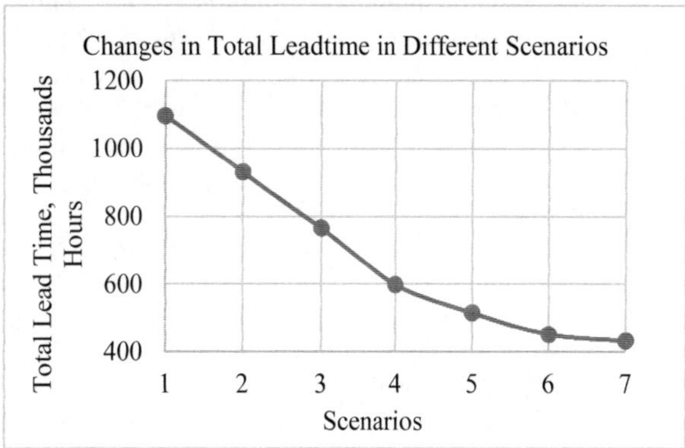

(b)

Figure 1: (a) Change in the total cost in each of the scenarios (b) Change in the total travel time in each of the scenarios.

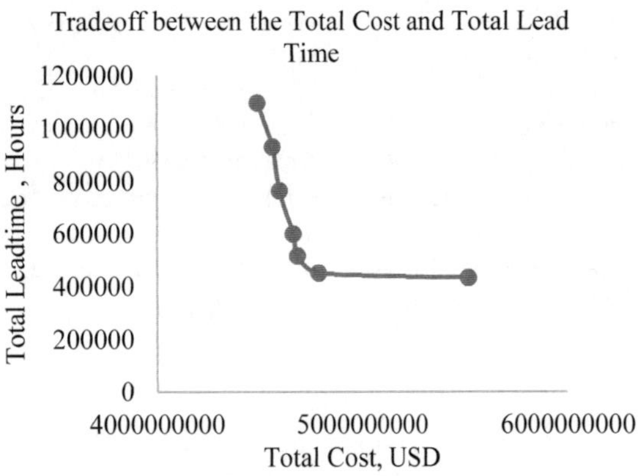

Figure 2: Tradeoff between the Total Cost and Total Lead Time

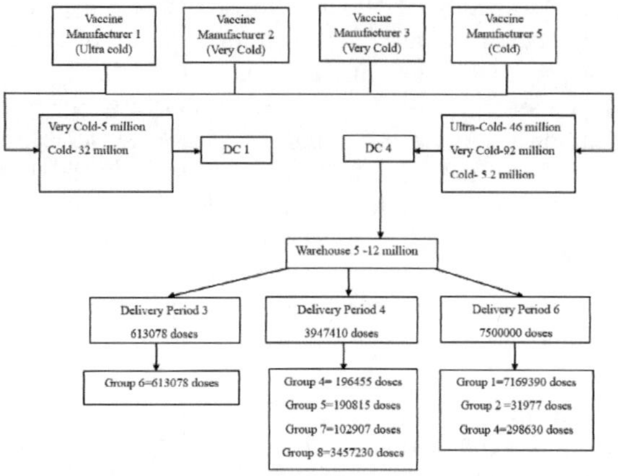

Figure 3: Vaccine Distribution from DC 4 to Warehouse 5 in Scenario 4

7. Conclusion

The model showed that total 179157000 doses of vaccines were ordered from manufacturers in each scenario. The total cost of distribution varied depending on factors such as the type of vaccine and the facilities available at the distribution centers, as well as the shipping delivery mode selected. The travel time also varied based on the shipping options, with faster delivery options increasing transportation costs and the overall cost of distribution. However, the model has some limitations, such as being cost-sensitive and not accounting for parameter uncertainties. The study was inspired by Tavana et al. [5] where we have applied multi-objective optimization techniques for the model formulation. Further study is required to eliminate these limitations of the study.

Acknowledgements

The paper is developed during a graduate course on quantitative modeling. AMPL teaching license was used to solve the models.

References

1. Duijzer, L.E., W. van Jaarsveld, and R. Dekker, *Literature review: The vaccine supply chain.* European Journal of Operational Research, 2018. **268**(1): p. 174-192.
2. Hovav, S. and D. Tsadikovich, *A network flow model for inventory management and distribution of influenza vaccines through a healthcare supply chain.* Operations Research for Health Care, 2015. **5**: p. 49-62.
3. Saif, A. and S. Elhedhli, *Cold supply chain design with environmental considerations: A simulation-optimization approach.* European Journal of Operational Research, 2016. **251**(1): p. 274-287.
4. Zandkarimkhani, S., et al., *A chance constrained fuzzy goal programming approach for perishable pharmaceutical supply chain network design.* Annals of Operations Research, 2020. **295**(1): p. 425-452.
5. Tavana, M., et al., *A mathematical programming approach for equitable COVID-19 vaccine distribution in developing countries.* Annals of Operations Research, 2021: p. 1-34.
6. Tang, L., et al., *Bi-objective optimization for a multi-period COVID-19 vaccination planning problem.* Omega, 2022. **110**: p. 102617.
7. Govindan, K., et al., *Medical waste management during coronavirus disease 2019 (COVID-19) outbreak: A mathematical programming model.* Computers & Industrial Engineering, 2021. **162**: p. 107668.
8. Mohammadi, M., et al., *Bi-objective optimization of a stochastic resilient vaccine distribution network in the context of the COVID-19 pandemic.* Omega, 2022. **113**: p. 102725.
9. Rastegar, M., et al., *An inventory-location optimization model for equitable influenza vaccine distribution in developing countries during the COVID-19 pandemic.* Vaccine, 2021. **39**(3): p. 495-504.

AUTHOR INDEX